Remote Sensing for Marine Environmental Disaster Response

Remote Sensing for Marine Environmental Disaster Response

Guest Editors

Zhixiang Fang
Quanyi Huang
Jaroslaw Tęgowski
Magaly Koch
Yukiharu Hisaki

MDPI

Basel • Beijing • Wuhan • Barcelona • Belgrade • Novi Sad • Cluj • Manchester

Guest Editors

Zhixiang Fang
State Key Laboratory for Information Engineering in Surveying, Mapping and Remote Sensing
Wuhan University
Wuhan
China

Quanyi Huang
Department of Engineering Physics
Tsinghua University
Beijing
China

Jaroslaw Tęgowski
Institute of Oceanography
University of Gdanska
Gdynia
Poland

Magaly Koch
Department of Earth & Environment
Boston University
Boston
USA

Yukiharu Hisaki
Department of Physics and Earth Sciences
University of the Ryukyus
Okinawa
Japan

Editorial Office
MDPI AG
Grosspeteranlage 5
4052 Basel, Switzerland

This is a reprint of the Special Issue, published open access by the journal *Remote Sensing* (ISSN 2072-4292), freely accessible at: https://www.mdpi.com/journal/remotesensing/special_issues/marine_disaster.

For citation purposes, cite each article independently as indicated on the article page online and as indicated below:

Lastname, A.A.; Lastname, B.B. Article Title. *Journal Name* **Year**, *Volume Number*, Page Range.

ISBN 978-3-7258-7532-0 (Hbk)
ISBN 978-3-7258-7533-7 (PDF)
https://doi.org/10.3390/books978-3-7258-7533-7

Contents

Preface

The vast ocean—the cradle of life and a crucial regulator of Earth's climate—is increasingly threatened by a variety of environmental disasters. From acute events like oil spills and devastating typhoons to chronic crises such as green tide outbreaks and coastal erosion, these hazards pose severe risks to ecological security, economic stability, and human life. In this context, the capacity for real-time, accurate, and large-scale dynamic monitoring has become a core requirement for disaster prevention, mitigation, and emergency response. Remote sensing technology, with its unparalleled spatiotemporal coverage, is evolving from a traditional observation tool into a key engine driving scientific understanding and intelligent decision making for marine disasters.

This reprint, *Remote Sensing for Marine Environmental Disaster Response*, aims to systematically present the latest advances and in-depth reflections in this field. It compiles fifteen cutting-edge studies published in the synonymous Special Issue of the journal *Remote Sensing*, which together form a comprehensive technical panorama spanning multi-source perception, intelligent identification, process analysis, and impact assessment. This volume not only showcases innovative applications of remote sensing in addressing various marine hazards but also delves into the complex coupling mechanisms among environmental factors, the integration of artificial intelligence with physical models, and how technology can serve practical disaster reduction efforts.

1. Theme, Scope, and Core Narrative

The reprint's theme focuses on utilizing space-borne, air-borne, and land-based remote sensing to address diverse marine environmental disasters. Its scope covers critical stages of the disaster lifecycle:

(1) Accurate Detection and Identification of Hazards: This is the first line of defense in emergency response. Several studies in the reprint are dedicated to enhancing the automated, high-precision identification of typical disaster targets.

(2) Dynamic Monitoring and Mechanistic Analysis of Disaster Processes: Moving beyond static identification, the reprint pays greater attention to the dynamic development of disasters and their interaction with environmental systems.

(3) Quantitative Impact Assessment and System Recovery: Post-disaster, quantifying losses and monitoring recovery are equally important.

(4) Early Risk Warning and Decision Optimization for Support Systems: Ultimately, all technologies must support risk prevention and decision support.

The common thread running through these studies is the integration of multi-source data, the combination of physical processes with artificial intelligence, and the connection between observation, mechanism, and application. From optical, radar, and infrared to hyperspectral sensing; from discrete image analysis to continuous time-series mining; and from phenomenological description to mechanistic interpretation, the works in this reprint represent a significant trend in current remote sensing research on marine environmental disasters, moving from "seeing" towards "understanding, explaining, and predicting accurately."

2. Motivation and Contemporary Value

Our motivation for compiling this reprint stems from a response to three pressing challenges: First, the combined effects of global climate change and human activities are increasing the frequency, intensity, and complexity of marine disasters, placing unprecedented demands on the timeliness and scientific rigor of monitoring and response.

Second, while remote sensing data acquisition capabilities have grown explosively, significant technical bottlenecks and management gaps remain in efficiently extracting disaster information from massive datasets, thoroughly understanding disaster processes, and translating this into actionable decision making knowledge.

Third, marine disaster response is inherently a systems engineering task requiring interdisciplinary collaboration and multi-sectoral coordination. However, effective communication bridges are often lacking among technical researchers, operational departments, and decision makers.

This reprint is born from the need to address these challenges. By compiling the latest findings from leading international teams in the field, it aims to (1) showcase cutting-edge technical methods to advance the intelligent development of data processing and information extraction; (2) reveal complex environmental mechanisms to deepen the scientific understanding of disaster genesis and evolution; and (3) provide examples of cross-domain application to facilitate the translation of remote sensing technology from research to operational and routine application. We hope that these works, embodying the wisdom of researchers, will provide solid technical support and decision making references for building a more resilient marine future.

3. Target Audience

This reprint addresses a broad and diverse readership:

(1) Researchers and Engineers: Scholars and technical developers in fields such as remote sensing, oceanography, environmental science, disaster science, and information science can find the latest research methods, technical challenges, and future directions herein.

(2) Operational Departments and Decision Makers: Practitioners in government agencies and industries related to marine environmental monitoring, maritime safety, emergency management, fisheries, and coastal zone planning can discover referable technical solutions, application cases, and assessment tools to aid daily monitoring, disaster emergency response, and long-term planning.

(3) Educators and Students: Senior undergraduate students, graduate students, and young faculty in related disciplines can use this reprint as a systematic reference reflecting the frontiers of the field, assisting in teaching, inspiring thinking, and exploring research topics. The ocean connects the world, and its protection is a shared responsibility of all humanity. We sincerely hope that this reprint will serve as a platform for knowledge convergence and intellectual exchange, inspiring more innovative research, closer interdisciplinary collaboration, and more effective international coordination. Together, we can advance the empowerment of remote sensing technology for marine environmental protection and disaster resilience, safeguarding this magnificent blue treasure.

Zhixiang Fang, Quanyi Huang, Jaroslaw Tęgowski, Magaly Koch, and Yukiharu Hisaki
Guest Editors

remote sensing

MDPI

Article

Analysis of Temporal Cumulative, Lagging Effects and Driving Mechanisms of Environmental Factors on Green Tide Outbreaks: A Case Study of the Ulva Prolifera Disaster in the South Yellow Sea, China

Zhen Tian [1,2,3], Jianhua Zhu [1,2,3,*], Huimin Zou [1], Zeen Lu [1], Yating Zhan [4,5,6], Weiwei Li [1], Bangping Deng [3,7], Lijia Liu [3,7] and Xiucheng Yu [3,7]

1 National Ocean Technology Center, Tianjin 300112, China; tjzhen451076@126.com (Z.T.)
2 Key Laboratory of Ocean Observation Technology, Ministry of Natural Resources, Tianjin 300112, China
3 Key Laboratory of Marine Ecological Monitoring and Restoration Technology, Ministry of Natural Resources, Shanghai 201206, China
4 Geological Survey of Jiangsu Province, Nanjing 210018, China
5 Natural Resources Satellite Application Technology Center of Jiangsu Province, Nanjing 210018, China
6 Key Laboratory of Satellite Remote Sensing Applications, Department of Natural Resources of Jiangsu Province, Nanjing 210018, China
7 East China Sea Ecological Center, Ministry of Natural Resources, Shanghai 201206, China
* Correspondence: besmile@263.net

Highlights

What are the main findings?

- Nutrients are identified as the dominant environmental drivers of Ulva prolifera green tide outbreaks in the South Yellow Sea, exerting strong cumulative and cross-seasonal lag effects that largely determine outbreak magnitude and timing.
- Light and thermal conditions play secondary, stage-dependent modulation roles, while the combined lag–accumulation mode represents the prevailing temporal response pattern across different outbreak stages.

What is the implication of the main finding?

- The results demonstrate that historical nutrient accumulation, rather than instantaneous environmental conditions, forms the material basis of large-scale green tide outbreaks, highlighting the necessity of incorporating temporal effects into predictive models.
- Quantifying nutrient-driven temporal windows provides a scientific foundation for early-warning systems and targeted management strategies focusing on nutrient source control and seasonal regulation.

Abstract

The Ulva prolifera green tide in the South Yellow Sea has erupted annually for many years, posing significant threats to coastal ecology, the economy, and society. While environmental factors are widely acknowledged as prerequisites for these outbreaks, the asynchrony and complex coupling between their variations and disaster events have challenged traditional studies that rely on instantaneous correlations to uncover the underlying dynamic mechanisms. This study focuses on the Ulva prolifera disaster in the South Yellow Sea, systematically analyzing its spatiotemporal distribution patterns, the temporal accumulation and lag effects of environmental factors, and the coupled driving mechanisms using the

Floating Algae Index (FAI). The results indicate that: (1) The disaster shows significant interannual variability, with 2019 experiencing the most severe outbreak. Monthly, the disaster begins offshore of Jiangsu in May, moves northward and peaks in June, expands northward with reduced scale in July, and largely dissipates in August. Years with large-scale outbreaks exhibit higher distribution frequency and broader spatial extent. (2) Environmental factors demonstrate significant accumulation and lag effects on Ulva prolifera disasters, with a mixed temporal mode of both accumulation and lag effects being dominant. Temporal parameters vary across different factors—nutrients generally have longer lag times, while light and temperature factors show longer accumulation times. These parameters change dynamically across disaster stages and display a clear inshore–offshore gradient, with shorter effects in coastal areas and longer durations in offshore waters, revealing significant spatiotemporal heterogeneity in temporal response patterns. (3) The driving mechanism of Ulva prolifera disasters follows a "nutrient-dominated, temporally relayed" pattern. Nutrient accumulation (PO_4, NO_3, SI) from the previous autumn and winter serves as the decisive factor, explaining 86.8% of interannual variation in disaster scale and 56.1% of the variation in first outbreak timing. Light and heat conditions play a secondary modulating role. A clear temporal relay occurs through three distinct stages: the initial outbreak triggered by nutrients, the peak outbreak governed by light–temperature–nutrient synergy, and the system decline characterized by the dissipation of all driving forces. These findings provide a mechanistic basis for developing predictive models and targeted control strategies.

Keywords: green tide; Ulva prolifera; time-accumulation effects; time-lag effects; FAI; driving mechanisms; XGBOOST; SHAP

1. Introduction

The rapid and dramatic growth of macroalgae in specific marine environments is known as a green tide disaster [1]. This phenomenon primarily affects mid-latitude coastal areas in the Northern Hemisphere, with the Asia–Pacific region, Europe, and North America being particularly impacted. Over the past few decades, green tide disasters have been documented in countries such as China, the United States, South Korea, France, and Japan [2]. These outbreaks deplete dissolved oxygen in the water, causing the suffocation and death of fish and other marine organisms, which severely disrupts the marine ecosystem's balance [3]. Additionally, the accumulation of green tide biomass can corrode fishing boats, block navigation channels, significantly affect fisheries and tourism, and threaten coastal economic development. Some algae also release toxic substances, which can kill other organisms, trigger other disasters like red tides, and pose health risks to humans through the food chain [4].

The Ulva prolifera disaster in China's Yellow Sea represents the world's largest green tide event. Since its initial occurrence in 2007, it has persisted annually for 19 years, inflicting substantial economic damage. Efforts like reducing raft aquaculture and implementing preemptive harvesting have somewhat mitigated the scale of these disasters, yet the core issue of prevention and control remains unresolved. In certain years, such as 2021, the problem has even intensified [5]. This persistent challenge is partly due to a lack of comprehensive understanding of the scientific mechanisms driving Ulva prolifera outbreaks and their interannual variations [6].

Current research on green tide disasters is concentrated in six primary areas [7]: (1) the use of satellite remote sensing technology for monitoring and predicting the occurrence

https://doi.org/10.3390/rs18020194

and progression of green tides [8,9]; (2) the role of eutrophication in the mechanisms behind green tide formation [10–12]; (3) phylogenetic analysis of green algae through molecular biology to investigate evolutionary relationships and genetic diversity [13,14]; (4) the effects of climate change on green tides, examining changes in frequency, intensity, and distribution [15–18]; (5) management and utilization strategies for green tides, emphasizing effective practices and potential economic and ecological benefits [19–24]; and (6) research on green tides in Chinese waters, with a focus on occurrence patterns and response strategies along China's coast. Eutrophication's impact on green tide formation and climate change's influence on green tides are considered studies on the driving factors of green tide growth and outbreaks, making them among the most popular research directions today [25]. Numerous studies indicate that the growth, proliferation, outbreak, and decline of Ulva prolifera disasters result from the interplay of various environmental factors, including nutrient enrichment (both inorganic and organic), sea surface temperature, solar radiation, and hydrodynamic conditions [26–29]. Sea temperature, effective photosynthetic radiation, and nutrients are essential for algal growth, while wind affects the development, dispersal, and drift direction of Ulva prolifera, with its influence varying at different stages.

While studies have identified key environmental factors influencing Ulva prolifera outbreaks, two major limitations remain in the mechanistic exploration, forming the core focus of this research. First, the temporal dimension effect is often overlooked. Most studies analyze the immediate correlation between environmental factors and green tide outbreaks, neglecting the inherent temporal dependence of ecological responses. Vegetation, as the primary producer in terrestrial ecosystems, shows significant cumulative and lagging effects in response to environmental factors [30–34]. Similarly, Ulva prolifera, an oceanic primary producer, is likely influenced by the continuous impact of environmental factors over time. For instance, nutrient inputs may need time to accumulate before triggering explosive algae proliferation, and changes in water temperature might affect growth rates with a lag of several days or weeks. Unfortunately, systematic quantitative research on these temporal effects (including cumulative and lagging effects) during Ulva prolifera outbreaks is lacking, limiting accurate predictions of critical outbreak points. Second, the relationship models are often oversimplified and lack interpretability. Current research frequently employs linear models to describe the relationship between single factors and Ulva prolifera growth, failing to capture the complex nonlinear relationships in the system. The impact of each environmental factor on Ulva prolifera growth may involve thresholds, saturation points, or optimal intervals, which linear models cannot adequately reveal.

Despite substantial progress in understanding Ulva prolifera green tides, existing studies still exhibit two major limitations. First, most analyses rely on instantaneous correlations between environmental variables and bloom dynamics, overlooking the cumulative and lagged responses that are intrinsic to ecological processes. The lack of a quantitative temporal-effect framework constrains mechanistic interpretation of the ecological time dependency underlying bloom initiation, expansion, and decline. Second, environmental drivers of green tides are commonly examined using linear or single-factor approaches, which are insufficient to capture nonlinear thresholds, saturation effects, and cross-factor coupling mechanisms. These limitations hinder the development of predictive models capable of representing the complex and stage-dependent interactions governing bloom evolution.

To address these knowledge gaps, the present study proposes an integrated framework that combines multi-source satellite remote sensing, temporal-effect modeling, and interpretable machine learning. FAI-based extraction is employed to characterize the long-term spatiotemporal patterns of Ulva prolifera blooms from 2008 to 2022. Four temporal-effect scenarios—no effect, accumulation-only, lag-only, and combined lag–accumulation—are

https://doi.org/10.3390/rs18020194

constructed to quantify the temporal dependence of major environmental variables across the Entire Period (EP), the Disaster Incident Period (DIP), and its internal stages, including the Initial Outbreak Period (IOP), Outbreak Peak Period (OP), and Decline Period (DOP). Based on pixel-level optimal temporal windows, an XGBoost–SHAP framework is further developed to identify dominant drivers, characterize nonlinear response behaviors, and reveal stage-wise evolution mechanisms. This framework enables the explicit integration of historical environmental influences with stage-specific ecological sensitivities.

The key novelties of this work are as follows: (i) it provides the first systematic quantification of both temporal accumulation and lag effects of environmental factors across multiple stages of Ulva prolifera bloom development; (ii) it reveals pronounced spatial gradients and stage-dependent heterogeneity in temporal responses; (iii) it identifies nonlinear driver–response relationships and ecological thresholds through an interpretable XGBoost–SHAP framework; and (iv) it proposes a unified "nutrient-dominated, temporally relayed" mechanism that links environmental history with outbreak-stage dynamics. These advances offer new mechanistic insights into green tide formation and provide a scientific foundation for time-explicit prediction and targeted management strategies.

2. Materials and Methods

2.1. Study Area

The South Yellow Sea is a semi-enclosed marginal sea of the western Pacific Ocean. For the purpose of this study, the geographical extent of the SYS is defined within the coordinates of 32°N to 37°N and 119°E to 124°E (Figure 1a), covering an area of approximately 309,000 km^2 with an average depth of 44 m [35]. This region features significant nutrient inputs from coastal rivers, as well as extensive Pyropia yezoensis aquaculture rafts along the Jiangsu coast, which provide initial substrates for macroalgae. These combined factors create a critical environment for the initiation and development of Ulva prolifera blooms, which this study focuses on [36]. The hydrodynamic environment of the Yellow Sea is shaped by several factors: the Yellow Sea Warm Current, coastal runoff from the Yangtze River, Sheyang River, and Jiangsu Northern Irrigation Main Channel, coastal upwelling along northern Jiangsu and Shandong, and the Yellow Sea Cold Water Mass (YSCWM) [37–40]. The region experiences a maritime monsoon climate, with annual precipitation ranging from 100 mm to 200 mm. During summer and autumn, sea surface temperatures vary between 10 °C and 28 °C [41,42]. These factors collectively contribute to the complex and varied hydrological and water quality parameters in the Yellow Sea.

2.2. Data Sources

2.2.1. Satellite Data Sources

This study obtained all MODIS-Terra and MODIS-Aqua satellite data for the study area from the NASA Goddard Space Flight Center Ocean Biology Processing Group (OBPG, http://oceancolor.gsfc.nasa.gov (accessed on 20 October 2024)), covering the period from 2003 to 2022. The data were acquired at the Level-0 processing stage. Following established methodologies [42], the SeaWiFS Data Analysis System software (SeaDAS, Version 9) was employed to preprocess each scene, generating Rayleigh-corrected reflectance (Rrc) products for specific bands: 469 nm, 555 nm, 645 nm, 859 nm, and 1240 nm. Each band was processed to a 250 m resolution and projected using a Cylindrical Equidistant (rectangular) projection. These bands facilitated image composition for green tide extraction. The composites included true-color images (R: 645 nm; G: 555 nm; B: 469 nm) and false-color images designed for identifying large floating algae (R: 645 nm; G: 859 nm; B: 469 nm). In these false-color images, water bodies with significant algal blooms appear green due to the high reflectance of macroalgae in the 859 nm band [43].

https://doi.org/10.3390/rs18020194

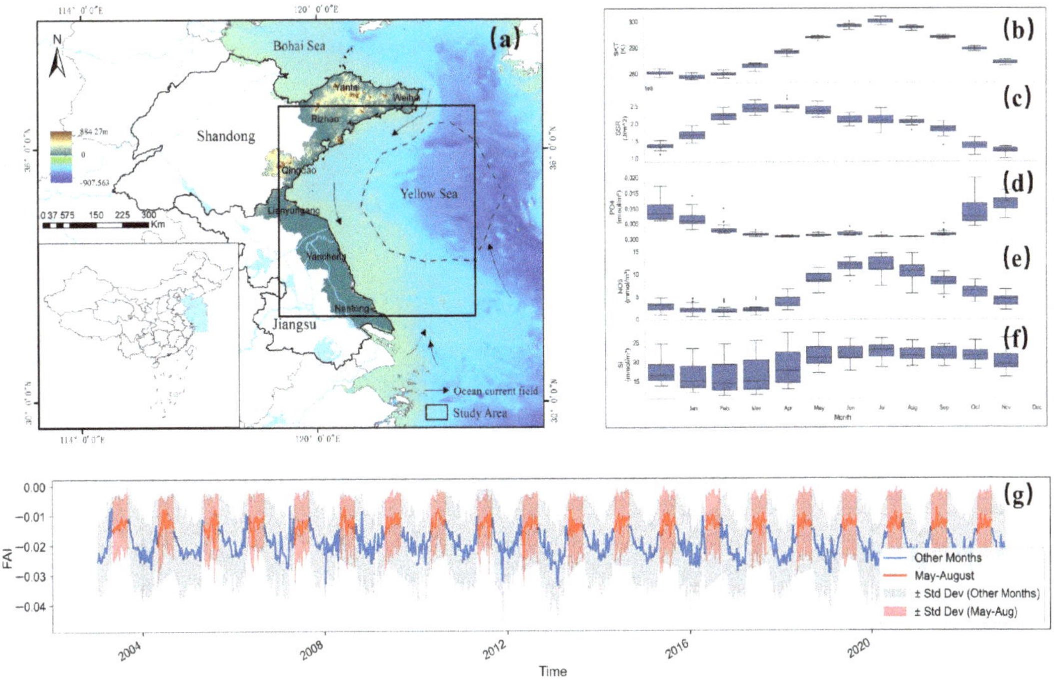

Figure 1. Geographical Extent of the Study Area and Temporal Distribution of Selected Environmental Factors (**a**) Geographical extent of the study area; (**b–f**) Monthly mean distributions (2008–2022) of key environmental factors within the study area: SKT, SSR, PO_4, NO_3, and SI, respectively; (**g**) Temporal distribution of FAI, where periods of disaster occurrence are highlighted in red, and other months are indicated in blue.

To further eliminate invalid information during the generation of the Rrc product, Level-2 product (Rrs) flag data were simultaneously produced using SeaDAS. Flags were used to mask invalid or anomalous information, such as clouds and sun glint. Notably, land masking was accomplished using land vector data in this study. The masked data were then used to generate the corresponding FAI data, which served as the basis for subsequent research.

To examine the temporal cumulative and lagged effects of environmental factors while ensuring data accuracy and enhancing data usability, the FAI data were resampled onto a grid with a resolution of 4 km. These standardized gridded data were then averaged into 8-day mean values to facilitate consistent temporal analysis.

2.2.2. Environmental Data Sources

Table 1 summarizes the environmental variables used in this study, including their definitions, units, data sources, and temporal coverage. The environmental factors employed in this study consist of Skin Temperature (SKT), Surface Solar Radiation (SSR), the U-component of 10 m wind (U10), the V-component of 10 m wind (V10), Mean Sea-Level Pressure (MSL), Mean Wave Period (MWP), and Total Precipitation (TP). These variables were obtained from ERA5, the fifth-generation atmospheric reanalysis dataset provided by the European Centre for Medium-Range Weather Forecasts (ECMWF, https://www.ecmwf.int/ (accessed on 5 March 2025)). The original dataset has a spatial resolution of 0.25° and spans the study area from 2000 to 2022. The data underwent processing through bilinear interpolation and spatiotemporal averaging, resulting in a final dataset characterized by a temporal resolution of 8-day means and a spatial resolution of 4 km [44].

Table 1. Summary of Environmental Factors.

Environmental Variables Name	Abbreviation	Spatial Resolution	Temporal Resolution	Unit
Skin Temperature	SKT	0.25°	Daily	K
Surface Solar Radiation	SSR	0.25°	Daily	J/m^2
U-component of 10 m wind	U10	0.25°	Daily	m/s
V-component of 10 m wind	V10	0.25°	Daily	m/s
Mean Sea-Level pressure	MSL	0.25°	Daily	hpa
Mean Wave Period	MWP	0.25°	Daily	s
Total Precipitation	TP	0.25°	Daily	mm
Mole concentration of nitrate in sea water	NO_3	0.25°	Daily	$mmol/m^3$
Mole concentration of phosphate in sea water	PO_4	0.25°	Daily	$mmol/m^3$
Mole concentration of silicate in sea water	SI	0.25°	Daily	$mmol/m^3$

Nutrient data were sourced from the biogeochemical hindcast for the global ocean provided by the Copernicus Marine Service (https://marine.copernicus.eu/ (accessed on 5 March 2025)), produced by Mercator-Ocean in Toulouse, France. This dataset offers 3D biogeochemical fields starting from 1993, featuring a 1/4° resolution across 75 vertical levels, using the PISCES biogeochemical model on the NEMO modeling platform. The selected nutrient data include the mole concentrations of nitrate (NO_3), phosphate (PO_4), and silicate (SI) in seawater. The data, with a spatial resolution of 0.25°, cover the study area from 2000 to 2022. Bilinear interpolation and spatiotemporal averaging were employed to achieve a final dataset with an 8-day temporal resolution and a 4 km spatial resolution.

2.3. Research Methods

2.3.1. Ulva Prolifera Extraction and Spatiotemporal Distribution Characteristic Analysis

In this study, the Floating Algae Index (FAI) was utilized for threshold segmentation to extract the Ulva prolifera disaster. The FAI's linear subtraction design offers greater resilience than the NDVI against disturbances from aerosols, sun glint, thin clouds, and foam [45]. Additionally, this design facilitates the linear unmixing of pixels when they contain only partial algal coverage. The formula for the FAI is as follows:

$$FAI = R_{rc}(859) - R_{rc}(859)'$$
$$R_{rc}(859)' = R_{rc}(645) + (R_{rc}(1240) - R_{rc}(645) \times \frac{\lambda_{859} - \lambda_{645}}{\lambda_{1240} - \lambda_{645}}) \quad (1)$$

$R_{rc}(645)$, $R_{rc}(859)$, and $R_{rc}(1240)$ denote the Rayleigh-corrected reflectance products at wavelengths of 645 nm, 859 nm, and 1240 nm, respectively. Meanwhile, λ_{645}, λ_{859}, and λ_{1240} signify the wavelengths corresponding to these specific bands.

FAI data were generated by processing each scene, with segmentation thresholds for each year determined through a combination of FAI calculations and visual interpretation. This approach enabled the extraction of Ulva prolifera results from each image. Results from individual images were then composited into daily maximum coverage maps, producing the final daily Ulva prolifera extraction results. The spatiotemporal distribution characteristics were analyzed based on these daily extraction results.

https://doi.org/10.3390/rs18020194

2.3.2. Analysis of Temporal Effects Between Ulva Prolifera Disaster and Environmental Variables

Utilizing techniques from cumulative and lagged analysis in terrestrial vegetation [32], this study employed linear regression to evaluate the temporal effects of each environmental factor on FAI. The regression formula is as follows:

$$FAI_t = k \times \sum_{j=0}^{P} X_{t-i-j} + b \tag{2}$$

In this analysis, k and b are regression coefficients, X represents a single environmental factor, t denotes the corresponding time point, i indicates the number of lag periods, and j signifies the number of accumulation periods (each unit is 8 days). Both i and j range from 0 to 46, covering roughly one year. The rationale for this range is explained in Section 4.1. Various combinations of i and j create four distinct temporal effect analysis modes: (i) $i = 0$, $j = 0$, indicating no temporal effect ('no'); (ii) $i = 0, j \neq 0$, considering only the temporal cumulative effect ('acc'); (iii) $i \neq 0, j = 0$, considering only the temporal lag effect ('lag'); and (iv) $i \neq 0, j \neq 0$, considering both lag and cumulative effects ('lagacc'). By comparing R^2 values across these combinations, the set that maximizes R^2 is chosen as the optimal temporal adjustment parameter set, identifying the temporal effect mode that best explains the variation in FAI.

Temporal effect mode analysis was conducted for five distinct periods: the entire period from 2008 to 2022 (Entire Period, EP), the disaster incident period within 2008–2022 (Disaster Incident Period, DIP), and three stages within the DIP: the Initial Outbreak Period (IOP), the Outbreak Period (OP), and the Decline Outbreak Period (DOP). Due to the ever-changing growth conditions and physical state of floating Ulva prolifera, accurately identifying the factors influencing algal development at specific stages is challenging. Drawing on the expertise of numerous scholars [30], this study defines 1 May–10 June as the IOP, 10 June–10 July as the OP, and 10 July–31 August as the DOP.

2.3.3. Analysis of Disaster Outbreak Driving Mechanisms

To quantitatively analyze the driving mechanisms behind the disaster outbreak and assess the significance of each environmental variable in influencing FAI changes, this section utilizes the Extreme Gradient Boosting (XGBoost) modeling method alongside Shapley Additive Explanations (SHAP) for model interpretation [45,46]. This integrated framework excels in managing high-dimensional nonlinear variables, capturing multivariate coupling relationships, and improving model interpretability [47].

Utilizing the optimal cumulative and lag adjustment parameters for each pixel, derived from the temporal effect analysis of the Ulva prolifera disaster, we adjusted the environmental variables from May to August annually from 2008 to 2022 on a pixel-by-pixel basis in their time series. Specifically, for each pixel, the optimal lag and accumulation periods determined from the search were used to identify the pre-adjusted environmental data corresponding to each FAI value timestamp. Instead of performing cumulative summation, the mean value of environmental data over the identified accumulation period preceding the lag-adjusted time was used to construct the temporally aligned dataset. These adjusted environmental variables were then used as input features for the nonlinear regression model. The fundamental predictive structure of the model is as follows:

$$\hat{y}_i = \sum_{k=1}^{k} f_k(x_i), f_k \in F(i = 1, 2, \ldots, n) \tag{3}$$

https://doi.org/10.3390/rs18020194

where $\hat{y}_i$ is the predicted value for surface sample i, x_i is the input feature vector, k represents the total number of regression trees, f_k is the k-th regression tree, and F represents the set of CART tree functions.

To enhance model interpretability, this study employed SHAP for quantification. SHAP values are based on the concept of Shapley values from cooperative game theory, quantifying the marginal contribution of each feature to the model output [48]. In this context, SHAP analysis accurately attributes the influence to each temporally adjusted environmental factor by computing the marginal contribution of each feature value to the prediction difference from the baseline. This approach effectively handles the time-aligned features by evaluating how much each adjusted environmental variable contributes to changing the model output from the expected value. For any given sample, the prediction can be expressed as the sum of the baseline model prediction and the SHAP values of all input features:

$$y_i = y_{base} + f(X_{i1}) + f(X_{i2}) + \cdots + f(X_{ij}) \quad (4)$$

where y_{base} represents the average prediction for all samples (baseline), and $f(X_{ij})$ represents the SHAP value of the j-th feature for the i-th sample, reflecting its marginal contribution to the model output. The sign of $f(X_{ij})$ indicates the direction of its influence on the response variable; a positive value suggests a positive promoting effect, while a negative value suggests an inhibitory effect.

The XGBoost model was utilized to examine the impact of environmental factors on the FAI. The hyperparameters were configured as follows: number of trees = 300, maximum depth = 6, learning rate = 0.05, and both subsample and column sample ratios = 0.8, with a random seed set to 42. The dataset was randomly divided into training and testing subsets in an 8:2 ratio. To achieve a balance between computational efficiency and statistical reliability, a maximum of 600,000 samples were employed during training, while up to 50,000 test samples were used for SHAP interpretation. Model performance was assessed using the coefficient of determination (R^2) and the mean absolute error (MAE) on the test set. All analyses were conducted in Python (version 3.9.23), utilizing the scikit-learn and shap libraries.

The SHAP analysis included generating feature importance rankings and SHAP summary plots for the five stages (EP, DIP, IOP, OP, DOP), as well as SHAP dependence plots for individual features during the DIP stage.

3. Results

3.1. Spatiotemporal Distribution Characteristics of Ulva Prolifera Green Tides

In 2007, the Ulva prolifera (green tide) disaster first occurred, but due to its limited area and brief duration [9], it is not examined in detail in this study. The disaster was identified using FAI threshold segmentation, and Figure 2 illustrates the annual daily maximum coverage area and annual distribution area. The trend in daily maximum coverage area showed distinct patterns over different periods: it generally declined from 2008 to 2011, rose from 2011 to 2016, dropped again to a relatively low level during 2017–2018, experienced a large-scale outbreak in 2019 peaking at 1732.75 km^2, decreased to a low level in 202, saw another major outbreak in 2021 with an area of 1385.69 km^2, and returned to a lower level in 2022. The trend in annual distribution area largely mirrored that of the daily maximum coverage area. In 2019, the largest distribution area was recorded at 6.12×10^4 km^2, followed by 2016 with approximately 5.76×10^4 km^2. The smallest distribution area occurred in 2001, measuring 1.95×10^4 km^2.

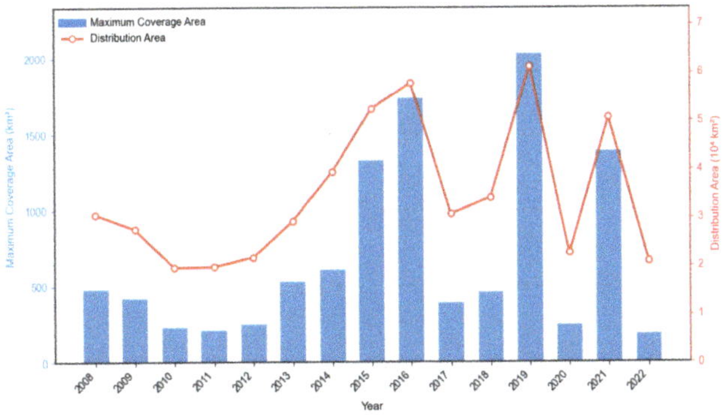

Figure 2. Annual Statistics of Ulva Prolifera Disasters in the South Yellow Sea. The blue bars depict the daily maximum coverage area, while the red line represents the annual distribution area.

Figure 3 highlights notable spatio-temporal variations in the distribution of Ulva prolifera in the South Yellow Sea (Figure 3a–o). From 2008 to 201, Ulva prolifera patches were mainly concentrated in the coastal waters of northern Jiangsu and the southern sea area of the Shandong Peninsula, within the coordinates 33°N–37°N, 119°E–122.5°E. The distribution area fluctuated annually. In 2016, 2019, and 2021, large-scale Ulva prolifera outbreaks occurred, marked by higher distribution frequency and wider spatial coverage compared to other years. To further explore the monthly temporal and spatial variations of the Ulva prolifera disaster, the study conducted two analyses: (1) plotting zonal cumulative distribution area line graphs for each year in Figure 3a–o, and (2) statistically analyzing the monthly disaster distribution and extracting the geometric center pixel values for the entire study period in Figure 3p. These analyses reveal that in most years, the scale of the Ulva prolifera disaster in May is relatively small and located near the Jiangsu coast. June typically marks the peak outbreak period, with the area generally surpassing that of other months, and the distribution center shifts northward. The disaster scale decreases in July, with the distribution range continuing to move north. By August, the disaster largely subsides, leaving only sporadic Ulva patches.

3.2. Temporal Effect Analysis of Environmental Factor

This study evaluated the global average R^2 for five periods (EP, DIP, IOP, OP, DOP) under four temporal effect scenarios (no, lag, acc, lagacc). As illustrated in Figure 4a, scenarios that included temporal effects (lag, acc, lagacc) yielded higher global average R^2 values than the scenario without temporal effects. Among these, the scenario combining both accumulation and lag effects (lagacc) consistently achieved the highest average R^2 across all stages.

The R^2 values under the optimal lagacc scenario displayed distinct variation patterns across different stages. The Entire Period (EP) demonstrated the highest explanatory power, with R^2 values ranging from 0.28 to 0.47 across various environmental factors. In contrast, the Disaster Occurrence Period (DIP) and its sub-stages exhibited significantly lower R^2 values (DIP: 0.16–0.23; IOP: 0.13–0.22; OP: 0.12–0.21; DOP: 0.12–0.22), with the lowest figures consistently during the Initial Outbreak Period (IOP). This reduction in explanatory power during active outbreak stages suggests that the rapid nutrient consumption and dynamic algal growth typical of these periods may weaken the immediate statistical relationships between environmental variables and FAI.

https://doi.org/10.3390/rs18020194

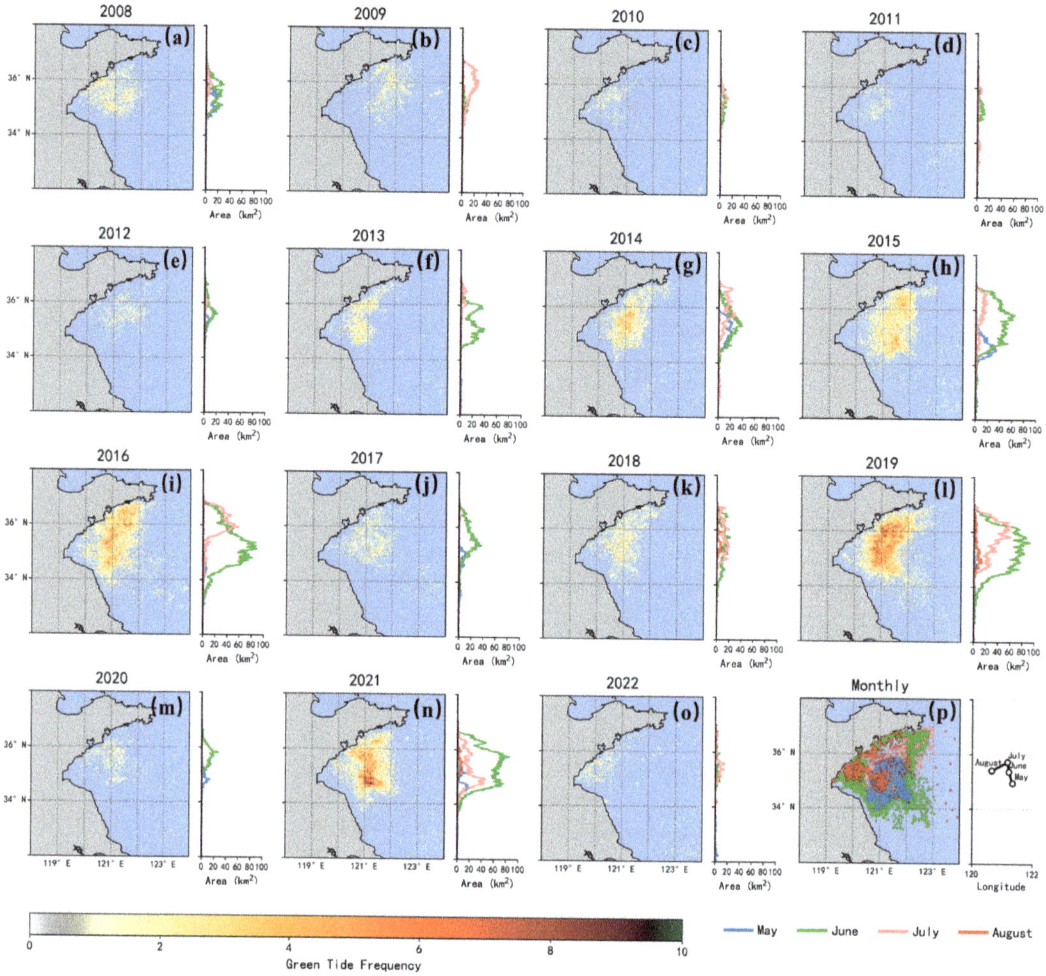

Figure 3. Spatio-temporal distribution statistics of the Ulva prolifera disaster (Subfigures (**a**–**o**) represent annual statistical results: left panels show annual Ulva occurrence frequency maps, right panels show monthly cumulative zonal distribution area line charts for each year—blue: May, green: June, yellow: July, red: August. Subfigure (**p**) represents multi-year monthly statistics: left panel shows monthly spatial distribution maps, right panel shows monthly center point trajectory map).

Figure 4b provides statistics on the proportion of pixels where each temporal effect mode was optimal, confirming that the scenario incorporating both lag and accumulation effects was the most common. The Optimal Temporal Scenario (OTS) determined for each pixel was then utilized in all subsequent analyses. These results bolster our initial hypothesis, underscoring significant temporal effects between Ulva prolifera outbreaks and environmental variables, with the mixed temporal scenario of accumulation and lag effects being the predominant form.

https://doi.org/10.3390/rs18020194

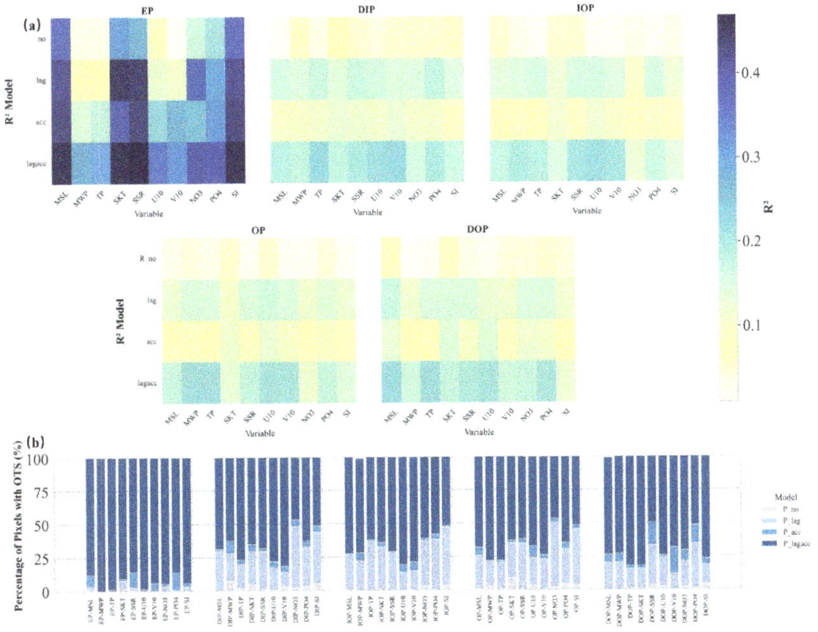

Figure 4. Assessment of Temporal Effects on the FAI-Environment Relationship: (a) Coefficient of Determination (R^2); (b) OTS Pixel Percentage per Variable and Stage.

After confirming the prevalence of temporal effects, this study quantified the temporal parameters—average lag and accumulation time—for each environmental factor at different disaster stages using the Optimal Temporal Scenario (OTS). For each pixel, the OTS was identified by comparing R^2 values across four temporal effect scenarios. The overall mean and spatial distribution of these OTS-derived parameters were then analyzed, highlighting systematic differences among factors and stages, as well as significant spatial heterogeneity. An examination of the spatial and temporal patterns (Figures 5 and A1–A5) reveals the following key findings:

(1) Temporal parameters varied significantly across different environmental factors. Nutrient factors (NO_3, PO_4, SI) generally displayed moderate to long lag times (typically 20–28 8-day periods) and highly variable accumulation times (ranging from 5.76 to 19.60 8-day periods). Regions with prolonged lag times for these nutrients were predominantly situated in the central and eastern parts of the study area. Particularly, PO_4 exhibited widespread high values (lag > 27 8-day periods) in the northern East China Sea during the OP and DOP stages. Among meteorological and hydrodynamic factors, U10 had the lengthiest average lag time (31.20 8-day periods) during EP compared to all factors, while V10's longest average lag time was observed during IOP (26.32 8-day periods). High-value regions for these factors were mainly found offshore and in the northern areas, with shorter lag times experienced in nearshore shelf areas. Sea Surface Temperature (SKT) and Surface Solar Radiation (SSR) showed their longest average accumulation times during DIP and its sub-stages. Specifically, SSR peaked in average accumulation during IOP (23.91 8-day periods), and SKT also exhibited significant accumulation during DIP (22.66 8-day periods). Areas with high accumulation for SSR (Accum > 23 8-day periods) were notably clustered in the southern study area and nearshore zones during the IOP and OP stages.

https://doi.org/10.3390/rs18020194

(2) Temporal parameters changed dynamically across disaster development stages, and their spatial distribution patterns exhibited distinct phase-specific characteristics. The global average R^2 reached its maximum during the EP. Spatially, in the EP, regions with high accumulation and long lag times for various factors tended to form patchy yet contiguous patterns. For example, MSL and MWP displayed greater lag times (Lag $\approx$ 24–31 8-day periods) in the Bohai Strait and the central Yellow Sea. In the IOP, the mean accumulation time peaked for several variables, including SSR (23.91 8-day periods) and NO_3 (19.60 8-day periods). Their high-value zones were concentrated offshore east of 123°E and in the seas south of Jiangsu, where values increased from the coast toward the open sea. During OP and DOP, some factors experienced relatively high lag times. For example, the average lag time for PO_4 was 27.31 and 28.28 8-day periods during OP and DOP, respectively, with high-value areas (Lag > 27 8-day periods) concentrated in the northeastern study area. Conversely, during DOP, the average accumulation time for SI fell to its lowest value among all stages, at 5.76 8-day periods, and its spatial distribution was relatively uniform, lacking significant high-value clusters. Spatially, the distribution of temporal parameters revealed a pronounced inshore–offshore gradient: nearshore and estuarine areas generally exhibited shorter lag and accumulation times, whereas offshore regions demonstrated longer temporal effects. This spatial pattern likely arises from fundamental physical and biogeochemical contrasts between dynamic coastal waters and the more stable open sea. Nearshore areas are characterized by stronger hydrodynamic mixing, pulsed terrestrial nutrient inputs, and active sediment–water interactions, which together lead to rapid environmental turnover and relatively short-lived ecological responses. In contrast, offshore waters tend to experience greater stratification, slower nutrient cycling dominated by water-column processes, and reduced direct anthropogenic disturbance. These conditions favor the gradual accumulation and prolonged influence of environmental drivers, manifesting as longer lag and accumulation times. This pronounced spatial heterogeneity indicates that the temporal response of Ulva prolifera outbreaks to environmental forcing is not spatially uniform but varies systematically across the seascape, highlighting the importance of incorporating spatially differentiated temporal parameters into targeted monitoring and prediction frameworks.

3.3. Coupled Driving Mechanisms During the DIP

For each pixel, the OTS (with the highest R^2) determined the accumulation and lag periods used to adjust the environmental variable time series. An XGBoost model was then fitted to the entire study area using this time-aligned dataset. SHAP analysis was employed to interpret the model results. During temporal adjustment, the optimal lag and accumulation periods identified in the search were used to match pre-adjusted environmental data with each FAI value timestamp. Instead of a cumulative summation, the mean value of environmental data over the identified accumulation period preceding the lag-adjusted time was used. This approach was chosen because the study's primary aim is to quantify the temporal effects (accumulation and lag) of environmental factors on green tide outbreaks. It involves constructing a temporally aligned dataset based on this quantification and using the XGBoost model and SHAP explainer to reveal each factor's relative contribution and non-linear influence during its key effective period.

https://doi.org/10.3390/rs18020194

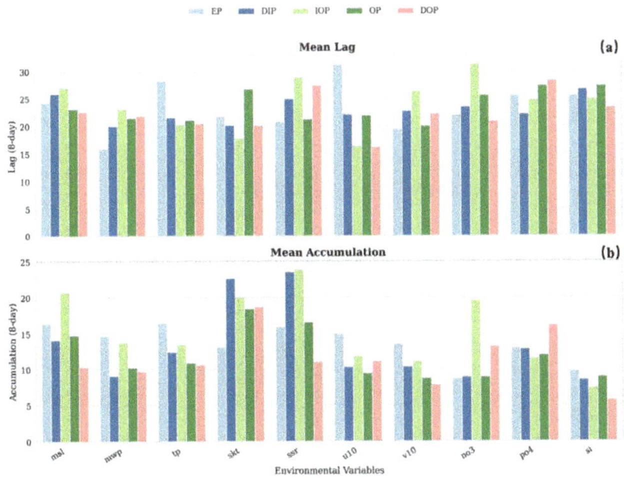

Figure 5. Histograms of average accumulation and lag periods for each environmental variable across five temporal phases (EP, DIP, IOP, OP, DOP) (Unit: 8-day; (**a**) distribution statistics of average accumulation, (**b**) distribution statistics of average lag).

Figure 6 offers a detailed SHAP-based analysis, highlighting both the influence direction and the relative significance of each environmental variable in the FAI estimation model during the DIP. In Figure 6a, the mean absolute SHAP values represent each variable's overall contribution to the model output, with longer bars signifying greater importance. Figure 6c shows the contribution percentages of each environmental variable. Among these, nutrients stood out: PO_4 had the highest mean absolute SHAP value, NO_3 was second, and SI ranked fourth, underscoring the crucial role of nutrient variables in influencing FAI variations. Figure 6b illustrates the directionality of feature effects, with point color indicating feature value (red for high, blue for low). A positive correlation is suggested when higher values (red) align with positive SHAP values and lower values (blue) align with negative SHAP values. Overall, the variables comprising the top 90% of cumulative importance included PO_4, NO_3, SKT, SI, SSR, and SI. SSR, SKT, and U10 positively influenced FAI prediction, while the nutrient variables PO_4, NO_3, and SI exhibited negative correlation characteristics.

SHAP dependence plots were used to quantitatively illustrate how individual environmental variables affect FAI (Figure 7). The single-factor dependence analysis with SHAP values revealed that each environmental variable's impact on the FAI displayed non-linear characteristics, threshold effects, and variations in direction and strength.

The SHAP dependence plots quantitatively elucidate the nonlinear influences of individual environmental variables on FAI during the DIP (Figure 7). These relationships reveal pronounced threshold effects and optimal response ranges, providing mechanistic insights into the proposed "nutrient-dominated, environmental synergy" driving pattern.

For nutrient variables, the SHAP responses clearly reflect the central role of the supply–consumption balance in green tide dynamics. Phosphate (PO_4; Figure 7i) exhibits a sharp positive response at very low concentrations (<0.05 mmol m^{-3}), indicating its potential role as a limiting nutrient during the initial development of biomass. Once this threshold is exceeded, the promoting effect rapidly saturates, consistent with the concept of nutrient sufficiency. Nitrate (NO_3; Figure 7h) and silicate (SI; Figure 7j) display similar transition behaviors, shifting from weak promotion to increasing suppression around ~10 mmol m^{-3} and ~20 mmol m^{-3}, respectively. These patterns suggest that moderate concentrations support algal growth, whereas excessively high levels may represent water masses associ-

https://doi.org/10.3390/rs18020194

ated with unfavorable physical conditions or enhanced ecological constraints. The overall negative correlation between nutrient-related SHAP values and FAI is consistent with the classical process of "rapid biomass growth followed by nutrient depletion," forming a key negative feedback mechanism during bloom evolution.

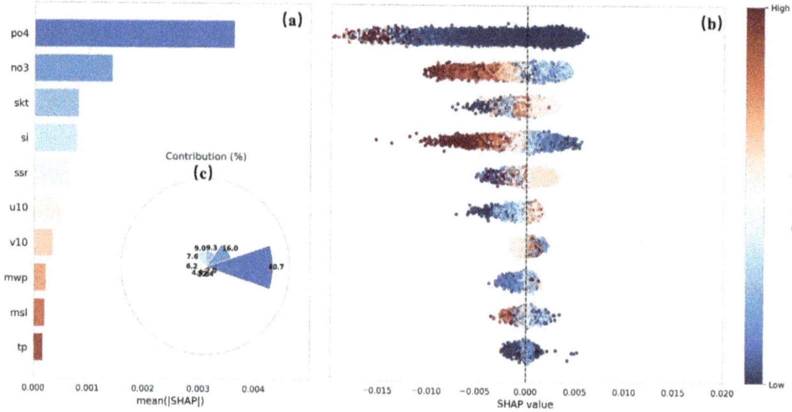

Figure 6. Importance Ranking and SHAP Plots of Driving Factors for FAI during the DIP Phase: (**a**) Bar plot illustrating feature importance, (**b**) SHAP summary plot, (**c**) Chart displaying feature contribution percentages.

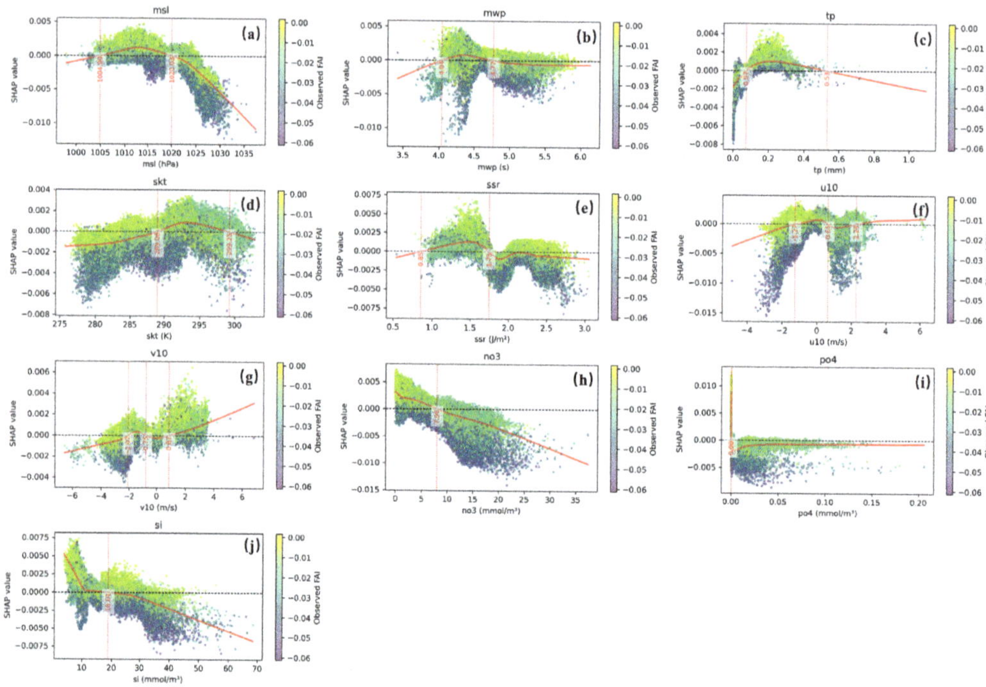

Figure 7. Presents SHAP Dependence Plots illustrating the influence of environmental factors on DIP: (**a**) MSL, (**b**) MWP, (**c**) TP, (**d**) SKT, (**e**) SSR, (**f**) U10, (**g**) V10, (**h**) NO_3, (**i**) PO_4, (**j**) SI.

For light and thermal conditions, the nonlinear responses closely align with the physiological tolerance windows of Ulva prolifera. Sea surface temperature (SKT; Figure 7d)

https://doi.org/10.3390/rs18020194

exhibits a clear unimodal relationship with FAI. The strong positive contribution within an optimal range (~289–299 K) reflects enhanced metabolic activity and growth efficiency, while deviations toward lower or higher temperatures lead to diminishing or negative effects, highlighting sensitivity to thermal stress. Surface solar radiation (SSR; Figure 7e) shows a "suppression at extremes and promotion at intermediate levels" pattern, with the most favorable range occurring at approximately 0.8–1.75 J m^{-2}. This response suggests that both light limitation under low irradiance and potential photoinhibition under excessive radiation can constrain biomass accumulation.

For meteorological and hydrodynamic factors (e.g., MSL, MWP, U10, and V10; Figure 7a–c,f,g), the observed nonlinear relationships are likely mediated indirectly through their effects on algal aggregation, dispersal pathways, and vertical mixing processes, which influence the spatial concentration and apparent surface abundance detected by FAI rather than directly regulating physiological growth.

Collectively, these nonlinear responses and identified thresholds are not merely statistical artifacts but ecologically meaningful indicators. They delineate the specific environmental windows that favor or inhibit Ulva prolifera proliferation. Incorporating such threshold-based insights into predictive models and management frameworks—such as early-warning indicators or nutrient reduction targets—can substantially enhance the accuracy and applicability of green tide forecasting and control strategies.

3.4. *Analysis of Stage-Wise Evolution Characteristics of the Disaster's Driving Mechanisms*

This study systematically uncovers the driving mechanisms by thoroughly comparing the importance rankings of environmental factors across the EP, the DIP, and its three internal sub-stages (IOP, OP, DOP) (Figure 8). The findings reveal that the driving mechanisms of the green tide display significant scale dependence and evolve distinctly at different stages.

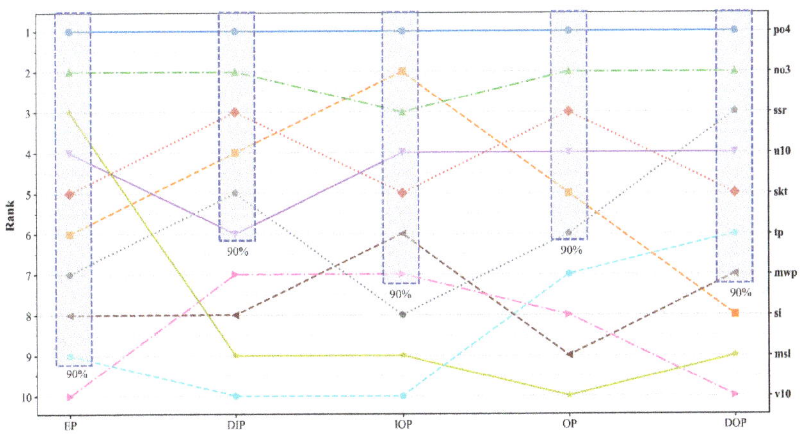

Figure 8. Feature Importance Rankings Across Different Phases (Elements within the gray rectangle indicate those with a cumulative importance of ≥90% in each phase).

3.4.1. Comparison of Driving Mechanisms Between EP and DIP

Comparing the SHAP importance rankings between EP and DIP (Figures 6, 8 and 9) reveals distinct driving modes for non-disaster and disaster periods. Although the composition of dominant factors varied, the core drivers remained consistent: nutrient factors (PO_4, NO_3) ranked first and second in both periods, underscoring their role as the most stable drivers of Ulva outbreaks. Sea Surface Temperature (SKT) also retained stable importance,

https://doi.org/10.3390/rs18020194

ranking fifth in EP and third in DIP, indicating its persistent influence. In contrast, Mean Sea Level Pressure (MSL) experienced the most significant shift, dropping from third in EP to last in DIP, suggesting its correlation with biomass is largely due to non-disaster processes rather than direct driving effects. Additionally, the concentration of contributing factors increased significantly during DIP, with only six factors (compared to nine in EP) accounting for $\geq$90% of cumulative importance, highlighting the enhanced dominance of core drivers during outbreak periods.

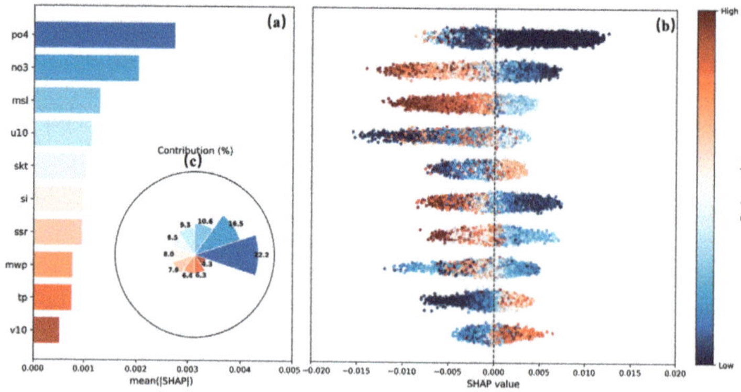

Figure 9. Importance Ranking and SHAP Plots of Factors Influencing FAI during the EP. (**a**) Bar plot illustrating feature importance, (**b**) SHAP summary plot, (**c**) Chart depicting feature contribution percentages.

3.4.2. Comparison of Driving Mechanisms Between IOP, OP and DOP

A comprehensive stage-specific analysis of the three sub-periods—IOP, OP, and DOP—was conducted based on the overall driving pattern identified during DIP. The significance of environmental factors in each stage (Figure 10) reveals a distinct "temporal relay" of driving influences.

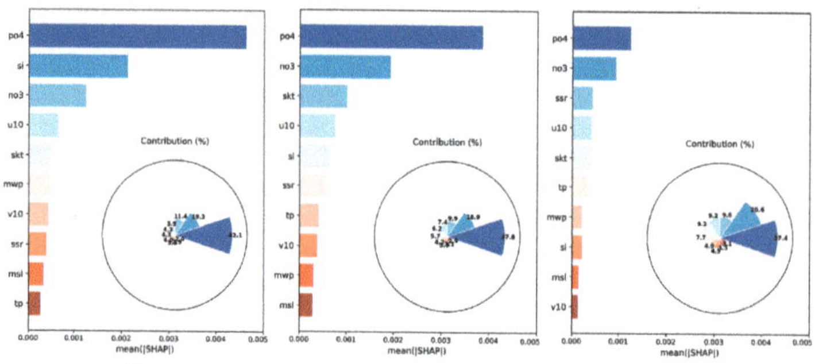

Figure 10. Feature Importance and Contribution Percentage in IOP, OP, and DOP: (**a**) IOP, (**b**) OP, (**c**) DOP.

During the IOP, the Ulva prolifera disaster exhibited a strong reliance on nutrients, establishing a "Nutrient-Initiation Mode" where nutrient factors were dominant. Phosphate (PO_4) had the highest mean absolute SHAP value (0.0045) among all factors, with SI and NO_3 following in second and third place, respectively. Along with PO_4, these factors were the primary drivers in this initial stage, while other environmental factors contributed less significantly. Factors with cumulative importance $\geq$ 90% during this stage included U10, SKT, MWP, and V10.

https://doi.org/10.3390/rs18020194

In the OP, the driving mechanism shifted from a single nutrient focus to a more diversified co-driving pattern, evolving into a "Light–Temperature–Nutrient Synergistic Mode." Although nutrients remained influential, their dominance decreased. PO_4 (0.0040) and NO_3 (0.0018) continued as the top two dominant factors, but their importance values declined compared to the IOP stage. Meanwhile, Sea Surface Temperature (SKT) gained prominence, with its importance rising substantially from fifth place (0.0005) in IOP to third (0.0010), making it a key driving factor in this stage. The importance of Solar Radiation (SSR) also increased significantly, joining the main driving factors. The factors with cumulative importance $\geq 90\%$ included PO_4, NO_3, SKT, U10, SI, and SSR.

During DOP, the system entered "System Decline Mode," marked by a significant reduction in the driving system's intensity and another round of adjustments in the main driving factors. The influence of core drivers notably decreased, with importance metrics for all factors dropping by an order of magnitude. The most crucial factor, PO_4, fell to 0.0013, less than one-third of its value during the IOP stage. Nutrients retained their core driving position, albeit with diminished impact; PO_4 and NO_3 still ranked as the top two factors, but their absolute controlling power was significantly reduced. The roles of light and wind fields remained relatively stable: U10 held its third-place ranking, while SSR's importance rose to fourth place, suggesting that the influence of transient environmental conditions became more prominent amid the overall biomass decline. The factors contributing to cumulative importance of $\geq 90\%$ were PO_4, NO_3, SSR, U10, SKT, TP, and MWP.

4. Discussion

4.1. Determination of Impact Periods and Comparison with Spatio-Temporal Background Regions

4.1.1. Determination of the Impact Period of Various Factors on Ulva Prolifera Disasters

Determining suitable accumulation and lag periods is crucial for analyzing the temporal effects on Ulva prolifera disasters. This study assessed these effects separately, using the maximum period length from each as the upper limit for the final analysis range, with the global average R^2 serving as the evaluation metric. Specifically, this study examined the average R^2 under scenarios of accumulation-only and lag-only within the Ulva disaster area from 2008 to 2022, covering a 5-year span (~230 8-day periods). This analysis aimed to identify the parameter ranges for i and j based on R^2 variation patterns.

Figure 11 illustrates the average R^2 of cumulative effects for environmental factors over periods ranging from 0 to 230. As the accumulation period extends, R^2 typically declines, with most peak values appearing within the first year. Consequently, the accumulation period j was established at 46 (approximately 1 year, k = 46).

Figure 12 demonstrates the lag effect, revealing distinct periodic characteristics. Although the interannual R^2 fluctuations follow a consistent trend, their amplitudes differ. To evaluate trends, the annual maximum R^2 was fitted: all factors, except for PO_4, which decreased, showed a slow increase (<0.01/yr). Despite SI having the highest growth rate (0.0136/yr), its overall level remained low. Consequently, the lag period i was set to 0–46 periods (approximately 1 year).

4.1.2. Comparison with Temporal Effects in Spatio-Temporal Background Regions

To pinpoint specific temporal effects during green tide outbreaks, this study established spatio-temporal background fields for comparison with the Disaster Occurrence Period (DIP). The Spatial Reference Area (SRA) included regions free from disasters during the same timeframe, while the Temporal Reference Period (TRP) encompassed historical non-outbreak months (May–August, 2002–2007). By comparing temporal parameters, such as optimal lag/accumulation periods and R^2, between the DIP and these reference fields, we aimed to uncover effect specificity and evolution. This analysis supports further investigation into the mechanisms driving these disasters.

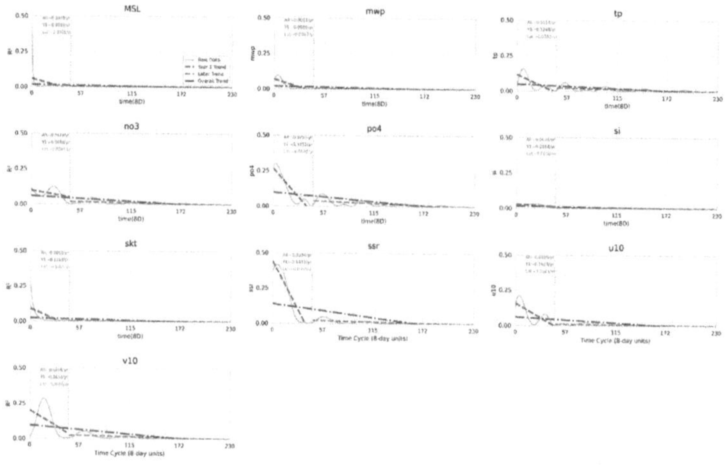

Figure 11. Analysis of Average R^2 under Different Temporal Accumulation Periods.

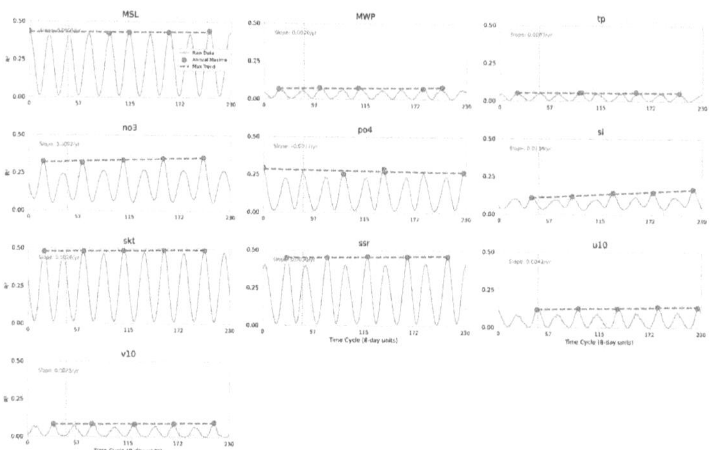

Figure 12. Analysis of Average R^2 under Different Temporal Lag Periods.

As shown in Figure 13, the temporal comparison reveals that TRP has a significantly higher explanatory power (R^2: 0.43–0.59) compared to DIP (R^2: 0.17–0.24), along with longer average lag and accumulation times. Notably, nutrients such as NO_3, PO_4, and SI exhibit the weakest explanatory power in DIP but the strongest in TRP. This suggests that rapid nutrient consumption and dynamic algal growth during outbreaks diminish the instantaneous correlations between nutrients and biomass.

https://doi.org/10.3390/rs18020194

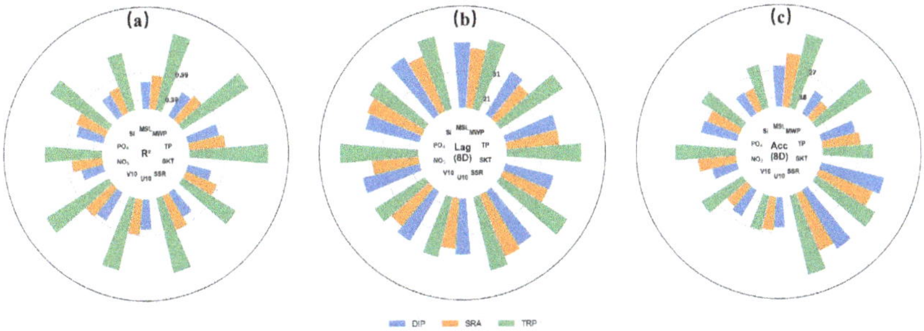

Figure 13. Comparison of Temporal Effects in Spatio-Temporal Background Fields: (**a**) R^2, (**b**) Lag, (**c**) Acc.

While the spatial differences in R^2 between DIP and SRA are minimal, the key temporal parameters show clear distinctions. In DIP, SKT and SSR have the longest accumulation times, at 22.66 and 23.52 8-day periods, respectively, surpassing those in SRA. Additionally, nutrient accumulation times in DIP are generally longer, such as for PO_4, which is 12.84 8-day periods compared to 11.93 8-day periods in SRA. This suggests that the sustained presence of green tides relies on prolonged environmental stability.

In summary, the comparison with the SRA demonstrates that green tide outbreaks require not only sustained light and heat conditions but also a continuous nutrient supply, as evidenced by the extended accumulation periods of these factors in the disaster area. Meanwhile, the comparison with the TRP reveals that the outbreak process itself disrupts the pre-existing ecological equilibrium in the region, leading to a substantial weakening of the statistical association between environmental parameters and FAI. This is clearly demonstrated by the significantly lower explanatory power of environmental factors during the DIP compared to the reference periods.

4.2. Comprehensive Analysis of Ulva Prolifera Outbreaks Using SHAP: From Environmental History to Temporal Relay

Drawing on the importance ranking of environmental factors (Figure 8), SHAP dependence plots (Figures 9 and A6–A8), and the monthly variation patterns of environmental factors (Figure 1b–f), this study develops an integrated model. This model combines "environmental history" and "temporal relay" to systematically explain the full ecological mechanism behind Ulva prolifera disasters in the South Yellow Sea.

4.2.1. DIP Overall Phase: Nutrient-Dominated Environmental Integration Driving

The comprehensive analysis of the DIP reveals a core pattern in its driving mechanism: Ulva prolifera outbreaks are primarily driven by nutrient accumulation from earlier periods, with environmental factors serving as secondary modulators. Nutrients from these earlier periods establish the foundational conditions for disaster occurrence through "delayed effects." Meanwhile, the Ulva population responds to these historical environmental signals via its physiological traits, which ultimately manifest as the biomass dynamics we observe.

The significant lag effects of nutrients (PO_4, NO_3, SI), averaged over 20 8-day periods, reveal that the scale and timing of disasters from May to August are heavily influenced by nutrient conditions from the preceding autumn and winter (Table 2). Multiple regression analysis shows that nutrient variables alone account for 86.8% of the variation in disaster scale and 56.1% in the timing of the first outbreak. Adding light and heat factors slightly boosts the explanatory power to 87.1% and 59.6%, respectively (Figure 14). These findings confirm a "nutrient-dominated, light-heat fine-tuning" mechanism, where nutrient accu-

https://doi.org/10.3390/rs18020194

mulation is nearly sufficient for disaster initiation, while light and thermal conditions play a secondary but noticeable role in adjusting the exact timing of outbreaks.

Table 2. Comparison of Key Environmental Factors Before and After Temporal Adjustment in DIP, IOP, OP, and DOP.

Phase	Dominant Factor	Temporal Window Before Adjustment	Effective Historical Time Window After Adjustment
DIP	PO_4	1 May–31 August	10 December (Previous Year)–8 April
	NO_3	1 May–31 August	29 November (Previous Year)–8 May
	SI	1 May–31 August	15 November (Previous Year)–8 May
	SKT	1 May–31 August	9 December (Previous Year)–9 April
	SSR	1 May–31 August	22 December (Previous Year)–29 April
IOP	PO_4	1 May–10 June	6 July–6 October (Previous Year)
	NO_3	1 May–10 June	29 May–14 September (Previous Year)
	SSR	1 May–10 June	10 September (Previous Year)–22 February
OP	PO_4	10 June–10 July	5 August–2 November (Previous Year)
	SKT	10 June–10 July	28 February–9 April
	SSR	10 June–10 July	5 January–29 March
DOP	PO_4	10 July–31 August	7 October–31 December (Previous Year)
	NO_3	10 July–31 August	25 December–10 March
	SI	10 July–31 August	2 December–28 December (Previous Year)

Note: This table contrasts nominal disaster periods with the actual influential time windows of environmental factors, taking temporal effects into account. The adjustment indicates that ecological impacts during disaster stages are mainly influenced by environmental conditions from earlier historical periods, rather than by concurrent conditions.

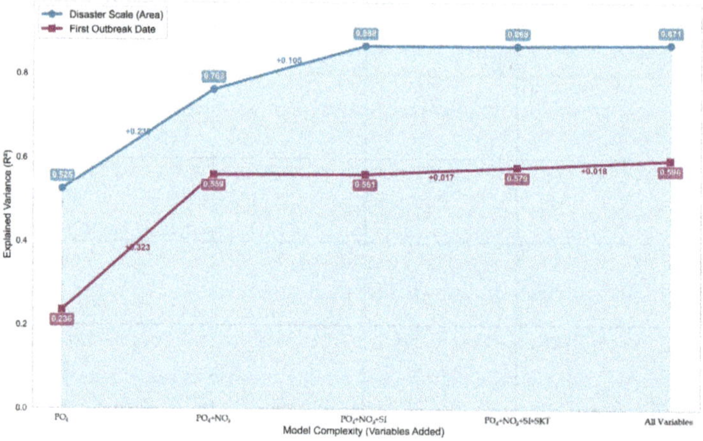

Figure 14. Changes in the explanatory power of core DIP environmental variables concerning the scale and timing of Ulva prolifera outbreaks. A statistical analysis was performed using multiple linear regression to assess changes in the coefficient of determination, R^2, between the annual distribution area/first occurrence date and the monthly mean values of environmental variables within the pre-adjusted temporal windows. The first occurrence date information was sourced from previous research [9].

The "nutrient-dominated, delayed-effect" mechanism identified in this study is highly consistent with, yet extends, existing research on nutrient sources and transport pathways in the South Yellow Sea by explicitly introducing a temporal dimension to the source–sink framework. Previous studies have demonstrated that nutrient release and associated micro-propagules from Pyropia aquaculture activities in the Subei Shoal constitute a major initial

https://doi.org/10.3390/rs18020194

source of green tides in the Yellow Sea [49,50], and that intensified coastal human activities and early raft recycling have contributed to the increasing frequency and magnitude of recent outbreaks [50]. By quantitatively revealing significant cross-seasonal lag effects of nutrients—averaging approximately 20 eight-day periods—this study provides direct evidence that these land-based and aquaculture-derived nutrients accumulate in the water column or sediments during autumn and winter, and are subsequently resuspended, transported, or biologically activated through spring hydrodynamic processes [51], thereby exerting a decisive influence on the scale of spring green tide outbreaks. This finding complements existing studies that primarily emphasize spatial source–sink correspondence, but lack explicit quantification of temporal linkages. Furthermore, the secondary modulating role of light and thermal conditions (SST and SSR) identified here is consistent with previous remote sensing–based observations, which suggest that sea surface temperature and solar radiation define favorable growth windows rather than primary material constraints [51,52]. Collectively, these results not only statistically validate the hypothesis that historical nutrient accumulation forms the material basis of large-scale Ulva prolifera outbreaks, but also precisely quantify its effective lag period and contribution weight. This advancement elevates the nutrient-driven mechanism from qualitative inference to a time-explicit, quantitatively constrained framework, providing a direct theoretical foundation for early-warning models based on historical nutrient loading.

The monthly variation diagram of key environmental factors (Figure 1b–f) indicates that nutrient levels remain high in autumn and continue through winter, forming the material basis for their delayed effects. While seasonal changes in SKT and SSR align with algal physiological rhythms, their direct impact on the scale of disasters is limited. The recovery of SKT and SSR beginning in January primarily offers favorable growth conditions rather than acting as crucial energy sources.

The nonlinear relationships among various factors highlight the ecological threshold characteristics of their driving effects (Figure 7). The negative correlation between nutrients and FAI supports the classic process of "rapid consumption leading to concentration depletion." This consumption-inhibition pattern forms the core negative feedback mechanism in disaster development. The "unimodal" curve of SKT and the "intermediate promotion" pattern of SSR illustrate the physiological adaptation window of Ulva. However, these environmental modulation effects become secondary when nutrients are abundant. These nonlinear responses align with the monthly variation patterns of environmental factors: during the DIP, SKT remains within the optimal range, SSR falls within the growth-promoting range, and nutrients maintain a dynamic balance between previous accumulation and current consumption. This matches the "nutrient-dominated, environmental synergy" signal revealed by SHAP analysis.

4.2.2. Internal DIP Phases: Temporal Relay of Driving Factors

Within the broader DIP framework, the three sub-phases clearly demonstrate "temporal relay" characteristics (Table 2). The IOP functions as a "nutrient initiation mode," with PO_4, SI, and NO_3 identified as the three most critical factors (Figure A6). The relevant environmental data from the preceding summer and autumn suggest that nutrient conditions during this time are crucial precursors to triggering disasters.

During the OP, the system transitions to a "light–temperature–nutrient synergy mode," where the significance of SKT and SSR markedly increases. The influence of SKT changes to a negative linear relationship (Figure A7), indicating stress from high temperatures, with an effective window from late February to early April. Simultaneously, SSR data from January to March underscore the critical role of winter-spring radiation accumulation in driving scale expansion.

https://doi.org/10.3390/rs18020194

In the DOP, the system enters a "system decline mode," characterized by the general decay of all driving factors. Nutrients, previously negatively correlated, now become indicators of decomposition and re-release. This phase highlights significant differences in the influence of various factors, underscoring the complexity of the system collapse phase.

4.3. *Implications for Green Tide Prediction and Management Practices*

This study systematically uncovers the substantial accumulation and lag effects of environmental factors on green tide outbreaks, identifying the distinctive "scale separation" and "stage relay" in the driving mechanisms. These findings offer crucial insights for enhancing existing prediction models and optimizing control strategies, thus promoting a paradigm shift in management from passive response to proactive foresight and from experience-based to mechanism-driven approaches.

4.3.1. Implications for Prediction and Early Warning Systems

The analysis indicates the necessity of establishing a medium to long-term risk warning mechanism, based on the critical role of nutrients during the early stages of disaster development. Findings from the DIP analysis reveal that nutrient accumulation from the previous autumn and winter serves as a crucial prerequisite for outbreaks. It is thus recommended that a monthly scale early warning model be developed, focusing on the temporal accumulation effects of key nutrients, particularly phosphate (PO_4) and nitrate (NO_3). By monitoring whether cumulative nutrient levels during specific historical periods exceed ecological thresholds, early identification of outbreak risks can be achieved, allowing for timely preventive measures. Additionally, a stage-adaptive prediction framework should be implemented as the disaster evolves. Given the distinct dominant drivers across different disaster phases, dynamic forecasting models capable of automatically prioritizing predictors based on the target lead time are essential. For medium- to long-term forecasts, models should emphasize the cumulative effects of key nutrients originating from the previous autumn and winter. For short-term forecasts approaching the outbreak season, models should further integrate real-time environmental factors such as sea surface temperature and solar radiation to improve predictions of outbreak timing and intensity. This temporally adaptive framework is expected to significantly enhance forecasting accuracy.

4.3.2. Implications for Whole-Process Targeted Control Strategies

The findings can also inform the optimization of targeted control strategies throughout the disaster cycle. During the IOP, enforcing land–sea integrated source control strategies is crucial. Given the significant role of nutrients, particularly phosphate (PO_4), efforts must focus on reducing phosphorus pollutants within their effective influence period. Precision emission reduction in river basins and strict control of land-based nutrient inputs are essential to fundamentally mitigate the core drivers. As the event progresses to the OP, prediction-based precision salvage operations are advised. Management resources should then focus on efficient physical removal. By utilizing real-time monitoring and short-term forecasting models that account for both nutrient depletion status and environmental fields, accurate prediction of Ulva aggregation areas is possible. This enables the dynamic deployment of salvage resources to target and clear high-biomass core areas, thereby controlling the disaster's impact at minimal cost. Finally, during the DOP, the focus should shift to activating secondary disaster risk prevention plans. The priority transitions from controlling the green tide itself to preventing subsequent ecological disasters, such as bottom water hypoxia and hydrogen sulfide production. This is achieved through early warnings for algal accumulation zones and timely cleanup of decaying algae on shores and beaches to ensure public environmental and ecological security.

https://doi.org/10.3390/rs18020194

5. Conclusions

The long-term spatiotemporal evolution of Ulva prolifera green tides in the South Yellow Sea exhibits pronounced interannual and seasonal variability. Both outbreak intensity and spatial extent differ markedly among years, with 2019 showing the largest affected area (6.12×10^4 km^2). Seasonally, the green tide follows a coherent progression: algal patches initially emerge offshore of Jiangsu Province in May, expand northward and reach their maximum extent between June and July, and subsequently weaken and dissipate in August. This spatiotemporal evolution reflects the combined influences of environmental conditions and regional hydrodynamic processes.

This study demonstrates that environmental drivers exert substantial temporal cumulative and lagged effects on Ulva prolifera development, with a coexisting lag–accumulation mode representing the dominant temporal response pattern. Optimal temporal parameters vary among environmental variables and outbreak stages. Nutrient factors generally exhibit longer lag periods, whereas temperature and solar radiation tend to require longer accumulation durations. These temporal responses also display clear spatial heterogeneity, characterized by shorter effective durations in nearshore waters and progressively longer ones offshore. Such spatiotemporal differences indicate that Ulva prolifera responds to environmental forcing in a stage-dependent and geographically structured manner, rather than instantaneously.

By integrating temporally adjusted environmental factors, this study reveals a nutrient-dominated driving mechanism, modulated by light and thermal conditions, that operates through a distinct temporal relay across outbreak stages. Nutrient accumulation during the preceding autumn–winter period explains much of the interannual variability in outbreak magnitude and timing, forming the fundamental ecological basis for bloom initiation. During the initial outbreak stage, nutrients act as the primary drivers, whereas the peak stage is jointly regulated by nutrients, temperature, and solar radiation. In the decline stage, the influence of all drivers weakens. This mechanistic understanding provides a solid theoretical foundation for improving predictive modeling and developing targeted, stage-specific management strategies for green tide disasters.

Author Contributions: Conceptualization, Z.T. and J.Z.; methodology, Z.T.; software, H.Z.; validation, W.L. and B.D.; formal analysis, Z.T.; investigation, Z.L.; resources, Y.Z. and H.Z.; data curation, L.L.; writing—original draft preparation, Z.T.; writing—review and editing, J.Z.; visualization, Z.T. and Z.L.; supervision, J.Z.; project administration, L.L.; funding acquisition, X.Y. All authors have read and agreed to the published version of the manuscript.

Funding: This research was funded by Key Laboratory of Marine Ecological Monitoring and Restoration Technology, Ministry of Natural Resources, grant number: MEMRT202313; Hainan Province's Key Research and Development Project, grant number: ZDYF2023GXJS023; National Key R&D Program of China, grant number: 2024YFC3109000; Hainan Province Science and Technology Special Fund, grant number: SOLZSKY2025008.

Data Availability Statement: The original contributions presented in the study are included in the article, further inquiries can be directed to the corresponding author.

Acknowledgments: This study utilized multiple data sources, including MODIS-Terra and MODIS-Aqua satellite data spanning 2003–2022 obtained from the NASA Goddard Space Flight Center Ocean Biology Processing Group, as well as environmental and nutrient variables derived from the ERA5 reanalysis dataset provided by the European Centre for Medium-Range Weather Forecasts (ECMWF) and the global ocean biogeochemical hindcast from the Copernicus Marine Service.

Conflicts of Interest: The authors declare no conflicts of interest.

https://doi.org/10.3390/rs18020194

Appendix A

Appendix A presents the spatial distribution of temporal parameters for ten environmental factors across five phases: EP, DIP, IOP, OP, and DOP.

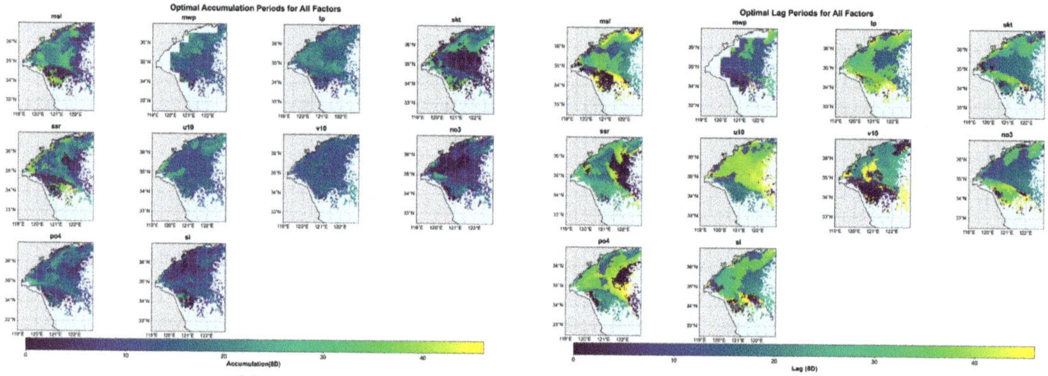

(a) (b)

Figure A1. Illustrates the spatial distribution of temporal parameters for environmental factors during the EP: (**a**) shows the average accumulation periods, and (**b**) depicts the average lag periods. Both periods are measured in 8-day intervals (8D).

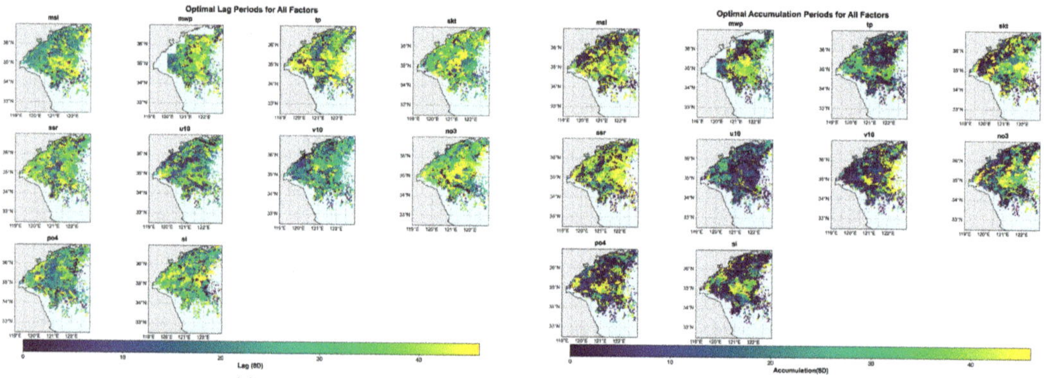

(a) (b)

Figure A2. Illustrates the spatial distribution of temporal parameters for environmental factors during the DIP: (**a**) depicts the average accumulation periods, and (**b**) shows the average lag periods. Both periods are measured in 8-day intervals (8D).

https://doi.org/10.3390/rs18020194

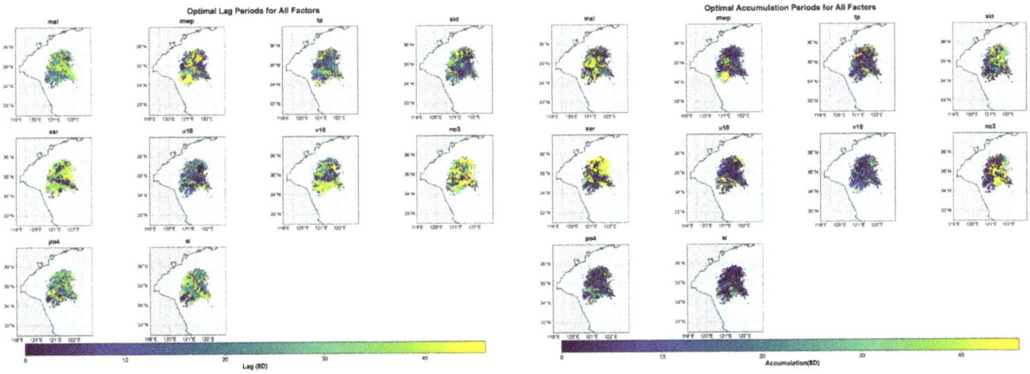

(a) (b)

Figure A3. Illustrates the spatial distribution of temporal parameters for environmental factors during the IOP. Panel (**a**) shows the average accumulation periods, while panel (**b**) depicts the average lag periods. Both periods are measured in 8-day intervals (8D).

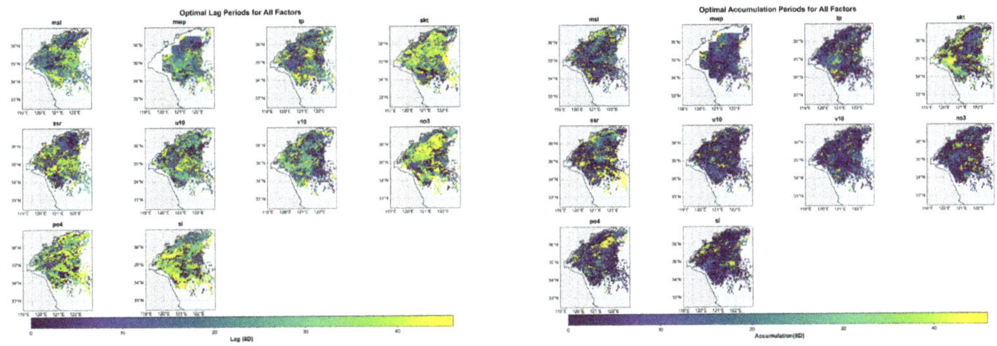

(a) (b)

Figure A4. Illustrates the spatial distribution of temporal parameters for environmental factors during the OP: (**a**) average accumulation periods and (**b**) average lag periods. Both periods are measured in 8-day intervals (8D).

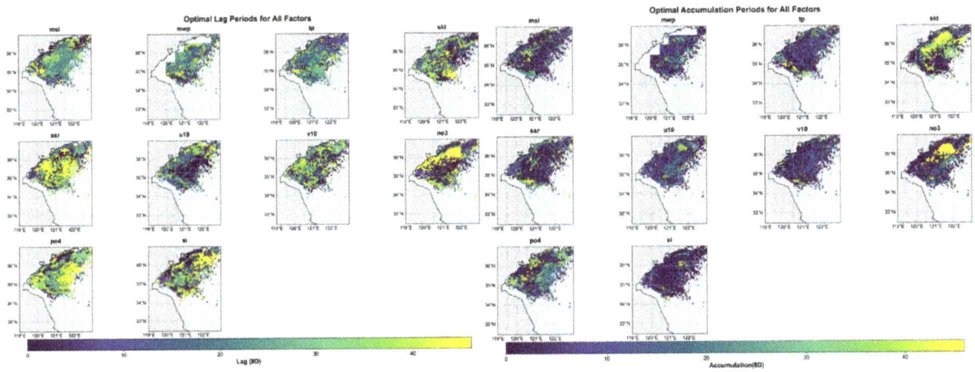

(a) (b)

Figure A5. Illustrates the spatial distribution of temporal parameters for environmental factors during the DOP: (**a**) shows the average accumulation periods, and (**b**) depicts the average lag periods. Both periods are measured in 8-day intervals (8D).

https://doi.org/10.3390/rs18020194

Appendix B

Appendix B presents the SHAP dependence plots for various environmental factors during the three phases: IOP, OP, and DOP.

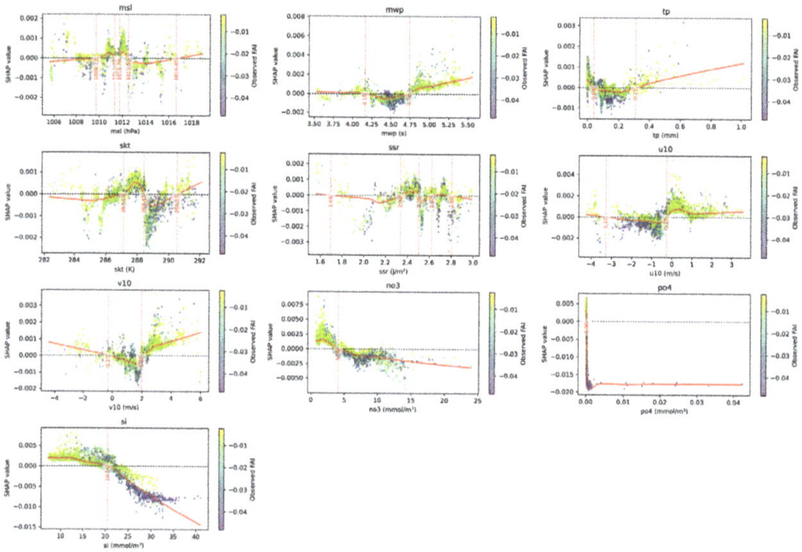

Figure A6. SHAP Dependence Plots for Environmental Factors in IOP.

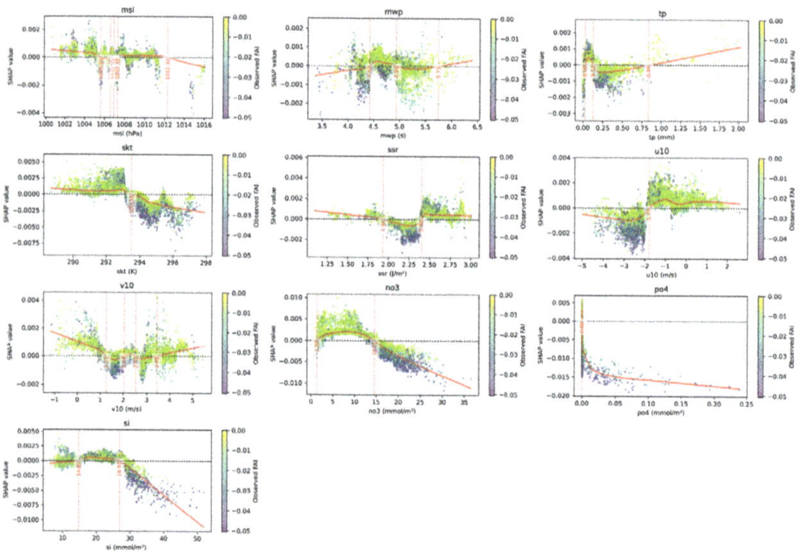

Figure A7. SHAP Dependence Plots for Environmental Factors in OP.

https://doi.org/10.3390/rs18020194

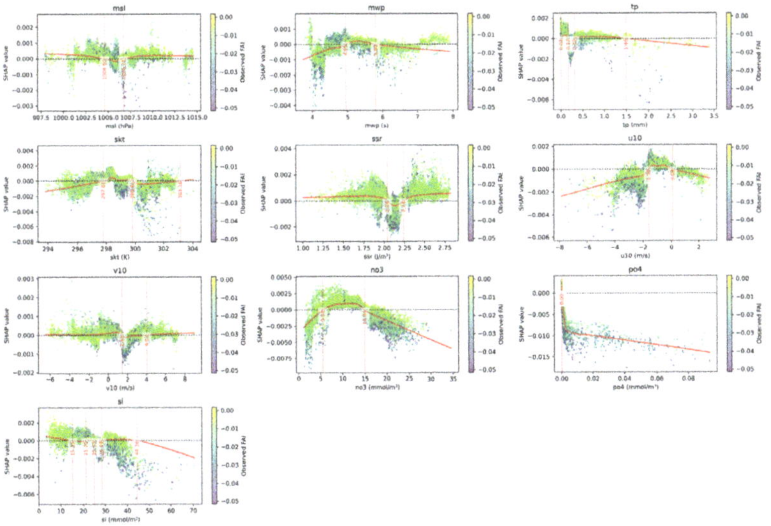

Figure A8. SHAP Dependence Plots for Environmental Factors in DOP.

References

1. Zeng, Y.; Yang, X.; Xia, Z.; Chen, R.; He, F.; Zhang, J.; He, P. Review of Allelopathy in Green Tides: The Case of Ulva prolifera in the South Yellow Sea. *Biology* **2024**, *13*, 456. [CrossRef]
2. Ye, N.; Zhang, X.; Mao, Y.; Liang, C.; Xu, D.; Zou, J.; Zhuang, Z.; Wang, Q. 'Green tides' are overwhelming the coastline of our blue planet: Taking the world's largest example. *Ecol. Res.* **2011**, *26*, 477–485. [CrossRef]
3. Li, H.; Feng, X.; Xiong, T.; He, C.; Wu, W.; Shi, Q.; Jiao, N.; Zhang, Y. Green Tides Significantly Alter the Molecular Composition and Properties of Coastal DOC and Perform Dissolved Carbon Sequestration. *Environ. Sci. Technol.* **2022**, *57*, 770–779. [CrossRef]
4. Qu, T.; Zhao, X.; Hao, Y.; Zhong, Y.; Guan, C.; Hou, C.; Tang, X.; Wang, Y. Ecological effects of Ulva prolifera green tide on bacterial community structure in Qingdao offshore environment. *Chemosphere* **2020**, *244*, 125477. [CrossRef]
5. Guo, X.; Zhu, A.; Chen, R. China's algal bloom suffocates marine life. *Science* **2021**, *373*, 751. [CrossRef]
6. Pan, X.; Cao, M.; Zheng, L.; Xiao, Y.; Qi, L.; Xing, Q.; Kim, K.; Sun, D.; Wang, N.; Guo, M.; et al. Remote Sensing of Ulva Prolifera Green Tide in the Yellow Sea Using Multisource Satellite Data: Progress and prospects. *IEEE Geosci. Remote Sens. Mag.* **2024**, *12*, 110–131. [CrossRef]
7. Feng, G.; Zeng, Y.; Wang, J.; Dai, W.; Bi, F.; He, P.; Zhang, J. A bibliometric review of Green Tide research between 1995–2023. *Mar. Pollut. Bull.* **2024**, *208*, 116941. [CrossRef]
8. Men, Y.; Liu, Y.; Ma, Y.; Wong, K.P.; Tsou, J.Y.; Zhang, Y. Remote sensing monitoring of green tide disaster using MODIS and GF-1 data: A case study in the Yellow Sea. *J. Mar. Sci. Eng.* **2023**, *11*, 2212. [CrossRef]
9. Zhan, Y.; Qiu, Z.; Wang, Y.; Su, Y.; Li, Y.; Cui, Y.; Qu, S.; Wang, P.; Rong, X. Long-term spatiotemporal characteristics of Ulva prolifera green tide and effects of environmental drivers on its monitoring by satellites: A case study in the yellow sea, China, from 2008 to 2023. *J. Mar. Sci. Eng.* **2024**, *12*, 630. [CrossRef]
10. Tang, J.; Wang, W.; Yang, L.; Qiu, Q.; Lin, M.; Cao, C.; Li, X. Seasonal variation and ecological risk assessment of dissolved organic matter in a peri-urban critical zone observatory watershed. *Sci. Total Environ.* **2020**, *707*, 136093. [CrossRef]
11. Zhang, H.; Su, R.; Shi, X.; Zhang, C.; Yin, H.; Zhou, Y.; Wang, G. Role of nutrients in the development of floating green tides in the Southern Yellow Sea, China, in 2017. *Mar. Pollut. Bull.* **2020**, *156*, 111197. [CrossRef]
12. Le Luherne, E.; Réveillac, E.; Ponsero, A.; Sturbois, A.; Ballu, S.; Perdriau, M.; Le Pape, O. Fish community responses to green tides in shallow estuarine and coastal areas. *Estuarine Coast. Shelf Sci.* **2016**, *175*, 79–92. [CrossRef]
13. Li, Y.; Huang, H.-J.; Li, H.; Liu, J.; Yang, W. Genetic diversity of Ulva prolifera population in Qingdao coastal water during the green algal blooms revealed by microsatellite. *Mar. Pollut. Bull.* **2016**, *111*, 237–246. [CrossRef]
14. Sun, Y.; Xia, Z.; Cao, X.; Tong, Y.; He, R.; Fu, M.; Sun, J.; Xu, H.; Xia, J.; Liu, J.; et al. A mixed acid treatment for the prevention of Ulva prolifera attachment to Neopyropia aquaculture rafts: Laboratory experimentation. *Mar. Pollut. Bull.* **2022**, *184*, 114134. [CrossRef]
15. Liu, X.; Wang, Z.; Fan, S.; Xiao, J.; Li, Y.; Zhang, X.; Li, R.; Wang, X. The current situations of green macroalgae and micro-propagules in Pyropia aquaculture of the Subei shoal in spring of 2013. *Acta Oceanol. Sin.* **2016**, *35*, 94–100. [CrossRef]

https://doi.org/10.3390/rs18020194

16. Gao, G.; Clare, A.S.; Rose, C.; Caldwell, G.S. Ulva rigida in the future ocean: Potential for carbon capture, bioremediation and biomethane production. *GCB Bioenergy* **2017**, *10*, 39–51. [CrossRef]
17. Kang, E.J.; Han, A.R.; Kim, J.H.; Kim, I.N.; Lee, S.; Min, J.O.; Nam, B.R.; Choi, Y.J.; Edwards, M.S.; Diaz-Pulido, G.; et al. Evaluating bloom potential of the green-tide forming alga Ulva ohnoi under ocean acidification and warming. *Sci. Total Environ.* **2021**, *769*, 144443. [CrossRef]
18. Zhao, X.; Zhong, Y.; Zhang, H.; Qu, T.; Jiang, Y.; Tang, X.; Wang, Y. Cooperation between photosynthetic and antioxidant systems: An important factor in the adaptation of Ulva prolifera to abiotic factors on the sea surface. *Front. Plant Sci.* **2019**, *10*, 648. [CrossRef]
19. Li, J.; Song, X.; Fan, X.; Yu, Z. Flocculation of Ulva microscopic propagules using modified clay: A mesocosm experiment. *J. Oceanol. Limnol.* **2020**, *38*, 1283–1291. [CrossRef]
20. Li, J.; Song, X.; Zhang, Y.; Pan, J.; Yu, Z. An investigation of the space distribution of Ulva microscopic propagules and ship-based experiment of mitigation using modified clay. *Mar. Pollut. Bull.* **2017**, *117*, 247–254. [CrossRef]
21. Rezayian, M.; Niknam, V.; Ebrahimzadeh, H. Oxidative damage and antioxidative system in algae. *Toxicol. Rep.* **2019**, *6*, 1309–1313. [CrossRef] [PubMed]
22. Fu, M.; Cao, S.; Li, J.; Zhao, S.; Liu, J.; Zhuang, M.; Qin, Y.; Gao, S.; Sun, Y.; Kim, J.K.; et al. Controlling the main source of green tides in the Yellow Sea through the method of biological competition. *Mar. Pollut. Bull.* **2022**, *177*, 113561. [CrossRef]
23. Li, C.; Tang, Y.; Sun, W.; Xia, J.; Xia, Z.; Zhang, J.; He, P.; Liu, J.; Zhao, S. Physiological responses of Ampithoe valida and its feeding potential on Ulva prolifera. *Mar. Environ. Res.* **2023**, *186*, 105942. [CrossRef] [PubMed]
24. Ruiz, J.M. Bivalves, tributyltin and Green tides: Ecosystem-level impact? *Mar. Ecol.* **1999**, *20*, 1–9. [CrossRef]
25. Geng, X.; Li, H.; Wang, L.; Sun, W.; Li, Y. A comprehensive review of remote sensing techniques for monitoring Ulva prolifera green tides. *Front. Mar. Sci.* **2025**, *12*, 1546289. [CrossRef]
26. Zhang, B.; Guo, J.; Li, Z.; Cheng, Y.; Zhao, Y.; Boota, M.W.; Zhang, Y.; Feng, L. Identifying the spatio-temporal variations of Ulva prolifera disasters in all life cycle. *J. Water Clim. Change* **2021**, *13*, 629–644. [CrossRef]
27. Zhang, G.; Wu, M.; Zhou, M.; Zhao, L. The seasonal dissipation of Ulva prolifera and its effects on environmental factors: Based on remote sensing images and field monitoring data. *Geocarto Int.* **2020**, *37*, 860–878. [CrossRef]
28. Xu, Q.; Zhang, H.; Ju, L.; Chen, M. Interannual variability of Ulva prolifera blooms in the Yellow Sea. *Int. J. Remote Sens.* **2014**, *35*, 4099–4113. [CrossRef]
29. Jin, S.; Liu, Y.; Sun, C.; Wei, X.; Li, H.; Han, Z. A study of the environmental factors influencing the growth phases of Ulva prolifera in the southern Yellow Sea, China. *Mar. Pollut. Bull.* **2018**, *135*, 1016–1025. [CrossRef]
30. Anderegg, W.R.L.; Schwalm, C.R.; Biondi, F.; Camarero, J.J.; Koch, G.W.; Litvak, M.; Ogle, K.; Shaw, J.D.; Shevliakova, E.; Williams, A.P.; et al. Pervasive drought legacies in forest ecosystems and their implications for carbon cycle models. *Science* **2015**, *349*, 528–532. [CrossRef]
31. Nemani, R.R.; Keeling, C.D.; Hashimoto, H.; Jolly, W.M.; Piper, S.C.; Tucker, C.J.; Myneni, R.B.; Running, S.W. Climate-driven increases in global terrestrial net primary production from 1982 to 1999. *Science* **2003**, *300*, 1560–1563. [CrossRef]
32. de Jong, R.; Schaepman, M.E.; Furrer, R.; De Bruin, S.; Verburg, P.H. Spatial relationship between climatologies and changes in global vegetation activity. *Glob. Change Biol.* **2013**, *19*, 1953–1964. [CrossRef]
33. Wen, Y.; Liu, X.; Pei, F.; Li, X.; Du, G. Non-uniform time-lag effects of terrestrial vegetation responses to asymmetric warming. *Agric. For. Meteorol.* **2018**, *252*, 130–143. [CrossRef]
34. Wu, D.H.; Zhao, X.; Liang, S.L.; Zhou, T.; Huang, K.C.; Tang, B.J.; Zhao, W.Q. Time-lag effects of global vegetation responses to climate change. *Glob. Change Biol.* **2015**, *21*, 3520–3531. [CrossRef]
35. Zhang, G.; Wu, M.; Wei, J.; He, Y.; Niu, L.; Li, H.; Xu, G. Adaptive threshold model in google earth engine: A case study of Ulva prolifera extraction in the south yellow sea, China. *Remote Sens.* **2021**, *13*, 3240. [CrossRef]
36. Zhang, H.; Wang, G.; Zhang, C.; Su, R.; Shi, X.; Wang, X. Characterization of the development stages and roles of nutrients and other environmental factors in green tides in the Southern Yellow Sea, China. *Harmful Algae* **2020**, *98*, 101893. [CrossRef] [PubMed]
37. Lie, H.J.; Cho, C.H.; Lee, S. Tongue-shaped frontal structure and warm water intrusion in the southern Yellow Sea in winter. *J. Geophys. Res. Ocean.* **2009**, *114*, C01003. [CrossRef]
38. Wei, Q.; Wang, B.; Yao, Q.; Fu, M.; Sun, J.; Xu, B.; Yu, Z. Hydro-biogeochemical processes and their implications for Ulva prolifera blooms and expansion in the world's largest green tide occurrence region (Yellow Sea, China). *Sci. Total Environ.* **2018**, *645*, 257–266. [CrossRef] [PubMed]
39. Yuan, D.; Li, Y.; Wang, B.; He, L.; Hirose, N. Coastal circulation in the southwestern Yellow Sea in the summers of 2008 and 2009. *Cont. Shelf Res.* **2017**, *143*, 101–117. [CrossRef]
40. Zhang, S.; Wang, Q.; Lü, Y.; Cui, H.; Yuan, Y. Observation of the seasonal evolution of the Yellow Sea Cold Water Mass in 1996–1998. *Cont. Shelf Res.* **2008**, *28*, 442–457. [CrossRef]

https://doi.org/10.3390/rs18020194

41. Naimie, C.E.; Blain, C.A.; Lynch, D.R. Seasonal mean circulation in the Yellow Sea—A model-generated climatology. *Cont. Shelf Res.* **2001**, *21*, 667–695. [CrossRef]
42. Lü, X.; Qiao, F.; Xia, C.; Wang, G.; Yuan, Y. Upwelling and surface cold patches in the Yellow Sea in summer: Effects of tidal mixing on the vertical circulation. *Cont. Shelf Res.* **2010**, *30*, 620–632. [CrossRef]
43. Qi, L.; Hu, C.; Xing, Q.; Shang, S. Long-term trend of Ulva prolifera blooms in the western Yellow Sea. *Harmful Algae* **2016**, *58*, 35–44. [CrossRef] [PubMed]
44. Ding, Y.; Li, Z.; Peng, S. Global analysis of time-lag and-accumulation effects of climate on vegetation growth. *Int. J. Appl. Earth Obs. Geoinf.* **2020**, *92*, 102179. [CrossRef]
45. Hu, C. A novel ocean color index to detect floating algae in the global oceans. *Remote Sens. Environ.* **2009**, *113*, 2118–2129. [CrossRef]
46. Xiao, Y.; Liu, G.; Hu, G.; Zou, D.; Li, R.; Du, E.; Wu, T.; Wu, X.; Zhao, G.; Zhao, Y.; et al. Mapping Mountain Permafrost via GPR-Augmented Machine Learning in the Northeastern Qinghai–Tibet Plateau. *Remote Sens.* **2025**, *17*, 2015. [CrossRef]
47. Xia, L.; Tan, H.; Zhang, J.; Yang, K.; Teng, C.; Huang, K.; Yang, J.; Cheng, T. Remote Sensing and Machine Learning Uncover Dominant Drivers of Carbon Sink Dynamics in Subtropical Mountain Ecosystems. *Remote Sens.* **2025**, *17*, 2843. [CrossRef]
48. Xue, Z.; Diao, S.; Yang, F.; Fei, L.; Wang, W.; Fang, L.; Liu, Y. Identifying Forest Drought Sensitivity Drivers in China Under Lagged and Accumulative Effects via XGBoost-SHAP. *Remote Sens.* **2025**, *17*, 2903. [CrossRef]
49. Xing, Q.; An, D.; Zheng, X.; Wei, Z.; Wang, X.; Li, L.; Tian, L.; Chen, J. Monitoring seaweed aquaculture in the Yellow Sea with multiple sensors for managing the disaster of macroalgal blooms. *Remote Sens. Environ.* **2019**, *231*, 111279. [CrossRef]
50. Fan, J.; Ma, Y.; Liang, T.; Cui, G. Multi-source remote sensing big data mining reveals cross-regional correlation between aquaculture and *Enteromorpha* disaster outbreaks. *Geo-Spat. Inf. Sci.* **2024**, *28*, 1072–1084. [CrossRef]
51. Pan, Y.; Ding, D.; Li, G.; Liu, X.; Liang, J.; Wang, X.; Liu, S.; Shi, J. Potential Temporal and Spatial Trends of Oceanographic Conditions with the Bloom of *Ulva Prolifera* in the West of the Southern Yellow Sea. *Remote Sens.* **2021**, *13*, 4406. [CrossRef]
52. Zheng, L.; Wu, M.; Cui, Y.; Tian, L.; Yang, P.; Zhao, L.; Xue, M.; Liu, J. What causes the great green tide disaster in the South Yellow Sea of China in 2021? *Ecol. Indic.* **2022**, *140*, 108988. [CrossRef]

https://doi.org/10.3390/rs18020194

remote sensing

MDPI

Article

Recovery of Water Quality and Detection of Algal Blooms in Lake Villarrica through Landsat Satellite Images and Monitoring Data

Lien Rodríguez-López [1,*], Iongel Duran-Llacer [2], Lisandra Bravo Alvarez [3], Andrea Lami [4] and Roberto Urrutia [5]

[1] Facultad de Ingeniería, Arquitectura y Diseño, Universidad San Sebastián, Lientur 1457, Concepción 4030000, Chile
[2] Hémera Centro de Observación de la Tierra, Facultad de Ciencias, Ingeniería y Tecnología, Universidad Mayor, Camino La Pirámide 5750, Santiago 8580745, Chile
[3] Department of Electrical Engineering, Universidad de Concepción, Edmundo Larenas 219, Concepción 4030000, Chile
[4] Institute of Water Research IRSA, 70132 Sezione di Verbania, Italy
[5] Facultad de Ciencias Ambientales, Universidad de Concepción, Concepción 4070386, Chile
* Correspondence: lien.rodriguez@uss.cl

Abstract: Phytoplankton is considered a strong predictor of the environmental quality of lakes, while Chlorophyll-a is an indicator of primary productivity. In this study, 25 LANDSAT images covering the 2014–2021 period were used to predict Chlorophyll-a in the Villarrica lacustrine system. A Chlorophyll-a recovery algorithm was calculated using two spectral indices (FAI and SABI). The indices that presented the best statistical indicators were the floating algal index (R^2 = 0.87) and surface algal bloom index (R^2 = 0.59). A multiparametric linear model for Chlorophyll-a estimation was constructed with the indices. Statistical indicators were used to validate the multiple linear regression model used to predict Chlorophyll-a by means of spectral indices, with the following results: a MBE of −0.136 μg/L, RMSE of 0.055 μg/L, and NRMSE of 0.019%. All results revealed the strength of the model. It is necessary to raise awareness among the population that carries out activities around the lake in order for them to take policy actions related to water resources in this Chilean lake. Furthermore, it is important to note that this study is the first to address the detection of algal blooms in this Chilean lake through remote sensing.

Keywords: water quality; spectral indices; lakes; remote sensing; Chlorophyll-a

1. Introduction

Inland freshwater bodies, such as rivers, lakes, wetlands, and streams, are important aquatic ecosystems [1]. The ecosystem services provided by inland water bodies are extensive, but people have not been able to safeguard them [2]. Human factors including nutrient loading, pollution from industry and agriculture, climate change, sewage discharge, invasive species incursion, and urbanization have triggered the rapid degradation of freshwater systems [3] and water quality [4,5]. Lakes are essential for meeting the necessities of life, including socio-cultural development, scientific advances, environmental balance, industrialization, agricultural activities, economic growth, protection of biological diversity, and finally, the well-being and health of organisms [4,6].

Traditional methods of monitoring water quality parameters, such as chemical, physical, and biological characteristics, consist of measurement by collecting samples in the field and then analyzing them using conventional techniques [7,8]. Even though in situ monitoring is very efficient, it is a time-consuming, costly, and technically deficient process that lacks the necessary human resources and provides trophic status on the day of sampling rather than in real-time [9,10]. In addition, traditional field sampling approaches do not easily detect temporal or spatial fluctuations in water quality, and doing so is essential for a proper evaluation and thorough management of water bodies [10–13].

Eutrophication is a major concern for the lakes around the world [14,15]. It is the biological reaction of lakes to nutrient oversaturation, developing algal productivity at large volumes [16]. The main problem is the increase in nitrogen and phosphorus load, which causes a high biomass of algae, an abundance of microalgae, and the loss of macroalgae [2,17]. Cyanobacteria require temperatures above 20 °C for growth rates to be competitive with eukaryotic phytoplankton taxa and above 25 °C for growth rates to be competitive with diatoms. In addition, they require relatively high irradiances to grow at maximal growth rates [18].

Thus, recently developed remote sensing technologies are increasingly used in the monitoring of various water quality parameters. One problem with detecting the concentration of Chlorophyll-a (Chl-a) in lakes from multispectral satellite data is the influence of atmospheric factors such as aerosols and clouds, which can affect the accuracy of the measurements [19]; another issue is the presence of other substances in the water that can interfere with the signal from Chlorophyll-a, making it difficult to accurately detect and quantify its concentration [20]. There are several approaches that researchers use to address the challenges of detecting Chlorophyll-a concentration from multispectral satellite data. Overall, the solution for detecting Chlorophyll-a concentration from multispectral satellite data depends on the specific challenges faced and the available data and resources, such as with collecting in situ measurements of Chlorophyll-a concentration to validate the accuracy of the satellite data and improve the calibration of the remote sensing algorithms. Remote sensing techniques have recently emerged as successful approaches to monitor algal blooms, allowing for the specific monitoring of phenomena affecting water quality [21]. Different spectral indices have been used to monitor blooms in inland water bodies and are usually based on spectral differences between aquatic vegetation and the bottom in the spectrum of visible radiation and infrared radiation, including near-infrared radiation [22]. Several spectral indices are available, including the normalized difference vegetation index [12], normalized difference turbidity index and green normalized difference vegetation index [23], emergent vegetation spectral index [22], floating algal index [24], and surface algal bloom index [25,26], which have been developed and successfully used in research and the mapping of aquatic plants and monitoring of vegetation or algae growth [12,25]. In this study, we will include most of these indices.

Furthermore, most of the remote sensing research on phenomena such as algal blooms have been conducted in the northern hemisphere, with less work conducted in the southern hemisphere. Investigations of unspecified Chlorophyll-a estimates of algal groups have been conducted with MODIS imagery to analyze the time series in the Great Lakes [27], and Landsat and Sentinel hyperspectral imagery was used in [28,29] to study the spatial variability of harmful algal blooms (cyanobacteria) in the western basin of Lake Erie, while other works such as [30,31] through band combinations detected cyanobacterial blooms in the Baltic Sea.

Lake Villarrica in Chile has attracted considerable attention due to its severe pollution issues from increased nutrients in the lake and algal blooms that have been recorded since 2016 [32]. Lake Villarrica is classified as an oligotrophic lake with a transition to a higher trophic state (mesotrophic). Therefore, the Ministry of Environment of Chile, in compliance with the obligations imposed by the Law on General Bases of the Environment, established a secondary environmental standards quality (hereinafter NSCA) for Lake Villarrica [33] to monitor changes in water quality parameters and the phenomenon of algal blooms. The predominant group of the mentioned blooms is Cyanophyceae, and the most repeated species is *Dolichospermum*. It should be noted that this microalga is toxic and has formed blooms in other Chilean aquatic systems such as the one reported in Lake Laja in the summer of 2018 [12]. In Lake Villarrica, these events have been taking place for more than a decade. The activities that take place in the basin determine the acceleration of natural processes such as the succession of trophic stages or eutrophication and are perhaps a consequence of urban expansion due to tourism [34–36]. The present research aims to: (i) analyze local conditions and water quality through the physical–chemical and biological parameters prevailing in Lake Villarrica during the 2014–2021 period, (ii) use spectral indices to estimate the concentration of Chlorophyll-a and detect algal blooms; and (iii) evaluate the prediction using statistical indicators and estimation maps.

2. Materials and Methods

2.1. Research Area

Villarrica or Mallolafquén (its name in Mapudungun, the native language of Chile's local native communities) is an Andean Lake located approximately at 39°18′S latitude and 72°05′W longitude in the lake district of the Araucanía Region, in Cautín Province [37]. The origin of the lake, like most of the southern lakes, can be traced to a glacial context, with it resulting from damming by a moraine in the last glaciation [38]. The coastal cities of Villarrica and Pucón are considered the main tourist hubs of the country due to the temperature of the water in the summer, from 19 °C to 22 °C, allowing for multiple sports and tourist activities (sport fishing, sailing, kayaking, water skiing, and swimming) [34,35]. The thermal amplitude increases towards the interior and there is rainfall during all seasons of the year, with an average annual rainfall of 1465 mm. Winter frosts are frequent and these cities host tourist attractions associated with snow-related activities due to their proximity to the volcano, which is a major attraction [34,35].

2.2. Field Data Collection

The study period covers the years 2014–2021. The data were collected from the database of the General Water Directorate (DGA) of Chile, as a system is in place for Lake Villarrica in which monitoring has been carried out twice a year in summer and spring from 1986 to the present [39]. Through the redefinition of the minimum network of lakes, Lake Villarrica has been one of the most studied lakes due to its great economic and touristic importance for the country and was one of the first to be included in it. The monitoring includes physical–chemical and biological parameters in 7 spatially distributed stations covering the surveillance zones defined according to the morphometric characteristics of Lake Villarrica (see Table 1). The monitored parameters were surface temperature (thermometry 2250 B standard methods 22 Ed, method used as a reference for temperature analysis according to the NCh 2313 compendium), surface Chlorophyll-a (fluorometric method), determination of total phosphorus (4500 P B Standard Methods 22 Edit. EAM), determination of total nitrogen (4500-N C Standard Methods 22 Edit-EAM), and transparency (measured using SDD). At each of the lake stations, water samples were collected at 5 depths using a 5 L Niskin sampling bottle. Samples were stored and transported in thermally insulated boxes and properly cooled with ice at a temperature of approximately 5 °C for subsequent analysis. Chemical analyses were performed at the chemistry laboratory of the General Water Directorate of Chile (DGA). This laboratory is accredited by the Instituto Nacional de Normalización (National Standards Institute) for the Chilean Standard NCh ISO 17.025 of 2005.

2.3. Raw Satellite Images

A total of 25 multispectral Landsat 8 Collection 2 Level 1 Operational Land Imager images (L8/OLI) were used (see Table 2). Landsat 8 was developed in collaboration with NASA and the United States Geological Survey (USGS) Earth-Explorer (https://earthexplorer.usgs.gov/, accessed on 15 November 2022). This site was used to download the images used in this study, which have a 30-m spatial representation and were accessed on 8 September 2022. The study area of interest (Roi) was covered by paths 232/233 and row 87 of the images. The chosen images were selected according to the following criteria: availability, a low percentage of cloud cover (less than 12%), and closeness to the sampling date (with an interval of ±9 days between satellite overpasses). To mask clouds, shadows, and cirrus clouds, quality assessment (QA) bands were used and analyzed via visual inspection. Only the data extracted from the areas without cloudiness were used.

Table 1. Morphometric characteristics of Lake Villarrica.

Parameter	Unit	Villarrica
Latitude	°	39°11′–39°18′S
Longitude	°	72°05′–72°15′W
Altitude	m.a.s.l.	250
Max. length	km	23
Max. width	km	11
Avg. width	km	7.6
Perimeter	km	71
Surface area	km^2	176
Max. depth	m	165
Average depth	m	120
Volume	km^3	20.9
Drainage area	km^2	2884.15
Avg. drainage/surface area		16.4
Renewal time	years	4

Table 2. Landsat image characteristics and dates of in situ measurements.

L8 Image ID	Path/Row	Year	In Situ Date	Image Date	Days Differences	N Samples
LC08_L1TP_233087_20140210_20200912_02_T1		2014	3, 4 Feb.	10 Feb.	6, 7	14
LC08_L1TP_233087_20141008_20200910_02_T1			6, 7 Oct.	8 Oct.	1, 2	14
LC08_L1TP_233087_20150128_20200909_02_T1		2015	26, 27 Jan.	28 Jan.	0, 1	14
LC08_L1TP_233087_20151011_20200908_02_T1	233/87		19, 20 Oct.	11 Oct.	8, 9	14
LC08_L1TP_233087_20160303_20200907_02_T1		2016	1, 2 Mar.	3 Mar.	1, 2	14
LC08_L1TP_233087_20161013_20200905_02_T1			18, 19 Oct.	13 Oct.	5, 6	14
LC08_L1TP_233087_20170306_20200905_02_T1		2017	1 Mar.	6 Mar.	5	7
LC08_L1TP_232087_20171025_20200902_02_T1	232/87		17, 19 Oct.	25 Oct.	7, 8	14
LC08_L1TP_233087_20180221_20200902_02_T1	233/87		27, 28 Feb.	21 Feb.	6, 7	14
LC08_L1TP_232087_20180302_20200902_02_T1	232/87	2018	27, 28 Feb.	2 Mar.	2, 3	14
LC08_L1TP_233087_20181019_20200830_02_T1			23, 24 Oct.	19 Oct.	4, 5	14
LC08_L1TP_233087_20190123_20200830_02_T1			28, 29 Jan.	23 Jan.	5, 6	14
LC08_L1TP_233087_20190224_20200829_02_T1		2019	26, 27 Feb.	24 Feb.	2, 3	14
LC08_L1TP_233087_20191123_20200825_02_T1			19, 20 Nov.	23 Nov.	3, 4	14
LC08_L1TP_233087_20191209_20200824_02_T1	233/87		3, 4 Dec.	9 Dec.	5, 6	14
LC08_L1TP_233087_20200126_20200823_02_T1			28, 29 Jan.	26 Jan.	2, 3	14
LC08_L1TP_233087_20200227_20200822_02_T1			24–26, 27 Feb.	27 Feb.	0, 1, 3	28
LC08_L1TP_233087_20200314_20200822_02_T1		2020	14 Mar.	14 Mar.	0	7
LC08_L1TP_233087_20201109_20210317_02_T1			10–12 Nov.	9 Nov.	1, 2, 3	21
LC08_L1TP_232087_20201118_20210315_02_T1	232/87		24 Nov.	18 Nov.	6	7
LC08_L1TP_232087_20201204_20210313_02_T1			26 Nov.	4 Dec.	8	7
LC08_L1TP_233087_20210301_20210311_02_T1	233/87		2, 3 Mar.	1 Mar.	1, 2	14
LC08_L1TP_233087_20211027_20211104_02_T1		2021	18–20 Oct.	27 Oct.	7, 8, 9	21
LC08_L1TP_232087_20211105_20211116_02_T1	232/87		8, 9 Nov.	5 Nov.	3, 4	14
LC08_L1TP_233087_20211128_20211208_02_T1	233/87		29, 30 Nov.	28 Nov.	1, 2	14

Subsequently, atmospheric corrections were performed with ACOLITE program (version 20211124.0) (https://github.com/acolite, accessed on 20 September 2022). ACOLITE brings together the atmospheric correction protocols and processing software developed at RBINS for aquatic remote sensing applications [40]. The default atmospheric correction using the "Dark Spectrum Fitting" DSF algorithm approach [41–43] and the older "Exponential extrapolation" or EXP algorithm [44–46] were utilized in the ACOLITE [36]. The resulting values represent the surface level reflectance (ρs) for L8 and with the values of the extracted bands at 3 × 3-pixel matrix, the spectral indices were calculated, in accordance with [10,47]. The lake contour was established using geospatial information obtained from the General Water Directorate of Chile (DGA, for its acronym in Spanish), and only the body of water was considered for the analysis [48].

2.4. Spectral Indices

Spectral indices are commonly used in remote sensing to quantify specific properties of vegetation, water, and other land cover types. In the context of the Chlorophyll-a algorithm, in some cases spectral indices are used as an intermediate stage to estimate the concentration of Chlorophyll-a in water bodies [49,50]. Some works have developed indices for the detection of algal blooms and due to the excellent results, they are being widely used by the scientific community [10,24,25,51]. The main motives for using spectral indices in the Chlorophyll-a algorithm can be calibrated and validated using field measurements, which can improve the accuracy of Chlorophyll-a concentration estimates. So, a total of nine spectral indices that have previously been used to detect the presence of aquatic vegetation through remote sensing were included in this study. Table 3 shows the algorithms of each index and the corresponding references.

Table 3. The list of indices compared in this study.

Indices	Formulae	Reference
Floating algal index (FAI)	FAI = NIR − 'NIR 'NIR = R + (SWIR − R) × (λNIR − λR)/(λSWIR − λR)	[24,52]
Green normalized difference vegetation index (GNDVI)	(NIR − G)/(NIR + G)	[23,53]
Normalized difference turbidity index (NDTI)	(R − G)/(R + G)	[23]
Normalized difference vegetation index (NDVI)	(NIR − R)/(NIR + R)	[12]
Enhanced vegetation index (EVI)	G × ((NIR − R)/(NIR + C1 × R − C2 × B + L))	[51,54]
Surface algal bloom index (SABI)	(NIR − R)/(B + G)	[25,26]
Emergent vegetation spectral index (EVSI)	EVSI = (R − SWR)/(R + SWR)	[22]
Modified normalized different water index (MNDWI)	MNDWI = (G − SWIR)/(G + SWIR)	[55]
Green Chlorophyll index (GCI)	GCI = (NIR/G) − 1	[56,57]

The FAI was designed to map floating algae in different aquatic environments and, according to the consulted bibliography, it is a very efficient index for the detection of surface vegetation, and it is more suitable than other indices since it is more sensitive to changes in environmental conditions [24,52]. It is defined as a linear spread of reflectivity in the near infrared (NIR, 0.85–0.88 μm), red (R, 0.64–0.67 μm), and shortwave infrared regions (SWR, 1.57–1.65 μm) [24,52]. The GNDVI is an index resistant to atmospheric effects and is very sensitive to Chlorophyll-a concentrations, so its use can be very beneficial in the detection of Chlorophyll-a in oligotrophic systems, hence the justification for its use in different investigations worldwide [12,23,53]. It is calculated according to the equation shown in Table 3, where G is the reflectivity in the green band (G, 0.53–0.59 μm) [12,23,53]. The NDTI is an index developed to estimate turbidity via remote sensing and is an indicator of water quality and can be related to Chlorophyll-a in inland waters, so it could infer the identification of algae [23]. Based on this, some investigations have included it in monitoring studies of algae blooms [58] and aquatic mucilage [59]. On the other hand, the NDVI is widely used for the detection of both terrestrial and aquatic vegetation. The values vary between −1 and 1, where values below 0 reflect the absence or sparse vegetation [12,60]. The EVI is an index like the NDVI, but more sensitive in areas with dense vegetation; however, it has been used in different investigations to detect Chlorophyll-a in lakes [12]. The EVI has an L value (L = 1) to adjust the canopy background, C values (C1 = 6, C2 = 7.5) as atmospheric resistance coefficients, a gain factor G (G = 2.5), and spectral reflectance values of the R, INR, and blue (B, 0.45–0.51 μm) bands [51]. Both the SABI and the EVSI were proposed to delineate the spatial distributions of floating algae or emergent vegetation extraction in aquatic systems. These have been able to detect cyanobacteria and other algae [26]. The GCI is used to estimate the Chlorophyll-a content of the leaves since using the NIR and

G wavelengths provides a better measurement of the Chlorophyll-a content [56,57]. Due to these characteristics, it is possible that it reflects the Chlorophyll-a content in the lakes, which is why it was selected in this investigation and also in [12]. Finally, the MNDVI is an index that is calculated with the ranges of wavelengths of G and SWIR, and these lengths have been used in algorithms to detect or monitor Chlorophyll-a and algae [22–24,53], so it could detect changes in the quantity of water and present good correlations with Chlorophyll-a values in the lakes.

2.5. Statistical Analysis

Algorithms for Chlorophyll-a Estimation and Mapping

The statistical indicators used for the analysis of the results are described in this section. Nine spectral indices incorporating the most relevant bandwidths of the variable under study were selected (Table 3). Pearson correlations (r) were then performed as measures of linear dependence between the calculated indices and the dependent variable Chlorophyll-a. To estimate the Chl-a variable, empirical expressions were developed and data from the different selected images covering summer and spring in the 2014–2021 period were used. Then, a multiparametric regression analysis of the best-correlated indices was developed to obtain a predictive quantitative relationship of Chl-a. Based on the methodology developed by [61], data were selected for internal validation/training and external validation/prediction. The resulting data matrix was composed of the response variable Chl-a and the prediction indices defined based on each of the 25 Landsat images. The methodology consisted of randomly selecting 70% of the data as the training series for the calibration process and the remaining 30% as the data set for prediction. The statistical analysis was performed using Origin Pro 2022b, version 9.7.0.788 (Academic). The statistical indicators used in the analysis are shown in Table 4. Smaller MBE, RMSE, and NRSME values suggest better model performance. A positive MBE value denotes the average quantity of underestimation in the estimated value and vice versa. The RMSE is usually employed to compare the prediction errors of different models. The NRMSE can be understood as a fraction of the mean error with respect to the mean. In addition, the RMSE measures the dispersion between the simulated and measured data. Finally, the mapping of Chl-a concentrations was performed in ArcGIS 10.8.1.

Table 4. Statistical indicators.

Statistical Indicators	Formulae	Reference
Determination coefficient (R^2)	$R^2 = \left[\frac{\sum_{i=1}^{n}(O_i-\overline{O})(P_i-\overline{P})}{\sqrt{\sum_{i=1}^{n}(O_i-\overline{O})^2\sum_{i=1}^{n}(P_i-\overline{P})^2}}\right]^2$	[62]
Mean bias error (MBE)	$MBE = \frac{1}{n}\sum_{i=1}^{n}(P_i - O_i)$	[62]
Root-mean square error (RMSE)	$RMSE = \left[\frac{\sum_{i=1}^{n}(P_i-O_i)^2}{n}\right]^{\frac{1}{2}}$	[63]
Normalized root means square error (NMRSE)	$NMRSE = \frac{RMSE}{\left[\frac{\sum_{i=1}^{n}O_i}{n}\right]}$	[63]

Where Pi is the estimated value, Oi is the measured value, *n* is the number of observations.

3. Results

3.1. Water Quality Parameters

Figure 1 shows the limnological parameters analyzed in this study as those that could provide optimal conditions for the generation of an algal bloom in the lake.

The characteristics of the limnological parameters shown in Figure 1 are evidence of the trophic state of Lake Villarrica. It is classified as an oligotrophic lake in transition to a higher trophic state (mesotrophic). The average transparency behavior fluctuates between 6.46 m in spring and 9.69 m in summer, while the nutrients' total nitrogen and total phosphorus are inversely proportional to Chl-a values, as can be observed in Figure 1.

As for temperature, it is known that this physical parameter regulates many processes in aquatic systems and algal productivity [64,65]. Temperature values range from 22 °C in summer to 11 °C in spring. The blooms in the lake manifest themselves when the temperature exceeds 18 °C.

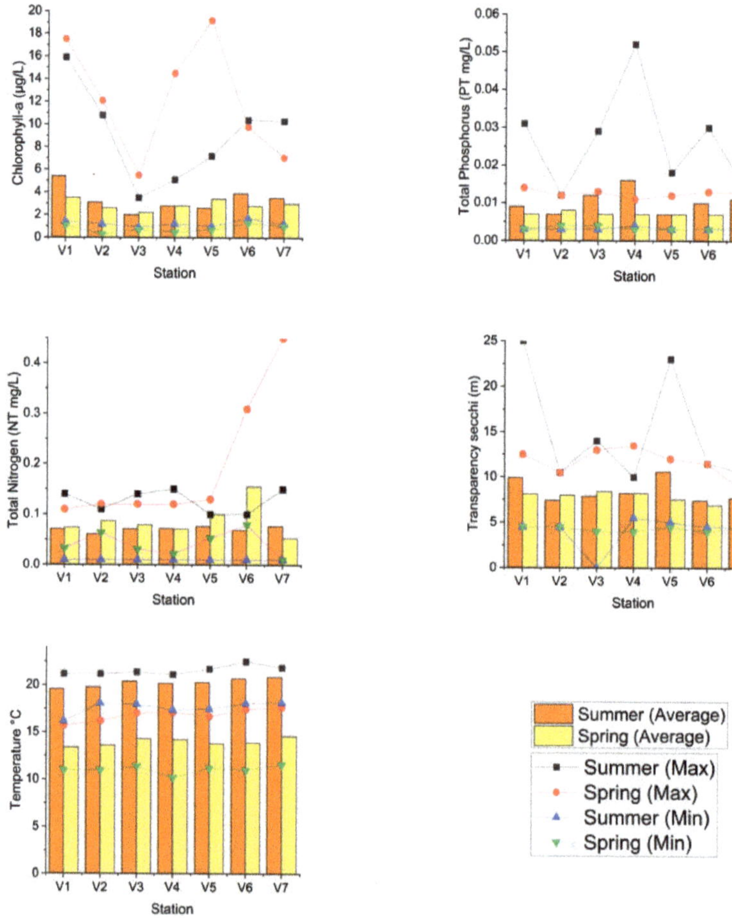

Figure 1. Behavior of limnological parameters in Lake Villarrica during summer and spring seasons. For the summer months (December, January, and February) and for the spring months (September, October, and November) were considered.

3.2. Estimation Model/Statistics

Pearson's correlation coefficient (r) was used to analyze the correlation between the Chlorophyll-a variable and the nine selected spectral indices. The different values obtained for the selected indices are shown in Table 5, and it can be observed that, of the nine indices calculated, three resulted in a high r (FAI, SABI, and GCI). This shows that the selected algorithms have a high degree of correlation. Meanwhile, indices such as EVSI and NDTI presented low values of Pearson's correlation coefficient; therefore, these spectral indices do not have a sufficient degree of correlation to estimate Chl-a in the lake. Several multiple linear regressions were performed combining the indices that presented the best correlation. The Chlorophyll-a recovery model between the FAI and SABI indices obtained statistical values of a MBE of −0.136 µg/L, RMSE of 0.055 µg/L, and NRMSE of 0.019%. On the other hand, the statistical results obtained from the combinations between the GCI with SABI and GCI with FAI indexes obtained MBE values of −0.56 µg/L and −0.50 µg/L, RMSE values of 0.057 µg/L and 0.055 µg/L, and NRMSE values of 0.030% and 0.023%, respectively. Therefore, the statistical results obtained demonstrate that the combination between FAI and SABI indices generates a multiparametric linear regression model that estimates Chl-a concentration in this lake more accurately. The resulting multiparametric regression model was obtained using the linear equation Chl-a = −6.26 × SABI − 10.34 × FAI + 4.72 (adjusted R^2 = 0.97). As can be seen, the adjusted coefficient of determination of the Chl-a estimation model was high. The model was validated by comparing samples of N = 52 with measured and estimated Chl-a concentrations (µg/L). For validation of the proposed Chl-a estimation model, 30% of the total data was used. Figure 2 shows the graphical comparison between the estimated and measured values, which present a high fit, represented by the red line. The results of the model show its robustness, but the amount of data used during the model validation stage remains a limitation due to the lack of real data.

Table 5. Pearson correlation coefficient between the spectral indices and in situ measured Chlorophyll-a concentration.

N°	Indices	r Pearson	Adjusted R^2
1	NDVI	0.76	0.56
2	NDTI	0.19	0.04
3	FAI	−0.78	0.59
4	SABI	−0.94	0.87
5	GNDVI	0.82	0.65
6	EVSI	−0.33	0.08
7	EVI	0.46	0.18
8	MNDWI	−0.69	0.46
9	GCI	0.81	0.65

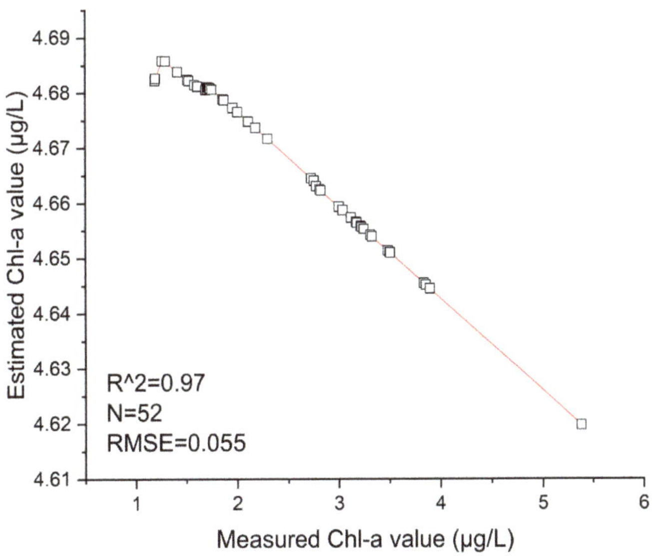

Figure 2. Measured Chl-a and Chl-a estimated using the obtained multiple linear regression model.

3.3. Bloom Estimation Maps

This section presents the spatial distribution maps based on the spectral indices and estimation model with the best results for Lake Villarrica. Figures 3 and 4 represent the indices, and Figure 5 shows the map resulting from the combination of the FAI and SABI indices reflecting the estimated Chlorophyll-a for Lake Villarrica.

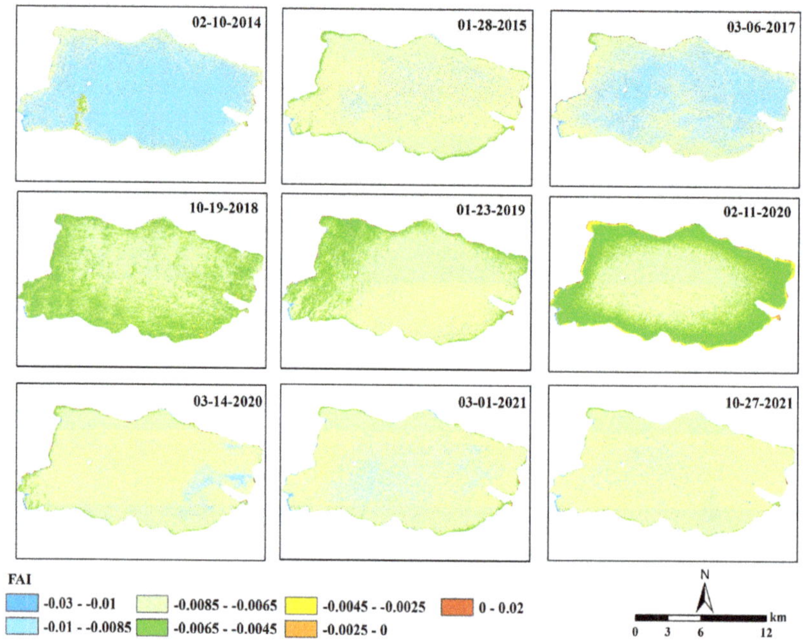

Figure 3. Spatial distribution of FAI in Lake Villarrica.

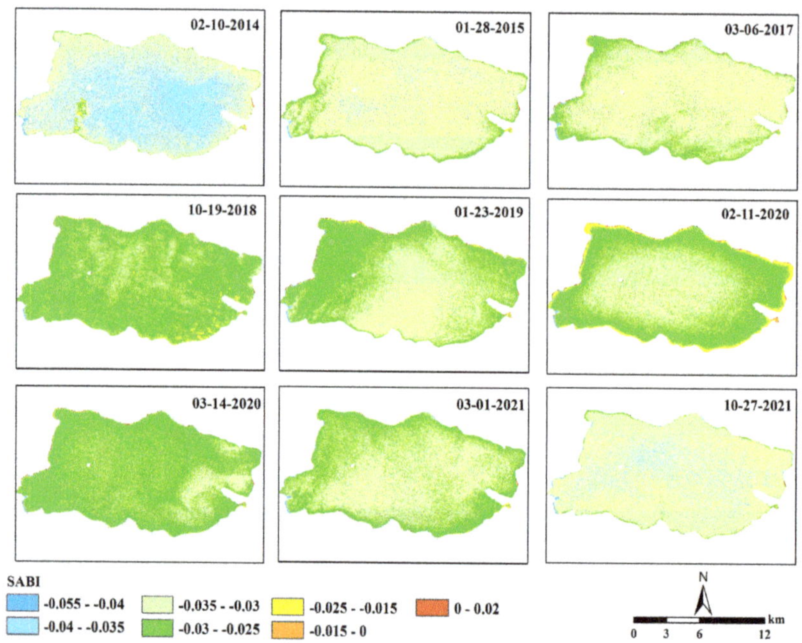

Figure 4. Spatial distribution of SABI in Lake Villarrica.

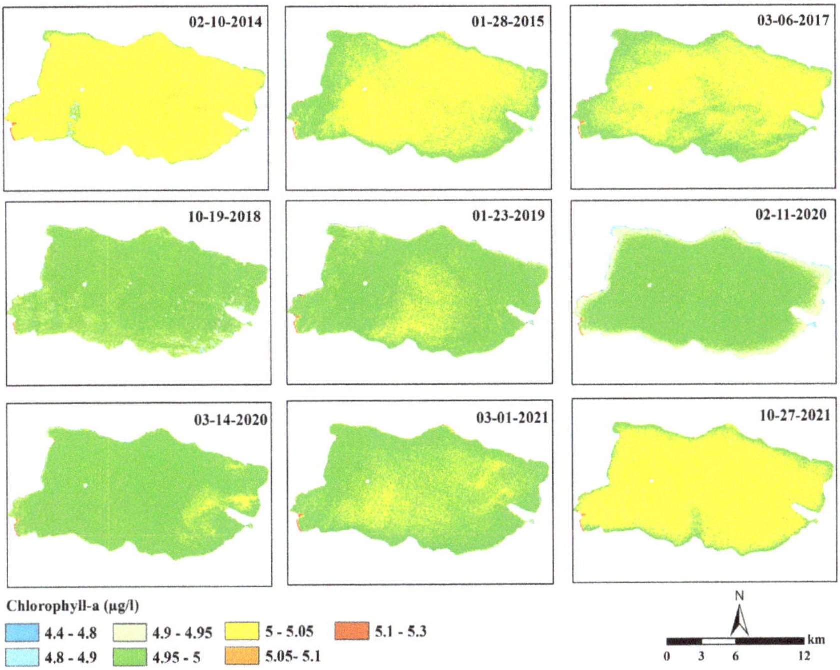

Figure 5. Spatial pattern of Chlorophyll-a arising from multivariate regression resulting from spectral indices through as seen Landsat imagery.

Figure 3 shows that the FAI has an excellent representation in the Landsat images, where the areas with the greatest predominance of vegetation can be observed spatially. Therefore, the FAI index can be applied to select surveillance sites with higher production values. The maps derived from the calculation of the FAI show a marked difference in the groups of primary producers in Lake Villarrica, which was confirmed by the results of the statistical correlation analysis. The satellite image of 11 February 2020 showed high values for the FAI index on the shores, suggesting the existence of some types of aquatic vegetation (positives values) or possible algae blooms. In addition, some positive values on the shore near population centers can suggest the existence of algal blooms. However, some negative values close to 0 (-0.0025) may indicate the presence or early initiation of an algal bloom.

Figure 4 shows spatial differences like those observed with the FAI, with high values near the shores being coincident with the image from 11 February 2020. However, in images such as that of 19 October 2018, the FAI more clearly reflects the vegetation at the surface level in Lake Villarrica than the SABI. This is why in the literature it is suggested that several indices be used to generate Chlorophyll-a estimation models for the lake [10,12,24,26,66].

The spatial behavior of Chlorophyll-a predicted using the optimal model for Lake Villarrica in the study seasons is shown in Figure 5. The Chl-a values range between 4.4 and 5.3 µg/L throughout the lake, with very low spatial variation observed, coinciding with the in situ values. This reaffirms that the lake has mesotrophic characteristics. However, as blooms are being reported in different parts of the lake, more frequent monitoring and studies should be carried out. Generally, the highest values are recorded in areas close to the shores, and most frequently for the shore of the city of Villarrica and Pucón.

3.4. Evolution of Algal Blooms in Lake Villarrica

Table 6 shows the blooms reported in Lake Villarrica in the last 28 years, as well as the species reported and the sector of the lake where the bloom developed.

Table 6. Algal blooms during the last 28 years in Lake Villarrica.

Year	Reported Blooms	Algae Species	Group	Lake Sector
1993	Summer February	*Microcystis aeruginosa*	Cyanophyceae	South, South-west
2005	Spring September 25	-	Cholophyceae	South Ribera
2008	Summer 12 January, Spring 6 November	*Fragilaria* sp.	Bacillariophyceae	North Ribera
2010	Summer 25 January, Autumn 26 April	*Dolichospermum* sp.	Cyanophyceae	Villarrica-Pucón shore
2011	Summer 10 January	*Dolichospermum* sp.	Cyanophyceae	La Poza and Pucón
2012	Summer 24 February	*Dolichospermum* sp.	Cyanophyceae	Pucón, La Poza
2014	Summer 25 January	*Anabaena spiroides*	Cyanophyceae	Pucón, La Poza
2015	Summer 19 January	*Dolichospermum* sp.	Cyanophyceae	Center
2016	Summer 2 January	*Dolichospermum* sp.	Cyanophyceae	South shore
2017	Summer 10 February	*Spirogyra* sp.	Charophyceae	South shore
2018	Summer 25 January, Autumn 25 May	*Dolichospermum* sp.	Cyanophyceae	South
2019	None reported	-	-	-
2020	Summer 19 January	*Dolichospermum* sp.	Cyanophyceae	Villarrica pelagial
2021	None reported	-	-	-

Based on information from reports generated by Vigilantes del Lago.

It can be seen from the information provided above that algal blooms in Lake Villarrica are occurring every summer in different parts of the lake (see Supplementary Materials). Finally, in situ measurements and algal reports were used to identify generic algal blooms occurring in summer and spring each year. This allowed us to investigate the spectral features of these blooms with respect to those of cyanobacteria and to verify the robustness of the conclusions reached in this work.

4. Discussion

Twenty-five Landsat multispectral images were selected to cover an 8-year time span (2014–2021) for which nine spectral indices were calculated. The FAI and SABI indices showed the best statistical results. In addition, in situ Chl-a sample data were available, covering the lake spatially across seven monitoring stations. Data for summer and spring were taken from the databases of the DGA. Chl-a ranged between minimum values of 0.21 μg/L and 0.24 μg/L in summer and spring and maximum values of 5.05 and 4.85 μg/L, respectively. Therefore, both seasons present high values of the study variable. Based on this data, we correlated the spectral indexes and Chl-a in the various sample areas. Using the resulting data, a linear Chl-a prediction model based on remote sensing data was constructed to estimate the Chl-a concentration in the lake by means of the spectral indices.

Based on remote sensing processing, the average maximum FAI value was found to be 0.067, and the average maximum SABI value was 0.048. According to Pearson's r, the FAI (−0.78) and SABI (−0.94) indices correlate well with Chlorophyll-a, presenting an inverse correlation (negative sign). Although in Boucher et.al [21] the relationship between the SABI algorithm and Chlorophyll-a was low, in this investigation it turned out to be quite high. The model constructed with the two indices estimated Chlorophyll-a concentration with high precision. The statistical indicators used to verify the linear multiparametric model of Chlorophyll-a estimation through the spectral indices were a MBE of −0.136 μg/L, RMSE of 0.055 μg/L, and NRMSE of 0.019%. Maps of vegetation indices (FAI, SABI) for Lake Villarrica were built, with the spatial distribution obtained from the processing of Landsat images using a range of established colors being shown, with negative values indicating the presence of water and values between 0 and 0.4 the presence of vegetation at the surface level. The FAI and SABI are spectral indices derived from satellite imagery and are used to quantify the amount and quality of light that is absorbed and reflected by water bodies. When you use these two indices in a regression equation to estimate Chl-a, the resulting equation provides a way to predict Chl-a concentrations based on the values of the FAI and SABI. This can be useful for monitoring and managing water quality in large bodies of water, where direct sampling may not be feasible or cost-effective. Chlorophyll-a is a pigment found in photosynthetic organisms such as plants, algae, and some bacteria. It is responsible for absorbing light energy during the process of photosynthesis. The FAI and SABI are two vegetation indices used to estimate the amount of photosynthetically active radiation absorbed by algae and algal leaf area, respectively. An inverse relationship, or negative correlation, between Chlorophyll-a and the FAI and SABI spectral indices means that as the amount of Chlorophyll-a decreases, the values of the FAI and SABI indices increase, and vice versa. Article [50] shows detailed examples of the application of the FAI index. From them, low values of the index correspond to clear water, and high values of the index correspond to the presence of macroalgae. The SABI index has been proposed to delineate the spatial distributions of floating microalgal species, such as cyanobacteria, or exposed intertidal vegetation, such as seagrasses [25]. Band ratio algorithms in the red and near-infrared spectral regions have great potential for remote Chlorophyll-a estimation. However, more research is needed to fully understand the complex relationships between Chlorophyll-a concentration and the FAI and SABI indices and identify the underlying mechanisms driving these relationships. It would be desirable in future research to include inherent optical properties of the lake system studied in order to further elucidate the optical signals used. The highest values of estimated Chlorophyll-a were observed in shore areas close to cities (Villarrica and Pucón), and this coincides with the highest values of the diffuse attenuation coefficient of solar radiation found in [36]. In addition, in some research [58] higher values near the shore have indicated the presence of algal blooms, and NDVI, NDWI, NDTI, and SABI indices were highly correlated with Chlorophyll-a. In the case of Villarrica, the NDTI had a low correlation with Chlorophyll-a, probably due to the variable turbidity of this lake, attributed, as with most of the Araucanian lakes, to mixing in the water column due to wind and frequent precipitation as well as sediment resuspension [47].

The greatest FAI values, as observed in Figure 3, occur in October 2018 and February 2020 (FAI > −0.0025), showing the presence of aquatic vegetation or algae bloom. In [67], the lake areas with positive FAI values (FAI > 0.003) were classified as algae blooms or aquatic vegetation, which is quite close to the results found and shown in this investigation. In the first season (summer) of 2018 and 2020, blooms were reported in Lake Villarrica at different points, and the cyanobacterial species was *Dolichorpermum* sp. The monitoring campaigns were carried out close to the dates of the satellite images (25 January 2018, and 19 January 2020). Both blooms can be captured through the indices, although another algal species (*Microcystis*) that has also been found in Lake Villarrica has been reported to form blooms, in other lake systems such as Lake Erie and Lake Taihu or Curonian Lagoon [29–31].

Algal blooms of cyanobacteria are becoming more frequent in freshwater systems, increasing due to the additional effects of climate change such as the gradual increase in surface water temperature, which makes it possible for algae to bloom for longer periods depending on how favorable the water temperature. In this study, it was not an objective to evaluate the behavior of temperature; however, temperature determines an increase in the productivity of phytoplankton communities; blooms in Lake Villarrica have manifested themselves when the temperature is higher than 18 °C.

Despite presenting good water quality in general, which is reaffirmed by the Chlorophyll-a estimation carried out in this work, the highest Chlorophyll-a values were recorded in the shore area of the lake, coinciding with the shore of Villarrica and Pucón city (see Figures 5 and 6). This indicates that in the shore waters of cities, algae blooms may occur more frequently, and an alert should be generated since they are very touristy areas. Therefore, it can be inferred that the main determining factor of the high values of Chlorophyll-a and algae blooms, considering only the land use (Figure 6), is the urban area. A deeper study should be carried out in other investigations to analyze the possible influence of agriculture. Of all the algal groups, cyanobacteria present advantages due to their environmental resistance to changing parameters related to either the climate or subaquatic column conditions. In addition, some of the described species present toxins that can cause intoxication problems for animals or skin irritation in humans. If these toxins are ingested at high concentrations, they can cause more serious health problems, including death, as stated by [68].

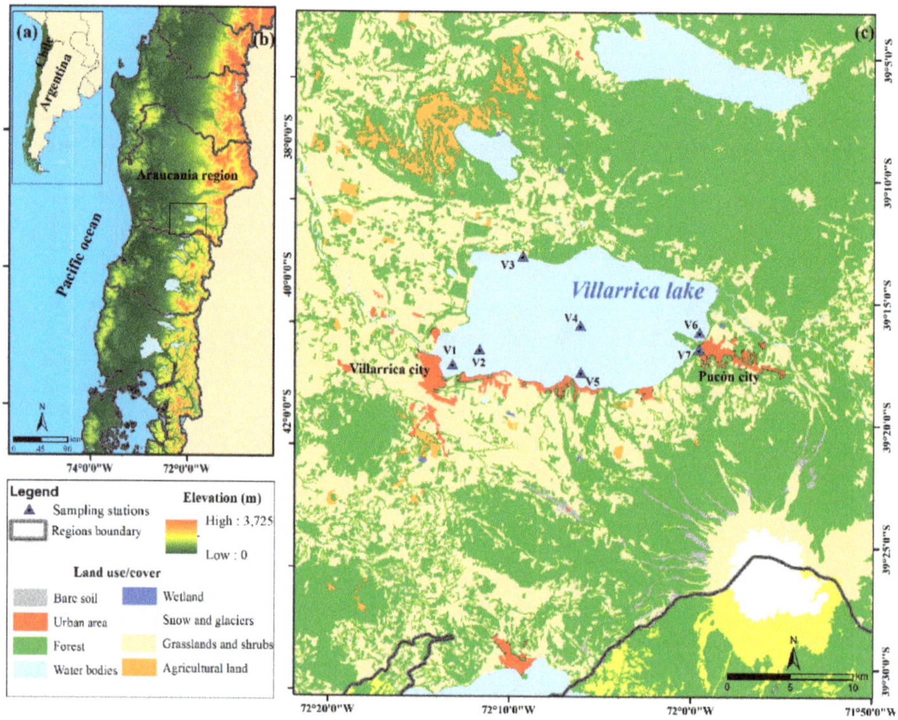

Figure 6. (**a**) Chile in South America; (**b**) Araucanía Region with topographic profiles, with the study area, including Lake Villarrica, outlined in black; and (**c**) the land uses of the Villarrica basin.

Phytoplankton is considered a good indicator of the environmental quality of lakes due to its tolerance and sensitivity to the increase in nutrients generated by anthropogenic pollution (associated with nitrates, nitrites, and phosphates, etc.); therefore, the measurement and analysis of the phytoplankton of a lake help determine its trophic state. It is important to mention that some phytoplankton are cyanobacteria which are associated with the algal blooms present in Lake Villarrica. Cyanobacteria can fix atmospheric nitrogen [69]; therefore, phosphorus functions as a limiting nutrient for blooms in most continental aquatic systems. In addition, blooms could be due to certain physicochemical conditions such as high concentrations of nutrients (mainly nitrogen and phosphorus), high temperatures, good light availability, low turbulence, absence of winds, and stratification of the water body, along with eutrophication processes that could also be influencing the blooms. In [70] the blooms are attributed to increases in the nutrients from small urban areas and agriculture. These same factors were the triggers for the appearance of cyanobacteria and other algae blooms in the investigations carried out by [31,68]. This phenomenon can be triggered in a few hours or several days and can disappear quickly or remain for long periods. In addition, wind is an important factor that can influence algal blooms in several ways. If there is mixing in the water column, wind-generated turbulence can mix surface water with deeper water, allowing nutrients such as nitrogen and phosphorus to rise to the surface [71–73]. These nutrients are essential for algal growth and reproduction. Wind can also transport algae to new locations, where they can thrive in favorable conditions, such as warmer temperatures, higher light levels, and greater nutrient availability. Sometimes wind can cause stratification of the water column, creating different layers of water with different temperatures and nutrient concentrations. This stratification can favor the growth of certain types of algae that prefer specific water conditions. On the other hand, wind-driven currents can also cause algae to accumulate in certain areas, such as near shorelines or in bays, where they can form dense mats or blooms. Therefore, remote sensing can provide an alternative to the spatial–temporal factors that are faced through research and surveillance programs of the lake ecosystems which are based mainly on costly on-site measurements. Furthermore, it is important to note that this study is the first to address the detection of primary producers in this Chilean lake using remote sensing techniques.

5. Conclusions

Using multispectral imagery is a reliable and efficient method for monitoring water quality in freshwater ecosystems. In Chile, this technique has gained popularity due to the limited research resources available. Although the General Water Directorate of Chile monitors many lakes, their efforts fall short in keeping up with changes in the system. The increasing occurrence of algal blooms poses a threat to the organisms and the scenery of the lake, and decreases income generated from tourism in the southern part of the country. Phytoplankton blooms are an indication of high nutrient concentrations that may result from contamination by organic and inorganic substances from urban and industrial effluents and crop fertilization.

This study employed two spectral indices algorithms to retrieve Chl-a data from Lake Villarrica. The statistical indicators of the model generated revealed that the SABI and FAI indices were the most precise in estimating Chl-a levels in Lake Villarrica. By detecting algae using satellite images, an early warning system can be implemented to anticipate the development of algal bloom events, inform the public, and execute policies that safeguard this valuable water resource.

Supplementary Materials: The following supporting information can be downloaded at: https://www.mdpi.com/article/10.3390/rs15071929/s1. Figure S1. Ninth bloom reported in Villarrica by Vigilantes del Lago (19 January 2015, *Dolichospermum* sp.) Sector Centro. Figura S2. Tenth bloom reported in Villarrica (2 January 2016, *Dolichospermum* sp.) Ribera Sur. Figure S3. Eleventh bloom reported in Villarrica (10 February 2017, *Spirogyra* sp.) South Bank. Figure S4. Twelfth bloom reported in Villarrica (25 January 2018, *Dolichospermum* sp.) South Sector. Figure S5. Last Bloom reported in Villarrica (19 January 2020, *Dolichospermum* sp.) Pelagial Villarrica.

Author Contributions: Conceptualization, L.R.-L.; methodology, L.R.-L.; software, I.D.-L.; validation, L.B.A. and L.R.-L.; formal analysis, L.R.-L.; investigation, L.R.-L.; resources, R.U.; data curation, L.R.-L.; writing—original draft preparation, L.R.-L.; writing—review and editing, L.R.-L. and I.D.-L.; visualization, L.B.A. and I.D.-L.; supervision, A.L.; project administration, R.U.; funding acquisition, L.R.-L. and R.U. All authors have read and agreed to the published version of the manuscript.

Funding: This research was funded by CRHIAM (ANID/FONDAP/15130015).

Data Availability Statement: The data presented in this study are available on request from the corresponding author.

Acknowledgments: L.R.-L. is grateful to the VRIDFAI21/10 project of the Universidad San Sebastian. Special thanks to the Centro de Recursos Hídricos para la Agricultura y la Minería (CRHIAM) (Project ANID/FONDAP/15130015).

Conflicts of Interest: The authors declare no conflict of interest.

References

1. Carpenter, S.R.; Stanley, E.H.; vander Zanden, M.J. State of the World's Freshwater Ecosystems: Physical, Chemical, and Biological Changes. *Annu. Rev. Environ. Resour.* **2011**, *36*, 75–99. [CrossRef]
2. Ho, L.T.; Goethals, P.L.M. Opportunities and Challenges for the Sustainability of Lakes and Reservoirs in Relation to the Sustainable Development Goals (SDGs). *Water* **2019**, *11*, 1462. [CrossRef]
3. Gozlan, R.E.; Karimov, B.K.; Zadereev, E.; Kuznetsova, D.; Brucet, S. Status, Trends, and Future Dynamics of Freshwater Ecosystems in Europe and Central Asia. *Inland Waters* **2019**, *9*, 78–94. [CrossRef]
4. Kumar, R.; Parvaze, S.; Huda, M.B.; Allaie, S.P. The Changing Water Quality of Lakes—A Case Study of Dal Lake, Kashmir Valley. *Environ. Monit. Assess.* **2022**, *194*, 228. [CrossRef] [PubMed]
5. Irannezhad, M.; Ahmadi, B.; Liu, J.; Chen, D.; Matthews, J.H. Global Water Security: A Shining Star in the Dark Sky of Achieving the Sustainable Development Goals. *Sustain. Horiz.* **2022**, *1*, 100005. [CrossRef]
6. Prado, R.; García, R.; Rioboo, C.; Herrero, C.; Cid, A. Suitability of Cytotoxicity Endpoints and Test Microalgal Species to Disclose the Toxic Effect of Common Aquatic Pollutants. *Ecotoxicol. Environ. Saf.* **2015**, *114*, 117–125. [CrossRef]
7. Park, J.; Kim, K.T.; Lee, W.H. Recent Advances in Information and Communications Technology (ICT) and Sensor Technology for Monitoring Water Quality. *Water* **2020**, *12*, 510. [CrossRef]
8. Ostroumov, S.A. Water Quality and Conditioning in Natural Ecosystems: Biomachinery Theory of Self-Purification of Water. *Russ. J. Gen. Chem.* **2017**, *87*, 3199–3204. [CrossRef]
9. Demetillo, A.T.; Taboada, E.B. Real-Time Water Quality Monitoring for Small Aquatic Area Using Unmanned Surface Vehicle. *Technol. Appl. Sci. Res.* **2019**, *9*, 3959–3964. [CrossRef]
10. Rodríguez-López, L.; González-Rodríguez, L.; Duran-Llacer, I.; Cardenas, R.; Urrutia, R. Spatio-Temporal Analysis of Chlorophyll in Six Araucanian Lakes of Central-South Chile from Landsat Imagery. *Ecol. Inform.* **2021**, *65*, 101431. [CrossRef]
11. Peterson, K.T.; Sagan, V.; Sidike, P.; Hasenmueller, E.A.; Sloan, J.J.; Knouft, J.H. Machine Learning-Based Ensemble Prediction of Water-Quality Variables Using Feature-Level and Decision-Level Fusion with Proximal Remote Sensing. *Photogramm. Eng. Remote Sens.* **2019**, *85*, 269–280. [CrossRef]
12. Rodríguez-López, L.; Duran-Llacer, I.; González-Rodríguez, L.; Abarca-del-Rio, R.; Cárdenas, R.; Parra, O.; Martínez-Retureta, R.; Urrutia, R. Spectral Analysis Using LANDSAT Images to Monitor the Chlorophyll-a Concentration in Lake Laja in Chile. *Ecol. Inform.* **2020**, *60*, 101183. [CrossRef]
13. Aranda, A.C.; Rivera-Ruiz, D.; Rodríguez-López, L.; Pedreros, P.; Arumí-Ribera, J.L.; Morales-Salinas, L.; Fuentes-Jaque, G.; Urrutia, R. Evidence of Climate Change Based on Lake Surface Temperature Trends in South Central Chile. *Remote Sens.* **2021**, *13*, 4535. [CrossRef]
14. Bhagowati, B.; Ahamad, K.U. A Review on Lake Eutrophication Dynamics and Recent Developments in Lake Modeling. *Ecohydrol. Hydrobiol.* **2019**, *19*, 155–166. [CrossRef]
15. Zou, R.; Wu, Z.; Zhao, L.; Elser, J.J.; Yu, Y.; Chen, Y.; Liu, Y. Seasonal Algal Blooms Support Sediment Release of Phosphorus via Positive Feedback in a Eutrophic Lake: Insights from a Nutrient Flux Tracking Modeling. *Ecol. Model.* **2020**, *416*, 108881. [CrossRef]
16. Harrison, J.W.; Smith, R.E.H. Environmental Effects of UV Radiation Effects of Ultraviolet Radiation on the Productivity and Composition of Freshwater Phytoplankton Communities. *Photochem. Photobiol. Sci.* **2009**, *8*, 1218–1232. [CrossRef]
17. Nazari-Sharabian, M.; Ahmad, S.; Karakouzian, M. Climate Change and Eutrophication: A Short Review. *Eng. Technol. Appl. Sci. Res.* **2018**, *8*, 3668–3672. [CrossRef]
18. Berg, M.; Sutula, M. Factors Affecting Growth of Cyanobacteria with Special Emphasis on the Sacramento-San Joaquin Delta EXECUTIVE SUMMARY. In *Southern California Coastal Water Research Project Technical Report*; Southern California Coastal Water Research Project: Costa Mesa, CA, USA, 2015; Volume 869, p. 100.
19. Li, S.; Song, K.; Wang, S.; Liu, G.; Wen, Z.; Shang, Y.; Lyu, L.; Chen, F.; Xu, S.; Tao, H.; et al. Quantification of Chlorophyll-a in Typical Lakes across China Using Sentinel-2 MSI Imagery with Machine Learning Algorithm. *Sci. Total Environ.* **2021**, *778*, 146271. [CrossRef]
20. Lillesand, T.; Kiefer, R.W.; Chipman, J. *Remote Sensing and Image Interpretation*; John Wiley & Sons.: Hoboken, NJ, USA, 2015; ISBN 978-1-118-34328-9.
21. Gholizadeh, M.H.; Melesse, A.M.; Reddi, L. A Comprehensive Review on Water Quality Parameters Estimation Using Remote Sensing Techniques. *Sensors* **2016**, *16*, 1298. [CrossRef]
22. Qing, S.; Runa, A.; Shun, B.; Zhao, W.; Bao, Y.; Hao, Y. Distinguishing and Mapping of Aquatic Vegetations and Yellow Algae Bloom with Landsat Satellite Data in a Complex Shallow Lake, China during 1986–2018. *Ecol. Indic.* **2020**, *112*, 106073. [CrossRef]
23. Elhag, M.; Gitas, I.; Othman, A.; Bahrawi, J.; Gikas, P. Assessment of Water Quality Parameters Using Temporal Remote Sensing Spectral Reflectance in Arid Environments, Saudi Arabia. *Water* **2019**, *11*, 556. [CrossRef]
24. Ma, J.; Jin, S.; Li, J.; He, Y.; Shang, W. Spatio-Temporal Variations and Driving Forces of Harmful Algal Blooms in Chaohu Lake: A Multi-Source Remote Sensing Approach. *Remote Sens.* **2021**, *13*, 427. [CrossRef]
25. Alawadi, F. Detection of Surface Algal Blooms Using the Newly Developed Algorithm Surface Algal Bloom Index (SABI). In *Remote Sensing of the Ocean, Sea Ice, and Large Water Regions 2010*; SPIE: St Bellingham, WA, USA, 2010; Volume 7825, p. 782506. [CrossRef]

26. Boucher, J.; Weathers, K.C.; Norouzi, H.; Steele, B. Assessing the Effectiveness of Landsat 8 Chlorophyll a Retrieval Algorithms for Regional Freshwater Monitoring. *Ecol. Appl.* **2018**, *28*, 1044–1054. [CrossRef]
27. Xu, J.; Liu, H.; Lin, J.; Lyu, H.; Dong, X.; Li, Y.; Guo, H.; Wang, H. Long-Term Monitoring Particulate Composition Change in the Great Lakes Using MODIS Data. *Water Res.* **2022**, *222*, 118932. [CrossRef]
28. Lekki, J.; Deutsch, E.; Sayers, M.; Bosse, K.; Anderson, R.; Tokars, R.; Sawtell, R. Determining Remote Sensing Spatial Resolution Requirements for the Monitoring of Harmful Algal Blooms in the Great Lakes. *J. Great Lakes Res.* **2019**, *45*, 434–443. [CrossRef]
29. Sayers, M.J.; Grimm, A.G.; Shuchman, R.A.; Bosse, K.R.; Fahnenstiel, G.L.; Ruberg, S.A.; Leshkevich, G.A. Satellite Monitoring of Harmful Algal Blooms in the Western Basin of Lake Erie: A 20-Year Time-Series. *J. Great Lakes Res.* **2019**, *45*, 508–521. [CrossRef]
30. Cazzaniga, I.; Zibordi, G.; Mélin, F. Spectral Features of Ocean Colour Radiometric Products in the Presence of Cyanobacteria Blooms in the Baltic Sea. *Remote Sens. Environ.* **2023**, *287*, 113464. [CrossRef]
31. Vaičiūtė, D.; Bučas, M.; Bresciani, M.; Dabulevičienė, T.; Gintauskas, J.; Mėžinė, J.; Tiškus, E.; Umgiesser, G.; Morkūnas, J.; de Santi, F.; et al. Hot Moments and Hotspots of Cyanobacteria Hyperblooms in the Curonian Lagoon (SE Baltic Sea) Revealed via Remote Sensing-Based Retrospective Analysis. *Sci. Total Environ.* **2021**, *769*, 145053. [CrossRef] [PubMed]
32. Nimptsch, J.; Woelfl, S.; Osorio, S.; Valenzuela, J.; Moreira, C.; Ramos, V.; Castelo-Branco, R.; Leão, P.N.; Vasconcelos, V. First Record of Toxins Associated with Cyanobacterial Blooms in Oligotrophic North Patagonian Lakes of Chile-a Genomic Approach. *Int. Rev. Hydrobiol.* **2016**, *101*, 57–68. [CrossRef]
33. Publicación, F.; Santiago, N. *19-Establece Normas Secundarias de Calidad Ambiental Para la Protección de Las Aguas Continentales Superficiales del lago Villarrica Ministerio del Medio Ambiente Establece Normas Secundarias de Calidad Ambiental Para la Protección de Las Aguas Continentales Superficiales del lago Villarrica*; MEDIO: Mar del Plata, Argentina, 2013; Volume 19.
34. Wojtiuk, J.R.; Otero, A.M. Condominiums: Indicators of Post-Tourism Change. The Case of Pucon, Chile. In *Realidad, Tendencias y Desafíos en Turismo*; Consejo de Decanos y Directores de Unidades Académicas Relacionadas con la Enseñanza del Turismo: Neuquén, Argentina, 2020; Volume 18, pp. 53–73.
35. Vergara, L.; Sánchez, C.; Zunino, H.M. Lifestyle Migration: Creating Diverse and Cohesive Communities? The Case of Los Riscos, Pucón, Chile. *Rev. Austral Cienc. Soc.* **2019**, *36*, 47–67. [CrossRef]
36. Rodríguez-López, L.; González-Rodríguez, L.; Duran-Llacer, I.; García, W.; Cardenas, R.; Urrutia, R. Assessment of the Diffuse Attenuation Coefficient of Photosynthetically Active Radiation in a Chilean Lake. *Remote Sens.* **2022**, *14*, 4568. [CrossRef]
37. Niemeyer, H. *Hoyas Hidrográficas De Chile Novena Región*; Dirección General de Aguas: Santiago, Chile, 1980.
38. Campos, H.; Steffen, W.; Roman, C.; Zúñiga, L.; Aguero, G. Limnological Studies in Lake Villarrica Morphometric, Physical, Chemical, Planktonical Factors and Primary Productivity. *Arch. Hydrobiologie. Supplementband. Monogr. Beiträge* **1983**, *65*, 371–406.
39. DGA Atlas Del Agua. *Atlas del Agua Chile 2016*; Ministerio de Obras Públicas: Santiago, Chile, 2016; Volume 1, pp. 1–24.
40. Ilori, C.O.; Pahlevan, N.; Knudby, A. Analyzing Performances of Different Atmospheric Correction Techniques for Landsat 8: Application for Coastal Remote Sensing. *Remote Sens.* **2019**, *11*, 469. [CrossRef]
41. Vanhellemont, Q.; Ruddick, K. Atmospheric Correction of Metre-Scale Optical Satellite Data for Inland and Coastal Water Applications. *Remote Sens. Environ.* **2018**, *216*, 586–597. [CrossRef]
42. Vanhellemont, Q. Adaptation of the Dark Spectrum Fitting Atmospheric Correction for Aquatic Applications of the Landsat and Sentinel-2 Archives. *Remote Sens. Environ.* **2019**, *225*, 175–192. [CrossRef]
43. Vanhellemont, Q. Sensitivity Analysis of the Dark Spectrum Fitting Atmospheric Correction for Metre- and Decametre-Scale Satellite Imagery Using Autonomous Hyperspectral Radiometry. *Opt. Express* **2020**, *28*, 29948. [CrossRef]
44. Vanhellemont, Q.; Ruddick, K. Turbid Wakes Associated with Offshore Wind Turbines Observed with Landsat 8. *Remote Sens. Environ.* **2014**, *145*, 105–115. [CrossRef]
45. Vanhellemont, Q.; Ruddick, K. Advantages of High Quality SWIR Bands for Ocean Colour Processing: Examples from Landsat-8. *Remote Sens. Environ.* **2015**, *161*, 89–106. [CrossRef]
46. Vanhellemont, Q.; Ruddick, K. Acolite for Sentinel-2: Aquatic Applications of MSI Imagery. In Proceedings of the 2016 ESA Living Planet Symposium, Prague, Czech Republic, 9–13 May 2016; Praga: Prague, Czechia, 2016; pp. 9–13.
47. Rodríguez-López, L.; Duran-Llacer, I.; González-Rodríguez, L.; Cardenas, R.; Urrutia, R. Retrieving Water Turbidity in Araucanian Lakes (South-Central Chile) Based on Multispectral Landsat Imagery. *Remote Sens.* **2021**, *13*, 3133. [CrossRef]
48. DGA Dirección General de Aguas de Chile. Available online: https://dga.mop.gob.cl/ (accessed on 2 March 2023).
49. Verstraete, M.M.; Pinty, B. Verstraete1996. *IEEE Trans. Geosci. Remote Sens.* **1996**, *34*, 1254–1265. [CrossRef]
50. Shi, K.; Zhang, Y.; Qin, B.; Zhou, B. Remote Sensing of Cyanobacterial Blooms in Inland Waters: Present Knowledge and Future Challenges. *Sci. Bull.* **2019**, *64*, 1540–1556. [CrossRef] [PubMed]
51. Huete, A.; Didan, K.; Miura, T.; Rodriguez, E.P.; Gao, X.; Ferreira, L.G. Overview of the Radiometric and Biophysical Performance of the MODIS Vegetation Indices. *Remote Sens. Environ.* **2002**, *83*, 195–213. [CrossRef]
52. Hu, C. A Novel Ocean Color Index to Detect Floating Algae in the Global Oceans. *Remote Sens. Environ.* **2009**, *113*, 2118–2129. [CrossRef]
53. Gitelson, A.A.; Kaufman, Y.J.; Merzlyak, M.N.; Blaustein, J. Use of a Green Channel in Remote Sensing of Global Vegetation from EOS-MODIS. *Remote Sens. Environ.* **1996**, *58*, 289–298. [CrossRef]
54. Duran-Llacer, I.; Arumí, J.L.; Arriagada, L.; Aguayo, M.; Rojas, O.; González-Rodríguez, L.; Rodríguez-López, L.; Martínez-Retureta, R.; Oyarzún, R.; Singh, S.K. A New Method to Map Groundwater-Dependent Ecosystem Zones in Semi-Arid Environments: A Case Study in Chile. *Sci. Total Environ.* **2022**, *816*, 151528. [CrossRef]

55. Xu, H. A Study on Information Extraction of Water Body with the Modified Normalized Difference Water Index (MNDWI). *J. Remote Sens.* **2005**, *9*, 589–595.
56. Hassan, M.A.; Yang, M.; Rasheed, A.; Jin, X.; Xia, X.; Xiao, Y.; He, Z. Time-Series Multispectral Indices from Unmanned Aerial Vehicle Imagery Reveal Senescence Rate in Bread Wheat. *Remote Sens.* **2018**, *10*, 809. [CrossRef]
57. Gitelson, A.A.; Viña, A.; Ciganda, V.; Rundquist, D.C.; Arkebauer, T.J. Remote Estimation of Canopy Chlorophyll Content in Crops. *Geophys. Res. Lett.* **2005**, *32*, L08403. [CrossRef]
58. Alharbi, B. Remote Sensing Techniques for Monitoring Algal Blooms in the Area between Jeddah and Rabigh on the Red Sea Coast. *Remote Sens. Appl.* **2023**, *30*, 100935. [CrossRef]
59. Kavzoglu, T.; Goral, M. Google Earth Engine for Monitoring Marine Mucilage: Izmit Bay in Spring 2021. *Hydrology* **2022**, *9*, 135. [CrossRef]
60. Ma, M.; Wang, X.; Veroustraete, F.; Dong, L. Change in Area of Ebinur Lake during the 1998-2005 Period. *Int. J. Remote Sens.* **2007**, *28*, 5523–5533. [CrossRef]
61. Bohn, V.Y.; Carmona, F.; Rivas, R.; Lagomarsino, L.; Diovisalvi, N.; Zagarese, H.E. The Egyptian Journal of Remote Sensing and Space Sciences Development of an Empirical Model for Chlorophyll-a and Secchi Disk Depth Estimation for a Pampean Shallow Lake (Argentina). *Egypt. J. Remote Sens. Space Sci.* **2018**, *21*, 183–191. [CrossRef]
62. Despotovic, M.; Nedic, V.; Despotovic, D.; Cvetanovic, S. Review and Statistical Analysis of Different Global Solar Radiation Sunshine Models. *Renew. Sustain. Energy Rev.* **2015**, *52*, 1869–1880. [CrossRef]
63. Li, Z.; Ye, L.; Zhao, Y.; Song, X.; Teng, J.; Jin, J. Short-Term Wind Power Prediction Based on Extreme Learning Machine with Error Correction. *Prot. Control Mod. Power Syst.* **2016**, *1*, 4–11. [CrossRef]
64. Kuefner, W.; Hofmann, A.M.; Geist, J.; Dubois, N.; Raeder, U. Algal Community Change in Mountain Lakes of the Alps Reveals Effects of Climate Warming and Shifting Treelines 1. *J. Phycol.* **2021**, *57*, 1266–1283. [CrossRef] [PubMed]
65. Caldwell, T.J.; Chandra, S.; Feher, K.; Simmons, J.B.; Hogan, Z. Ecosystem Response to Earlier Ice Break-up Date: Climate-Driven Changes to Water Temperature, Lake-Habitat-Specific Production, and Trout Habitat and Resource Use. *Glob. Chang. Biol.* **2020**, *26*, 5475–5491. [CrossRef]
66. Yin, Z.; Li, J.; Zhang, B.; Liu, Y.; Yan, K.; Gao, M.; Xie, Y.; Zhang, F.; Wang, S. Increase in Chlorophyll-a Concentration in Lake Taihu from 1984 to 2021 Based on Landsat Observations. *Sci. Total Environ.* **2023**, *873*, 162168. [CrossRef]
67. Pu, J.; Song, K.; Lv, Y.; Liu, G.; Fang, C.; Hou, J.; Wen, Z. Distinguishing Algal Blooms from Aquatic Vegetation in Chinese Lakes Using Sentinel 2 Image. *Remote Sens.* **2022**, *14*, 1988. [CrossRef]
68. Kislik, C.; Dronova, I.; Grantham, T.E.; Kelly, M. Mapping Algal Bloom Dynamics in Small Reservoirs Using Sentinel-2 Imagery in Google Earth Engine. *Ecol. Indic.* **2022**, *140*, 109041. [CrossRef]
69. Molot, L.A.; Higgins, S.N.; Schiff, S.L.; Venkiteswaran, J.J.; Paterson, M.J.; Baulch, H.M. Phosphorus-Only Fertilization Rapidly Initiates Large Nitrogen-Fixing Cyanobacteria Blooms in Two Oligotrophic Lakes. *Environ. Res. Lett.* **2021**, *16*, 064078. [CrossRef]
70. Matej-Lukowicz, K.; Wojciechowska, E.; Nawrot, N.; Dzierzbicka-Głowacka, L.A. Seasonal Contributions of Nutrients from Small Urban and Agricultural Watersheds in Northern Poland. *PeerJ* **2020**, *8*, e8381. [CrossRef] [PubMed]
71. Arfi, R.; Guiral, D.; Bouvy, M. Wind Induced Resuspension in a Shallow Tropical Lagoon. *Estuar. Coast. Shelf Sci.* **1993**, *36*, 587–604. [CrossRef]
72. Meyer, J.; Leonhardt, V.; Blindow, I. Sedimentation in a Shallow Brackish Water Lagoon Influenced by Wind-Induced Waves—A Methodical Study. *Estuar. Coast. Shelf Sci.* **2019**, *218*, 359–367. [CrossRef]
73. Zhu, Y.; Mulholland, M.R.; Macías Tapia, A.; Echevarría, M.A.; Pérez Vega, E.; Bernhardt, P. Cyanate Dynamics under Algal Blooms and Sediment Resuspension Events in a Shallow Micro-Tidal Estuary in the Lower Chesapeake Bay. *Estuar. Coast. Shelf Sci.* **2023**, *281*, 108188. [CrossRef]

remote sensing

MDPI

Article

Neural Gas Network Optimization Using Improved OAT Algorithm for Oil Spill Detection in Marine Radar Imagery

Baozhu Jia [1,2,3,4], Zekun Guo [1,2], Jin Xu [1,2,3,4,*], Peng Liu [5] and Bingxin Liu [5]

1 Shenzhen Institute of Guangdong Ocean University, Shenzhen 518116, China; jiabzh@gdou.edu.cn (B.J.); zkguo@stu.gdou.edu.cn (Z.G.)
2 Naval Architecture and Shipping College, Guangdong Ocean University, Zhanjiang 524091, China
3 Technical Research Center for Ship Intelligence and Safety Engineering of Guangdong Province, Zhanjiang 524088, China
4 Guangdong Provincial Key Laboratory of Intelligent Equipment for South China Sea Marine Ranching, Guangdong Ocean University, Zhanjiang 524088, China
5 Navigation College, Dalian Maritime University, Dalian 116026, China; liupeng@dlmu.edu.cn (P.L.); gisbingxin@dlmu.edu.cn (B.L.)
* Correspondence: jinxu@gdou.edu.cn; Tel.: +86-15842695866

Abstract: With the increasingly frequent exploitation and transportation of offshore oil, the threat of oil spill accidents to the marine ecological environment has become increasingly serious. It is urgent to develop efficient and reliable oil film monitoring technology. Based on the marine radar oil spill data, an innovative OAT-NGN hybrid strategy segmentation algorithm was proposed. By integrating the local feature learning ability of a Neural Gas Network (NGN) and the global search strategy of the Oat optimization algorithm (OAT), the proposed method effectively meets the challenges of traditional oil film segmentation methods in complex sea conditions. Firstly, the raw data of marine radar were preprocessed by using co-frequency interference and speckle noise suppression. Then, the OAT algorithm guided the updating of neural weights in the NGN on a global scale for the exploration of a more optimal solution space during the optimization process. Finally, the oil spill segmentation results were projected to the polar coordinate system through post-processing technology. The experimental results showed that this method effectively balanced the problem of false detection and missing detection. Compared with existing methods, OAT-NGN shown stronger adaptability in complex scenarios. In order to improve the segmentation performance, its innovative dynamic weight adjustment mechanism and spatial constraint design provide a new technical path.

Keywords: oil spill; marine radar; image segmentation; OAT-NGN

1. Introduction

With the rapid development of global marine economy, offshore oil transportation and exploitation activities are increasingly frequent [1–4]. However, this development trend has led to severe oil spill accidents, which pose a serious threat to the marine ecological environment and coastal economy [5–8]. In recent years, massive oil spill accidents have aroused worldwide attention. In October 2021, a large-scale oil spill accident occurred in the southern waters of California, the USA. It was estimated that more than 126,000 gallons of crude oil leaked into the Pacific Ocean, resulting in the forced closure of some beaches and the death of extensive wild animals [9]. In January 2022, a crude oil spill accident also occurred in the northern waters of Lima, Peru. The Italian oil tanker *Mare Doricum* leaked in the process of transporting crude oil to the submarine pipeline operated by Spanish oil company *Repsol*, resulting in the serious pollution of the coastline. The accident had brought far-reaching negative impacts on the local marine ecosystem and fisheries [10]. In June 2025, an oil tanker collision occurred in the Gulf of Oman. About 1500 hectares of the sea region were polluted by oil spills [11]. Therefore, the development of efficient and reliable oil spill detection technology is crucial to the effective response to oil spill emergency rescue.

The role of satellite remote sensing technology in the field of marine environment monitoring has been continuously enhanced [12–14]. HJ-1B CCD, an environmental disaster reduction satellite, and HY-1C/D CZI, an ocean color satellite, were independently developed by China [15–17]. The National Aeronautics and Space Administration (NASA) acquires the long-term stable observation data with the MODIS system and Landsat series satellites. It has provided important support for global marine pollution monitoring [18–23]. The Sentinel series satellites deployed by the European Space Agency through the Copernicus plan contribute to marine environmental surveys [24–27]. Li et al. [28] built a closed-loop system for training data generation, model enhancement and oil film detection. The problems of insufficient training data and the false detection of suspected oil films were solved. Its self-evolution mechanism significantly improved the generalization ability of the model [28]. Fan et al. [29] developed a feature fusion network (FMNet) for oil film recognition in Satellite images. The global features obtained by threshold segmentation were organically combined with the local features of U-Net to improve the accuracy of oil film recognition [29]. Chen et al. proposed a seminal image representation collaborative network (SRC-Net), which achieves the accurate segmentation of oil film boundary under complex sea conditions by means of the fusion of physical representation and the generation and discrimination framework [30]. Zhu et al. [31] proposed a contextual and boundary supervised detection network (CBD-Net). It further strengthens edge detection ability through the use of multi-scale feature fusion and attention mechanism [31]. Cheng et al. [32] proposed an improved J-M/k-means method by means of the perspective of feature optimization. The robust performance of dual polarimetric SAR oil film detection was significantly enhanced [32]. Song et al. [33] developed an adaptive network based on dynamic convolution and boundary constraints. Through the cooperation of conditional parametric convolution and two-dimensional discrete wavelet transform, high-precision detection in a complex oil pollution scene using full polarimetric SAR was realized [33].

Although significant progress has been made in the development of oil spill detection methods based on satellite image technology, these methods cannot completely satisfy real-time monitoring. With the breakthrough of remote sensing technology, shipborne and shore-based radar can effectively penetrate clouds and fog to achieve all-weather monitoring by virtue of its centimeter wavelength [34]. Its high-frequency scanning ability makes it possible to track the dynamic diffusion of oil films in real time. The multi-polarization channel provides an important physical basis for distinguishing oil films and sea clutter. Based on the characteristics of X-band radar, researchers have developed a series of efficient oil spill detection methods. Lau and Huang built an offshore oil spill monitoring framework that organically combined Mask R-CNN and a Geographic Information System (GIS). With the synergy of deep learning segmentation and GIS verification, the limitations of traditional on-site monitoring methods in terms of timeliness and dependence on artificial experience were effectively overcome [35]. In the field of marine radar oil spill detection, Chen et al. [36] proposed a BP neural network model based on Gray-Level Co-occurrence Matrix (GLCM) texture features and Principal Component Analysis (PCA) dimensionality reduction. This model successfully realized the high-precision extraction of oil films in an X-band maritime radar image by sliding window threshold segmentation technology [36]. Liu et al. [37] developed a semi-automatic X-band maritime radar oil spill detection method based on texture analysis, machine learning and adaptive threshold segmentation. The co-frequency interference and noise were firstly eliminated in this method. Then, the texture index was constructed to improve the extraction accuracy. It validly solved the problem of manually setting the threshold in the traditional method [37]. The study shown that X-band radar can availably identify the oil spill regions on the sea surface through a specific signal processing algorithm and the feature extraction method. However, this technology still faces several key challenges. Firstly, the radar echo signal has inherent complexity, and its scattering characteristics are affected by many factors. Secondly, oil leaks and other sea surface substances may show similar scattering characteristics in radar images, which increases the difficulty of identification. In addition, detection accuracy fluctuates under different sea conditions, especially in extreme sea conditions, and the performance may be significantly reduced.

This study proposes a hybrid segmentation algorithm based on a Neural Gas Network (NGN) and Wild Oat Optimization (OAT) to process oil spill images obtained from marine X-band radar, effectively solving the above-mentioned problems. To address the issue of incoherent complexity in radar echo signals, the local feature learning capability of the NGN is used to adaptively capture scattering characteristics. For the identification difficulties caused by the similarity in scattering characteristics between oil spills and other sea surface materials, a connectivity penalty term is introduced to enhance spatial consistency, and the global optimization of OAT is combined to achieve accurate differentiation. A collaborative optimization mechanism is adopted to maintain the stability of the data topology structure and address the issue of detection accuracy fluctuations under different sea conditions. The experimental results show that this hybrid strategy has stronger adaptability to maintain high accuracy compared to traditional methods, providing a more reliable technical solution for oil spill monitoring.

The structure of this paper is as follows: Section 2 describes the data and methods in detail. Section 3 shows the results. In Section 4, the algorithm is compared with other methods. Section 5 summarizes this study.

2. Materials and Methods

2.1. Dataset

During the crude oil spill accident caused by the oil pipeline explosion of the Dalian New Port in 2010, the experimental vessel *Yukun* of Dalian Maritime University obtained relevant data. The data collection adopted a shipborne X-band radar with a detection range of 0.75 nautical miles. The pixel resolution of radar images is 1024×1024. The data were gathered in horizontal polarization mode. The relevant equipment and original data in polar coordinates are shown in Figure 1.

The original data were preprocessed to facilitate the subsequent oil film detection and analysis. The pretreatment process is shown in Figure 2. Firstly, the radar image coordinate system was converted from polar coordinates to Cartesian coordinates. The Cartesian coordinate transformation significantly improves the interpretability, algorithm compatibility, and computational efficiency of radar data, and the converted image remains two-dimensional. This conversion provides a standardized input space for subsequent segmentation. The horizontal axis of the transformed Cartesian coordinate system is the azimuth angle, and the vertical axis is the detection distance. Secondly, a Laplace operator was used to process the image and accurately extract the co-frequency interference noise. Then, an average filter was used to smooth the extracted interference noise, which reduced the impact of co-frequency noise on image quality. Furthermore, through the precise calibration of gray-level and pixel-area thresholds, speckle noise was effectively isolated. Afterwards, a median filter was used to smooth the speckles again to further improve the clarity and stability of the image. Finally, with the help of gray correction and local contrast enhancement technology, the contrast between oil film targets and background was improved. This provided strong support for detection and analysis. Ground-truth labels for the processed images were meticulously generated via manual interpretation. The results of each step are shown in Figure 3.

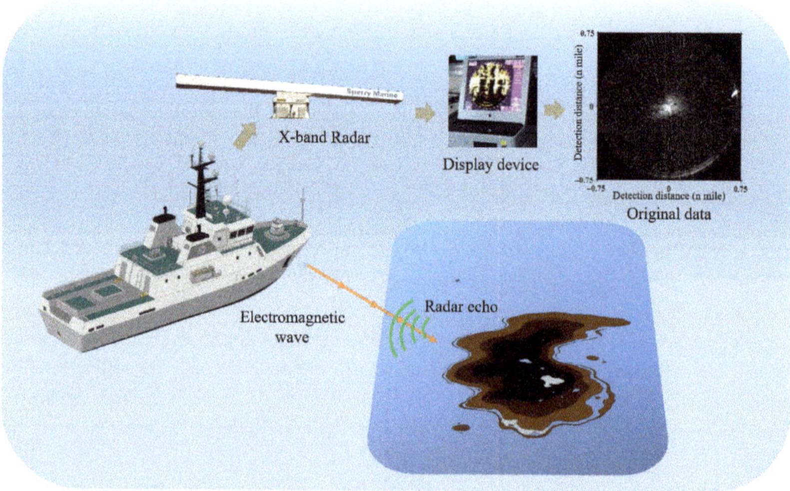

Figure 1. Data collection equipment and experimental sample.

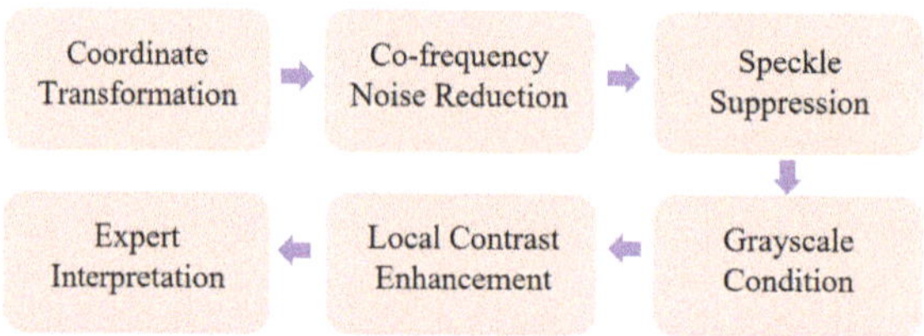

Figure 2. Data preprocessing process.

2.2. NGN-OAT Oil Film Segmentation Algorithm

In radar-acquired imagery, the grayscale difference between oil film regions and the background is small, with boundaries representing misidentification. Consequently, conventional segmentation methods frequently fail to achieve satisfactory results when processing such images. To address this challenge, this paper presented a hybrid segmentation framework integrating Neural Gas Network clustering with optimal adaptive thresholding.

2.2.1. The NGN Model

The NGN model is an unsupervised clustering method based on competitive learning, which approximates the probability distribution of input data by dynamically adjusting the weights of neurons. In the field of image segmentation, the NGN achieves pixel clustering by minimizing the intra class variance [38–42]. The objective function can be expressed as

$$E = \sum_{i=1}^{R} \sum_{j=1}^{C} (x_{ij} - w_{c(i,j)})^2 \quad (1)$$

where R and C represent the number of rows and columns of the image, respectively. x_{ij} represents the pixel value of the position (i,j) in the image. w represents the weight of the neuron. The detailed learning process of the NGN model includes six parts.

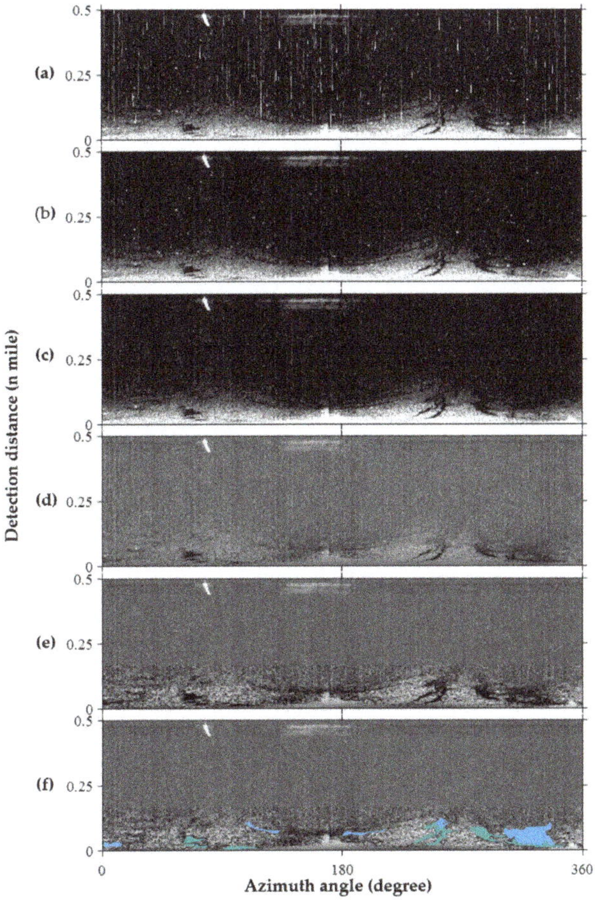

Figure 3. Date pretreatment. (**a**) Original data in Cartesian coordinate system. (**b**) Noise reduction by co-frequency interference. (**c**) Speckle noise suppression. (**d**) Grayscale adjustment. (**e**) Local contrast enhancement. (**f**) Expert interpretation (the suspected oil film regions are marked with a blue label, and the oil film regions are marked with cyan).

(1) Grid initialization: the random initialization of the weight vector:

$$w_i \sim \mathcal{U}(X_{min}, X_{max}), i = 1, \ldots, N \tag{2}$$

where N is the preset number of neurons.

(2) For each input sample x, calculate the distance from all neurons and sort

$$d_i = \| x - w_i \|_2, i = 1, \ldots, N \tag{3}$$

(3) Learning rate $\epsilon(t)$ and domain parameters $\lambda(t)$ decay with the number of iterations:

$$\epsilon(t) = \epsilon_{\text{initial}} \left({\epsilon_{\text{final}}}/{\epsilon_{\text{initial}}} \right)^{t/t_{\max}} \tag{4}$$

$$\lambda(t) = \lambda_{\text{initial}}\left(\lambda_{\text{final}}/\lambda_{\text{initial}}\right)^{t/t_{\max}} \tag{5}$$

$$T(t) = T_{\text{initial}}\left(T_{\text{final}}/T_{\text{initial}}\right)^{t/t_{\max}} \tag{6}$$

where $t_{\max}$ represents the total time step of the run, t is the current iteration progress, T_{final} represents the initial lifespan threshold of neuronal connections. T_{initial} is the threshold for the termination lifespan of neuronal connections.

(4) Update weights w_i of the sorted neurons:

$$w_i \leftarrow w_i + \epsilon(t)\cdot\exp\left(-k_i/\lambda(t)\right)\cdot(x - w_i) \tag{7}$$

where k_i is the sorting position of neuron i (starts from 0).

(5) In the training process of the network, the connection between the nearest two neurons is established by setting the connection matrix. Then, the aging process of neuronal connections is simulated by increasing the values of relevant elements in the connection matrix. Finally, if the age of a connection surpasses the predefined threshold, $T(t)$, the corresponding element in the connection matrix is set to 0, effectively pruning these aged connections. Dynamic synaptic pruning enhances data distribution adaptation. Topological neuron relationships boost clustering efficacy.

(6) After the final weights are re-quantified, the model will map each pixel to the nearest neuron to generate the segmented image.

2.2.2. OAT Optimization Algorithm

Although the NGN model demonstrates unique advantages in image segmentation, several critical limitations remain. Firstly, the convergence speed of the NGN algorithm is relatively slow, which leads to its low efficiency in processing complex images. It is difficult to meet application scenarios with high real-time requirements. Secondly, the algorithm can easily fall into the local optimal solution in the optimization process, which makes the final segmentation result inaccurate. Then, in the process of image segmentation, the NGN algorithm often easily ignores the spatial information. This may lead to discontinuous or inconsistent segmentation results, which will affect reliability and practicability. So as to overcome the limitations of the NGN algorithm, a global optimization algorithm inspired by plant growth called the OAT optimization algorithm was introduced here to improve the NGN model [43]. The complete process of the algorithm is as follows:

(1) Preprocess the image and normalize the image data:

$$I_{\text{norm}}(x,y) = I(x,y)/255 \tag{8}$$

where $I(x, y)$ represents the pixel value of the original gray image.

(2) Calculate the histogram peak of the normalized image:

$$\hat{t}_{\text{init}} = \arg \max_{b\in\{0,0.02,\ldots,1\}} \text{hist}(b) \tag{9}$$

where $hist(b)$ is the total number of pixel values falling in interval b.

(3) The fitness function, $F(t)$, is defined to evaluate the segmentation threshold, t. The function takes into account two key indicators: intra class variance and regional connectivity. Intra class variance is used to measure the uniformity of pixel values in each region after segmentation. Regional connectivity ensures the continuity and

consistency of segmentation results in space. Through threshold t, the normalized image is binarized to obtain $B_t(x,y)$:

$$B_t(x,y) = \begin{cases} 1, \text{if } I_{\text{norm}}(x,y) > t \\ 0, \text{otherwise} \end{cases} \tag{10}$$

Calculate the intra class variance $\sigma^2_{\text{within}}(t)$:

$$\sigma^2_{\text{within}}(t) = w_1(t)\sigma_1^2(t) + w_0(t)\sigma_0^2(t) \tag{11}$$

where

$$w_1(t) = \frac{1}{N}\sum\nolimits_{x,y} B_t(x,y) \tag{12}$$

$$w_0(t) = 1 - w_1(t) \tag{13}$$

$$\sigma_1^2(t) = \text{Var}(I_{\text{norm}}(x,y) \mid B_t(x,y) = 1) \tag{14}$$

$$\sigma_0^2(t) = \text{Var}(\{I_{\text{norm}}(x,y) \mid B_t(x,y) = 0\}) \tag{15}$$

The penalty term $P(t)$ is introduced to improve the regional consistency of the segmentation results:

$$P(t) = {}^{N_c(t)}/_{N} \tag{16}$$

where $Nc(t)$ is the number of connected regions of the binary image. This item can effectively suppress the fragmentation problem caused by oversegmentation.

(4) For each generation, generate candidate seeds.

$$t_i^{(k)} = t_{k-1} + \epsilon_i \tag{17}$$

where $\epsilon_i \sim \mathcal{U}(-0.1, 0.1), i = 1, \ldots, 10$.

Select the best seed and update the threshold:

$$t_{\text{best}}^{(k)} = \arg \underset{t \in \{t_i^{(k)}\}}{min} \mathcal{F}(t) \tag{18}$$

$$t_k = \alpha \cdot t_{\text{best}}^{(k)} \tag{19}$$

(5) Output the optimal threshold, t*, and segmentation results:

$$t^* = \arg \min_{t \in \{t_k\}_{k=1}^K} \mathcal{F}(t) \tag{20}$$

The algorithm determines the optimal segmentation threshold by minimizing the weighted sum of intra class variance and connected region complexity. In this process, the OAT algorithm, as the core optimization framework, conducts a directional random search in the threshold space to efficiently explore the potential optimal solution:

$$t_{\text{new}} = t_{\text{best}} \cdot \alpha + \epsilon, \epsilon \sim \mathcal{U}(-0.1, 0.1) \tag{21}$$

where t_{best} is the current optimal threshold. α is the attenuation factor. The multi-seed parallel evaluation and adaptive threshold attenuation characteristics of the OAT algorithm can effectively make up for the global search defects of the NGN model. The multi-seed parallel evaluation and adaptive threshold decay mechanisms of the OAT algorithm compensate for the global exploration deficiencies inherent in the NGN model.

2.2.3. OAT-NGN Hybrid Strategy

To synergistically leverage the complementary strengths of the two algorithms, this study developed a hybrid optimization framework that integrates NGN topological preservation with OAT global search capabilities. The structural framework of the algorithm is shown in Figure 4.

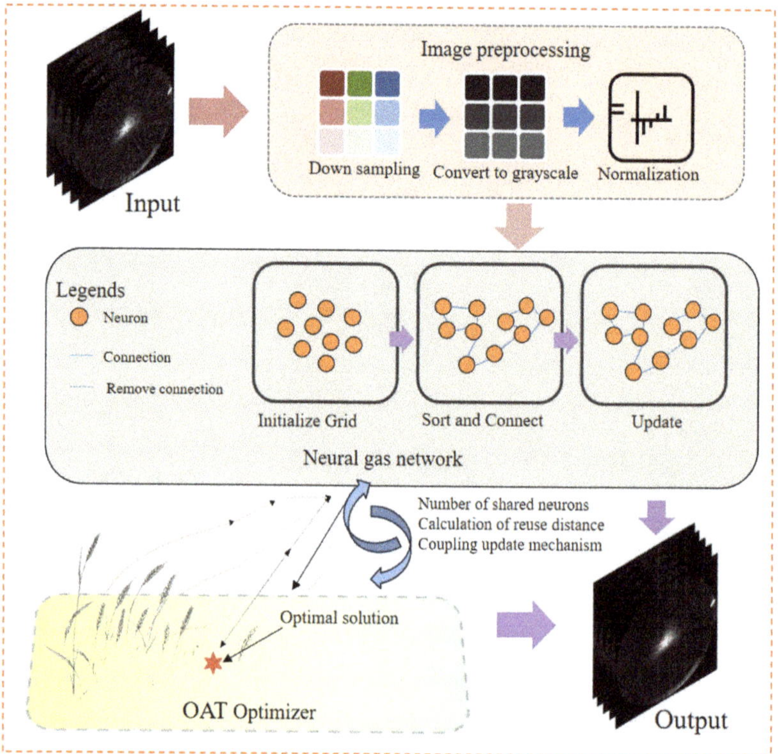

Figure 4. OAT-NGN hybrid strategy structure.

(1) Global and local collaborative optimization

The NGN model achieves pixel clustering with the help of a neuron competition mechanism, which can effectively preserve the topology of data. The local learning mechanism is as follows:

$$w_i \leftarrow w_i + \epsilon e^{-k_i/\lambda}(x - w_i) \quad (22)$$

The advantage of the NGN model is that it can adaptively learn local features, so as to better adapt to the distribution characteristics of data. However, this local learning mechanism also has some limitations. The OAT algorithm guides from a global perspective, generating candidate solutions and optimizing weights through periodic iterative processes. The formula for this mechanism is as follows:

$$w_{\text{new}} = w + (\text{seed} - \overline{w}), \quad (23)$$

$$\text{seed} \sim \mathcal{U}(t_{\text{best}} - 0.1, t_{\text{best}} + 0.1) \quad (24)$$

where seed is a random seed point sampled from a uniform distribution, $\mathcal{U}$, with a range fluctuating around the current optimal threshold, t_{best}. $\overline{w}$ is the mean of the neuron weight vector. w_{new} represents the new weight after adjusting the seed point.

The advantage of the OAT algorithm lies in its ability to effectively break free from the constraints of local optima. By implementing a global search strategy to avoid getting stuck in local optima, the likelihood of finding the optimal solution is increased. However, there are certain limitations when OAT is used alone, as it may overlook the spatial topology of the data. If this structure is ignored, it may lead to discontinuous or inconsistent segmentation results, thereby affecting the accuracy and reliability of the segmentation results.

During implementation, the OAT algorithm globally directs NGN synaptic weight updates, expanding the solution search space to identify optimal segmentation thresholds. The hybrid OAT-NGN algorithm synergistically combines the NGN model's local topology-preserving capability with the global optimization strength of the OAT algorithm, thereby maintaining data topology while circumventing local optima.

(2) Spatial information integration

The NGN model can achieve pixel clustering through its neuron competition mechanism in image segmentation and retain the topology of the data. However, this model will ignore the defects of data space topology when it is used alone. The fitness function of the OAT algorithm sets the spatial constraints by introducing the connectivity penalty term so as to effectively improve the consistency of segmentation. This mechanism can solve the problem of spatial correlation and then make up for the shortcomings of the NGN model. The penalty function F is as follows:

$$\mathcal{F} = \sigma^2_{\text{within}} + 0.1 \cdot P(t) \tag{25}$$

The NGN-OAT hybrid designed here aims to efficiently complete the task of oil pollution image segmentation. The algorithm is mainly composed of parameter initialization, the main segmentation process and multiple function functions. Its pseudo code Algorithm 1 is as follows:

2.3. Evaluation Indicators

The performance of the OAT-NGN oil film segmentation algorithm is evaluated by using several indicators, such as accuracy, precision, recall, the F1-score, IoU and the Dice coefficient. These indicators comprehensively reflect the performance of the algorithm in oil film segmentation, including classification accuracy, false positive suppression, missing feature detection and boundary fidelity.

In the calculation process, the segmentation results of the algorithm were compared with the binary true value image interpreted by experts at the pixel level. A complete confusion matrix was constructed as the basis for evaluation. Some possible interference factors were eliminated in the calculation, which was helpful to evaluate the performance of the algorithm more accurately. The four key elements of the confusion matrix include true positive (TP), false positive (FP), false negative (FN), and true negative (TN). TP represents the number of pixels correctly identified as oil film, reflecting the ability of the algorithm to capture the target area. FP reflects the quantity of pixels that misjudge the background as oil film, which is used to evaluate the false alarm of the algorithm. FN represents the number of real oil film pixels missed and displays the detection blind area. TN counts the quantity of pixels correctly identified as the background. The calculated indicators have different emphases. The definition and calculation formula of each indicator are as follows:

$$Accuracy = \frac{TP + TN}{TP + FP + FN + TN} \tag{26}$$

$$Precision = \frac{TP}{TP + FP} \quad (27)$$

$$\text{Recall} = \frac{TP}{TP + FN} \quad (28)$$

$$F1 = 2 \times \frac{\text{Precision} \times \text{Recall}}{\text{Precision} + \text{Recall}} \quad (29)$$

$$\text{IoU} = \frac{TP}{TP + FP + FN} \quad (30)$$

$$Dice = \frac{2TP}{2TP + FP + FN} \quad (31)$$

Algorithm 1. Pseudo code of the NGN-OAT Oil Segmentation algorithm

```
Input: Oil Spill Image (Org)
  Function: Initialize Parameters
    Calculate histogram of imgData
    Set initThresh to the most frequent intensity value
    N = min (20, max (5, round(numel(imgData)/10,000)) // Number of neurons
    Return N, initThresh
  Function: Main Segmentation
    NGNnetwork = GasNN-Oat (imgData, params)
    Return segmented image
  Core Function: GasNN-Oat (X, params)
    Extract parameters from params
    Shuffle input data X
    Initialize weights w from initThresh and N
    Repeat for it = 1 to MaxIt:
      Update parameters: epsilon, lambda, T
      If oat_active and mod (it, oat_updateInterval) == 0:
        Generate seeds and evaluate scores in parallel
        Update best weights if score improves
      Process data in batches to update weights w
    Return w
  Function: oatFitness (X, w)
    Minimize distance to weights w and compute internal variance
    Return fitness score
  Function: fastDist (X, Y)
    Calculate distance matrix D between X and Y
    Return D
  Output: Segmented Image (segmented)
```

3. Results

The image segmentation framework developed in this research runs in a hardware environment equipped with the GeForce GTX 1660 super graphics card (NVIDIA Corporation, Santa Clara, CA, USA). The processor is an Intel Core i5-12400f processor (Intel Corporation, Santa Clara, CA, USA) with main frequency 2.50 GHz and maximum turbo frequency 4.40 GHz. The algorithm is optimized with the help of a parallel computing toolbox and GPU acceleration, which significantly improves the processing speed of oil film segmentation. The parameter setting of this algorithm is based on the adaptive strategy of task characteristics combined with empirical values. The model undergoes 50 iterations to ensure full convergence while avoiding overfitting. The learning rate decays exponentially from an initial value of 0.5 to 0.01, balancing the need for rapid convergence in the early stages and fine adjustment in the later stages. The neighborhood range decays from 3.0 to 0.5, gradually narrowing the local influence range of neuron updates. The connection lifetime threshold is gradually reduced from 30 to 5, and the network topology is dynamically optimized. The number of neurons (5–20) is automatically calculated based on the amount of input data, while the seed number of the OAT optimizer is set to half of the number of neurons (not less than 20), with a threshold decay factor of 0.73. All attenuation processes are controlled by normalized progress ($t/t_{\max}$), where $t_{\max}$ ensures that the parameter attenuation is adapted to the data size. This setting preserves the local feature learning capability of the NGN while enhancing robustness through the global search of OAT.

3.1. Oil Film Segmentation Results

The result of oil film segmentation by the OAT-NGN algorithm is shown in Figure 5. The figure clearly shows the distribution of oil film regions. In Figure 5a, the processed oil film image is visually presented in color. A variety of colors reflect the different changes in oil film concentration. The upper part of the image is mainly dark red, indicating that there may be no-oil films in this region. The blue, green, and yellow colors in the lower area indicate the accumulation of oil film. Figure 5b was obtained after converting Figure 5a into a gray image. The dark regions may correspond to oil filmsor may be caused by wave induced fluctuations or interference reflection. Figure 5c was obtained by color system conversion for directly displaying the oil film targets. The overall effect was relatively clear, but there were still noise and background interference. Finally, the threshold was auto-selected from Figure 5c to obtain the preliminary oil film segmentation, as shown in Figure 5d.

To verify the effectiveness of this method in the emergency response for oil spill monitoring, another marine radar oil spill image was also processed by using the same steps. In order to better support actual emergency rescue, the experimental results were further projected to the polar coordinate system through speckle noise suppression, as shown in Figure 6. Subsequent discussions were focused on the data in Figure 6a.

The results shown that the shape and edge details of the oil film are clearly preserved. The proposed algorithm was highly effective in accurately identifying the characteristics of oil films. However, for the recognition of suspected oil film regions, although the algorithm had shown preliminary recognition ability, the performance was still imperfect. Due to the influence of ship wake or wind waves on some suspected oil film regions, the oil film recognition precision is not good enough.

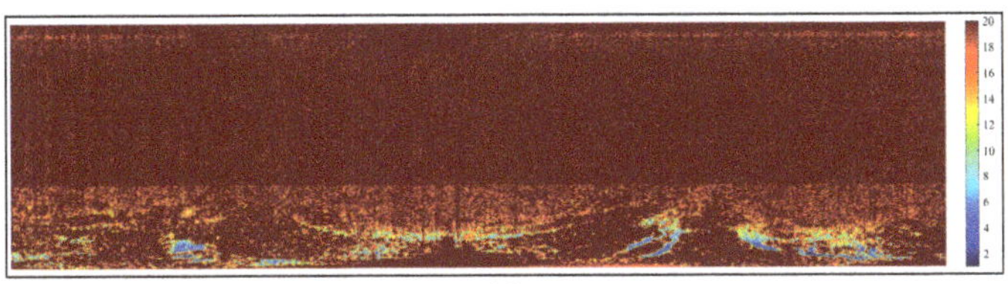

(**a**)

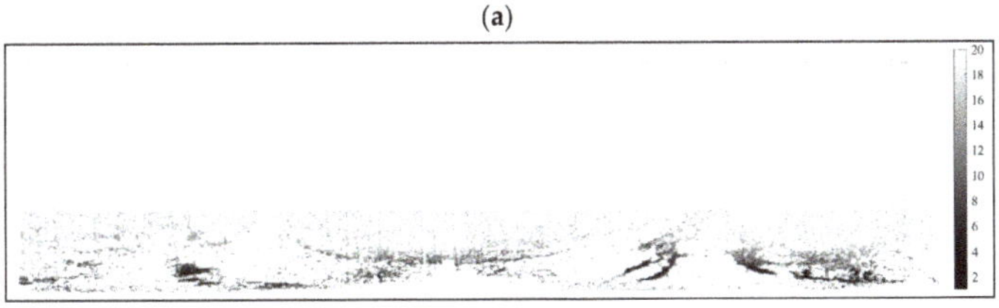

(**b**)

Figure 5. *Cont.*

(c)

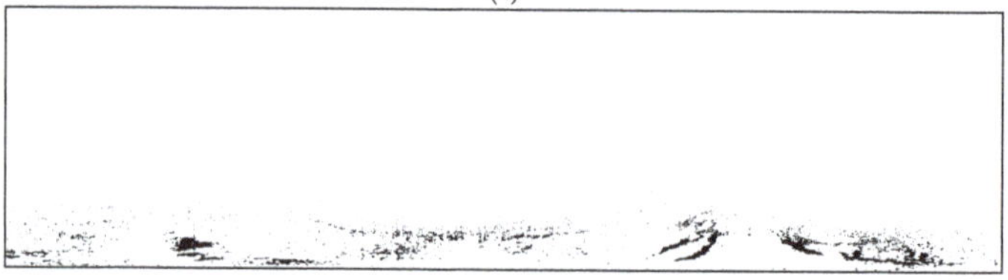

(d)

Figure 5. Preliminary oil film segmentation results. (**a**) Color quantization. (**b**) Grayscale display mode. (**c**) Auto threshold selection. (**d**) The preliminary segmentation.

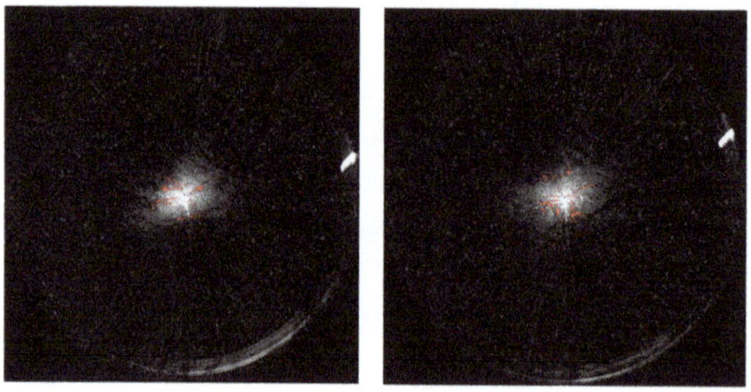

(**a**) (**b**)

Figure 6. The final results. (**a**) The current oil spill image. (**b**) Another oil spill image.

3.2. Evaluation of Experimental Results

The proposed oil film segmentation algorithm performs well in the evaluation, and each index meets practical needs for emergency rescue, as shown in Table 1. The accuracy achieves 94.5%, which shows that it can correctly classify the vast majority of pixels. The precision rate is up to 99.8%, indicating that there is almost no false detection in the segmentation results. The recall rate is 94.6%, which shows that the algorithm can capture most of the real oil film regions. Both the F1-score and Dice coefficient are 97.14%, and the IoU is 94.43%. These indicators further verify its good performance in balancing accuracy and missed detection.

Table 1. Evaluation index.

Indicators	Performance
Accuracy	94.50%
Precision	99.80%
Recall	94.61%
F1-score	97.14%
IoU	94.43%
Dice	97.14%

4. Discussion

4.1. Comparison with NGN Model

The results of the OAT-NGN algorithm were compared with those of the NGN model, as shown in Figure 7. The segmentation results of the OAT-NGN algorithm had less noise interference, and the target regions were more obvious. By contrast, the NGN model exhibited relatively significant noise interference. This not only blurred the boundaries of the target region but also potentially obscured some important details of the oil films, exerting a negative impact on the accuracy and integrity of the segmentation effect.

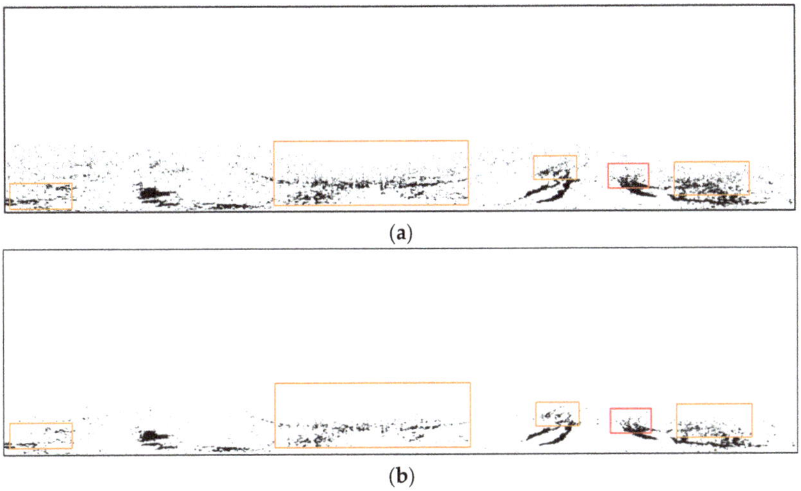

Figure 7. The comparison with the NGN model. (**a**) The NGN model. (**b**) The OAT-NGN algorithm.

The orange boundary areas in the above results were the suspected oil films and stern flow regions. The NGN model mistakenly identified a large number of regions similar to oil areas considered as oil films, resulting in a high misclassification rate. In contrast, the OAT-NGN algorithm performed better in identifying oil film regions. Some oil films in the red boundary areas were not identified. It was indicated that neither algorithm could achieve ideal recognition accuracy in complex marine environments such as ship wake or wind wave interference.

The performance indicators related to both algorithms are shown in Figure 8. The OAT-NGN algorithm exhibited superior performance in various key indicators, which was of great significance for oil film detection tasks. The accuracy of the OAT-NGN algorithm reached 94.50%, significantly higher than the 83.30% of the NGN. This proved that the OAT-NGN algorithm was more effective in distinguishing oil film regions from the backside in overall classification tasks. In terms of the recall rate, the OAT-NGN algorithm performed well with 94.61%, while the NGN model only achieved 83.11%. This means that the OAT-NGN algorithm could capture the true characteristics of the oil films more comprehensively. In terms of precision, the value of the NGN model was 99.95%, slightly higher than the 99.80% of the OAT-NGN algorithm. Both methods had extremely high precision, indicating that the probability of misjudgment in identified oil film regions was almost negligible. However, the slightly lower value of the OAT-NGN algorithm might reflect its greater emphasis on recall, sacrificing some accuracy. As for the F1-score and Dice, the OAT-NGN algorithm was once again in a leading position, reaching 97.14% and 90.75%, respectively. This means that the OAT-NGN algorithm could maintain high segmentation quality. In terms of IoU, the value of the OAT-NGN algorithm was 94.43%, far exceeding the 83.07%

of the NGN algorithm. Its high IoU value implied that the segmented areas had a higher value of overlap with the real labeled areas, ensuring accuracy.

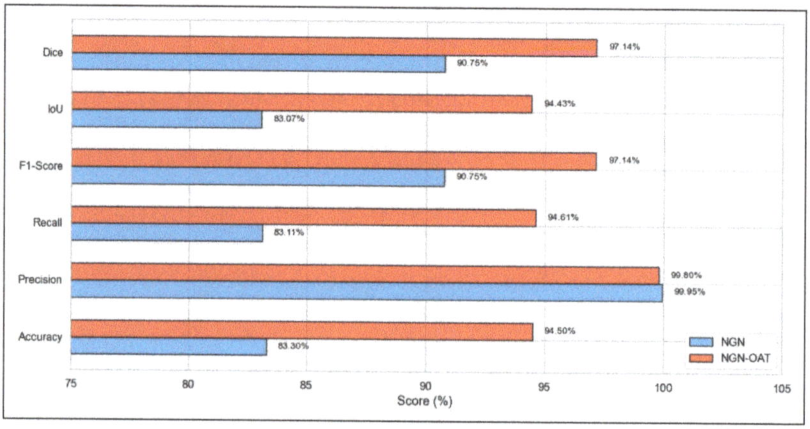

Figure 8. Performance comparison of NGN and OAT-NGN.

4.2. Comparison with Other Algorithms

4.2.1. Comparison with Global Adaptive Threshold Method

The threshold segmentation method was based on the minimum fuzzy metric proposed by Huang and Wang [44] (called Method 1 here) and the global Otsu adaptive threshold [45–47] (called Method 2 here). The segmentation results are shown in Figure 9. The method proposed was compared with these two global adaptive threshold methods.

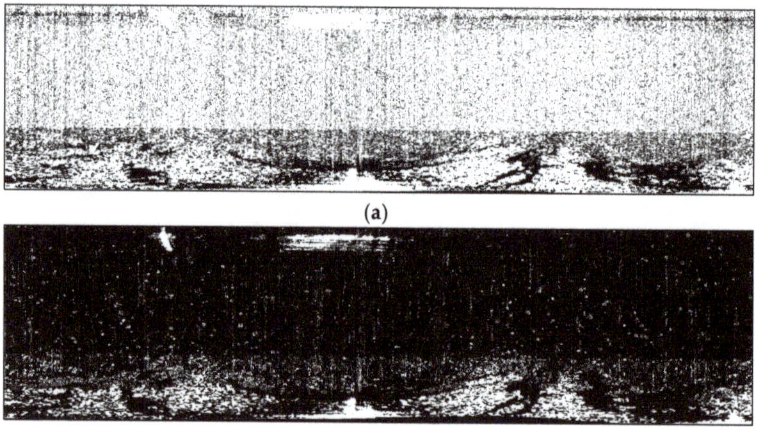

(a)

(b)

Figure 9. Global adaptive threshold algorithm segmentation results. (**a**) Method 1. (**b**) Method 2.

In terms of noise suppression in the oil film results, the two methods shown significant differences in noise processing. Although Method 1 had oil film recognition ability, there were intensive noises in the result. The residual interference noise still needed to be further optimized and suppressed to improve the accuracy of the segmentation results. For Method 2, the results were filled with dense noise pixels, which deeply mixed with the details of the oil films, seriously interfering with the target extraction. This resulted in the need to invest a significant amount of additional cost in denoising during subsequent processing. In contrast, little noise was present in the results of our method. Through adaptive background modeling and the iterative optimization of thresholds, our method could significantly reduce the interference of noise. This mechanism provided favorable conditions for the accurate segmentation of oil films. In terms of oil film recognition ability, our method presented the outline of the oil films more clearly. Although Method 1 could identify a large number of oil film regions, its feature discrimination mechanism had limitations. It was easy to mistake suspected oil film regions and non-oil regions as real targets. Method 2 found it difficult to effectively distinguish between oil films and noises due to excessive noise interference, which led to the inaccurate extraction of target boundaries, resulting in poor recognition performance. In terms of identifying suspected oil film regions, the proposed method effectively reduced the probability of misclassification by fusing multiple features for discrimination. This feature fusion mechanism effectively improved the accuracy of oil film detection. Method 2 involved a significant amount of noise interference, making it difficult to reliably distinguish between oil films and noise. Method 1 not only has the problem of false negatives due to incomplete oil film region segmentation but also results in some oil film regions being inaccurately identified. There were also false positive cases where the background was mistakenly labeled as oil films. Therefore, our method yielded better results.

4.2.2. Comparison with Local Adaptive Threshold Method

The local Otsu adaptive thresholding method (called Method 3 here) and Sauvola local thresholding method (called Method 4 here) [48–50] were compared here with our method, as shown in Figure 10.

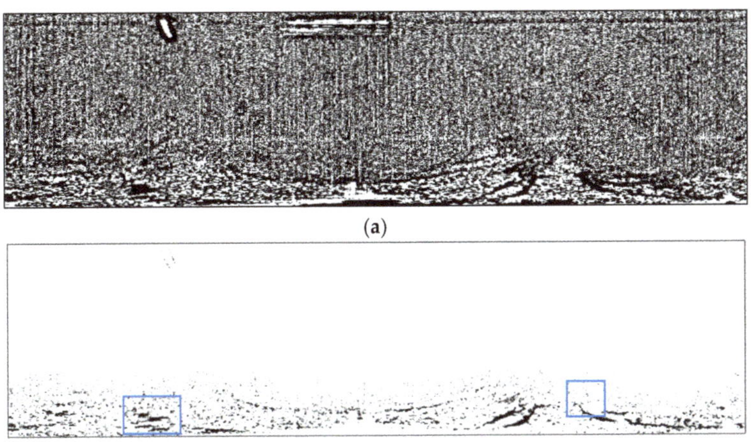

Figure 10. The segmentation results of the local adaptive algorithm. (**a**) Method 3. (**b**) Method 4.

In terms of noise suppression in oil film results, Method 3 relied on local grayscale statistics, resulting in a large amount of residual fine noise. The deep overlap between oil film and noise made it difficult to effectively separate them, which seriously affected the accurate extraction of oil film features. Method 4 adjusted the threshold by local mean and standard deviation, and its noise suppression effect was improved compared to Method 3. However, there were still some residual noise interferences. Our method could significantly reduce the noise interference on oil film detection. Our method integrated multi-dimensional features such as the grayscale gradient and area connectivity. The segmentation results could clearly and completely present the contour of the oil films. The results of Method 3 were remarkably affected by local noise, resulting in blurred oil film boundaries, leading to poor recognition performance. Part of the oil film regions were accurately identified by Method 4 without noise interference. However, the edges of the oil films and weak signal parts were prone to loss. To some extent, application effectiveness in oil spill detection was limited in Method 4. Method 3 used a single threshold, which led to excessive false positives and false negatives. This might lead to the misjudgment of oil film distribution. Method 4 had a better performance than Method 3 in terms of misclassification in suspected oil film regions. However, there were still issues with mislabeling regarding targets and the incomplete identification of oil film regions. So, our method could effectively reduce the probability of misclassification in non-oil film regions.

4.2.3. Comparison with Other Machine Learning Algorithms

Xu et al. proposed a shipborne radar oil spill detection method based on Local Binary Pattern (LBP) texture features and K-Means clustering analysis [51] (called Method 5 here). Li et al. proposed a method for offshore oil spill detection that integrates the Gray-Level Co-occurrence Matrix (GLCM), Support Vector Machine (SVM), and Fuzzy C-Means (FCM) [52] (called Method 6 here). The segmentation results of the two methods are shown in Figure 11.

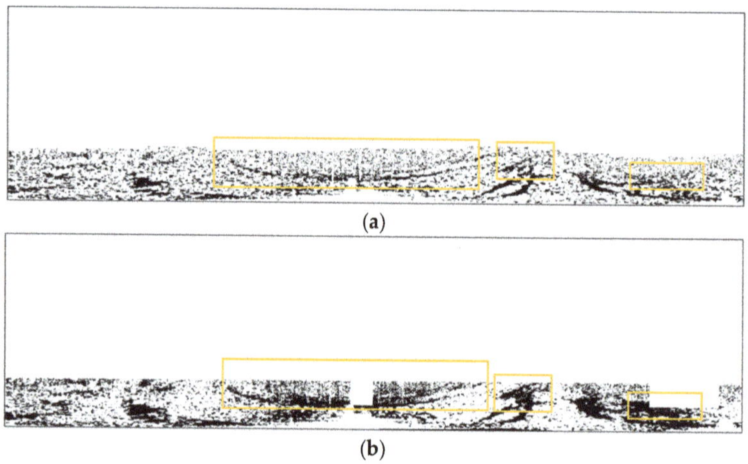

Figure 11. The segmentation results of machine learning algorithms. (**a**) Method 5. (**b**) Method 6.

The results of both methods were filled with small speckles. Both methods are severely affected by the wake of the ship, leading to the misjudgment of suspected oil film regions as actual oil film areas. However, compared with the other two methods, Method 6 still performed well in identifying certain oil film regions. It could better determine the oil film region. For the identification of suspected oil films, our method and Method 5 perform well in accurately distinguishing suspected oil film areas. Method 6 had a weak ability to identify suspected oil films, resulting in a sum of false alarms. The suspected oil films were mistakenly classified as actual oil films.

4.2.4. Quantitative Comparative Analysis

The comparative experimental results clearly demonstrated that our proposed method was superior, as shown in Table 2. Specifically, our method significantly outperformed the other comparative methods in classification metrics such as accuracy, recall, and the F1-score. This outstanding performance not only confirmed the accuracy of the model in object recognition and classification but also reflected its excellent detection coverage. Although some comparison methods achieved near-perfect performance in accuracy metrics, these methods often come at the cost of sacrificing recall. Our method significantly improved recall and spatial accuracy while maintaining high precision, fully verifying its excellent performance in image recognition and segmentation tasks.

Table 2. Comparison of evaluation indicators.

	Accuracy (%)	Precision (%)	Recall (%)	F1-Score (%)	IoU (%)	Dice (%)
Our Method	94.50	99.80	94.61	97.14	94.43	97.14
Method 1	88.54	99.91	88.46	93.84	88.39	93.84
Method 2	58.43	98.41	58.81	73.62	58.25	73.62
Method 3	52.80	99.75	52.28	68.60	52.21	68.60
Method 4	77.79	99.91	77.57	87.34	77.52	87.34
Method 5	75.89	99.92	75.55	86.08	75.55	86.08
Method 6	79.63	99.93	79.35	88.49	79.35	88.49

4.3. Limitations of Marine Radar Oil Spill Monitoring Technology

The detection effect of marine radar oil spill monitoring technology is significantly affected by the conditions of waves, which are directly constrained by weather conditions. There are two key limitations to the practical application of this technology. Firstly, in exceptionally calm sea conditions, the insufficient roughness of the sea surface leads to low wave echo signal strength, making it difficult to effectively identify oil film features. This low-energy echo environment will significantly reduce the detection sensitivity of the system. Secondly, under strong wind conditions, there will be a dual-interference effect: (a) Intense wind and waves will promote the mixing of oil spills and seawater, weakening the inhibitory effect of oil films on sea surface roughness. (b) Enhanced sea clutter will significantly increase the background noise level and increase the difficulty of extracting oil film features from complex wave information. More seriously, excessive wind force may also cause the physical dispersion and dilution of the oil film, fundamentally reducing its detectability.

4.4. Potentiality of NGN-OAT Model in High-Frequency Ground Wave Radar Data

The NGN-OAT segmentation strategy proposed in this study performs well in oil spill detection tasks and also demonstrates potential application value in the field of target detection using high-resolution-range Doppler images. This algorithm can effectively distinguish target signals from sea clutter interference through adaptive background modeling and a multi-feature fusion mechanism, which is highly compatible with the technical requirements of weak target detection in high-frequency ground wave radar (HFSWR) systems [53–56]. Specifically, the dynamic topology optimization mechanism based on NGNs can effectively adapt to the non-stationary distribution characteristics of targets and clutter in the time–frequency domain in the Doppler distance images, while the global search strategy of the OAT model provides a new approach for solving the detection problem of Doppler extended targets. It should be pointed out that the current research mainly focuses on the verification of marine radar oil spill segmentation tasks. In the future, specialized radar datasets will be needed to further validate the cross-domain applicability of this model in HFSWR systems.

5. Conclusions

In this study, an innovative OAT-NGN algorithm based on oil spill data collected by shipborne X-band radar is proposed. This algorithm has successfully achieved the high-precision detection and segmentation of oil film regions in the radar image. The results show that this algorithm exhibits prominent advantages in multiple performance indicators. Compared to the NGN model, the OAT-NGN algorithm has improved accuracy by 11.2%, the recall rate by 11.5%, and IoU by 11.36%. This method effectively overcomes the drawbacks of traditional methods that are prone to falling into local optima and ignoring spatial information. The OAT-NGN algorithm achieves a breakthrough mechanism. Firstly, the NGN module utilizes a neuron competition mechanism to adaptively learn local topology structures. Thus, the spatial distribution characteristics of the oil films are effectively preserved. Secondly, the OAT module introduces multiple seeds for a parallel search and connection penalty terms, which improves the robustness of the algorithm in complex scenarios such as ship wake and wind wave interference. Thirdly, the collaborative optimization mechanism of the NGN module and the OAT module achieves the organic integration of local feature learning and a global threshold search. The algorithm is enhanced to balance computational efficiency while ensuring segmentation accuracy. Compared with the other six methods, the superiority of our algorithm is further verified. The OAT-NGN algorithm can provide an efficient solution for detecting oil spills in shipborne radar datasets. Future work will focus on enhancing the detection ability of weak target signals in complex sea conditions by introducing deep learning methods and further improving recognition accuracy.

Author Contributions: Conceptualization, Z.G. and J.X.; methodology, Z.G.; software, Z.G. and J.X.; validation, Z.G., P.L., Z.G. and J.X.; resources, J.X. and B.J.; data curation, J.X.; writing—original draft preparation, Z.G.; writing—review and editing, J.X. and B.J.; supervision, J.X. and B.L.; project administration, J.X.; funding acquisition, J.X. All authors have read and agreed to the published version of the manuscript.

Funding: This research was supported by the Guangdong Basic and Applied Basic Research Foundation, grant numbers 2025A1515010886 and 2023A1515011212, the National Natural Science Foundation of China, grant number 52271359, the Special Projects in Key Fields of Ordinary Universities in Guangdong Province, grant number 2022ZDZX3005, the Shenzhen Science and Technology Program, grant number JCYJ20220530162200001, the Postgraduate Education Innovation Project of Guangdong Ocean University, grant numbers 202421, 202539 and 202551, and the Guangdong Provincial Key Laboratory of Intelligent Equipment for South China Sea Marine Ranching, grant number 080508132401.

Data Availability Statement: The data collection department did not agree to share the analysis data.

Conflicts of Interest: The authors declare no conflicts of interest.

References

1. Yin, C. International Law Regulation of Offshore Oil and Gas Exploitation. *Environ. Impact. Assess. Rev.* **2021**, *88*, 106551. [CrossRef]
2. Zhou, X.; Ruan, X.; Wang, H.; Zhou, G. Exploring Spatial Patterns and Environmental Risk Factors for Global Maritime Accidents: A 20-Year Analysis. *Ocean Eng.* **2023**, *286*, 115628. [CrossRef]
3. Novaglio, C.; Bax, N.; Boschetti, F.; Emad, G.R.; Frusher, S.; Fullbrook, L.; Hemer, M.; Jennings, S.; van Putten, I.; Robinson, L.M.; et al. Deep Aspirations: Towards a Sustainable Offshore Blue Economy. *Rev. Fish Biol. Fish.* **2022**, *32*, 209–230. [CrossRef] [PubMed]
4. Yu, J.; Zhou, D.; Yu, M.; Yang, J.; Li, Y.; Guan, B.; Wang, X.; Zhan, C.; Wang, Z.; Qu, F. Environmental Threats Induced Heavy Ecological Burdens on the Coastal Zone of the Bohai Sea, China. *Sci. Total Environ.* **2021**, *765*, 142694. [CrossRef] [PubMed]
5. Chen, J.; Di, Z.; Shi, J.; Shu, Y.; Wan, Z.; Song, L.; Zhang, W. Marine Oil Spill Pollution Causes and Governance: A Case Study of Sanchi Tanker Collision and Explosion. *J. Clean. Prod.* **2020**, *273*, 122978. [CrossRef]
6. Zhang, W.; Li, C.; Chen, J.; Wan, Z.; Shu, Y.; Song, L.; Xu, L.; Di, Z. Governance of Global Vessel-Source Marine Oil Spills: Characteristics and Refreshed Strategies. *Ocean Coast. Manag.* **2021**, *213*, 105874. [CrossRef]
7. Ma, X.; Xu, J.; Pan, J.; Yang, J.; Wu, P.; Meng, X. Detection of Marine Oil Spills from Radar Satellite Images for the Coastal Ecological Risk Assessment. *J. Environ. Manag.* **2023**, *325*, 116637. [CrossRef]
8. Wang, L.; Niu, W.; Ding, L.; Zhao, Z. A Bibliometric Review on Marine Economy: Current Status, Development and Future Directions. *Mar. Policy* **2023**, *155*, 105705. [CrossRef]
9. Fernández-Macho, J. Forecasting Marine Spill Risk along the U.S. Pacific Coasts. *Mar. Pollut. Bull* **2024**, *198*, 115826. [CrossRef]

10. Mogollón, R.; Arellano, C.; Villegas, P.; Espinoza-Morriberón, D.; Tam, J. REPSOL Oil Spill off Central Perú in January 2022: A Modeling Case Study. *Mar. Pollut. Bull* **2023**, *194*, 115282. [CrossRef]
11. Oil Tanker Collision and Fire in Gulf of Oman Caused by 'Navigational Misjudgment'. Available online: https://www.thenationalnews.com/news/uae/2025/06/17/uae-coast-guard-rescues-24-crew-members-after-oil-tanker-collision-in-sea-of-oman (accessed on 5 July 2025).
12. Zhang, C.; Zhu, J.; Mai, W.; Chen, Z.; Xie, Y.; Fu, S.; Xia, D.; Cai, C.; Zheng, W.; Liu, J.; et al. Satellite Remote Sensing-Implemented Nontargeted Screening of Emerging Contaminant Fingerprints in a River-to-Ocean Continuum through Interpretable Machine Learning: The Pivotal Intermediary Role of Dissolved Organic Matter. *Environ. Sci. Technol.* **2025**, *59*, 8714–8726. [CrossRef] [PubMed]
13. Mokarram, M.; Pham, T.M. Environmental Monitoring of Oil Pollution in the Marine Waters Using Machine Learning and Remote Sensing. *Adv. Space Res.* **2025**, *75*, 6173–6187. [CrossRef]
14. El Mahrad, B.; Newton, A.; Icely, J.D.; Kacimi, I.; Abalansa, S.; Snoussi, M. Contribution of Remote Sensing Technologies to a Holistic Coastal and Marine Environmental Management Framework: A Review. *Remote Sens.* **2020**, *12*, 2313. [CrossRef]
15. Hengkai, L.; Feng, X.; Qin, L. Remote Sensing Monitoring of Land Damage and Restoration in Rare Earth Mining Areas in 6 Counties in Southern Jiangxi Based on Multisource Sequential Images. *J. Environ. Manag.* **2020**, *267*, 110653. [CrossRef] [PubMed]
16. Shi, D.; Zou, J.; Liu, L.; Huang, F.; Li, L. Research Progress and Prospect of Super-Large Field-of-View Infrared Imaging Technology. *Laser Optoelectron. Prog.* **2025**, *62*, 1000004. [CrossRef]
17. Sun, K.; Jiang, G.; Wang, N.; Yu, D.; Teng, J.; Gao, S.; Huang, J.; Zhao, Z.; Song, Y.; Xin, L.; et al. A Semantic Segmentation Network for Red Tide Detection Based on Enhanced Spectral Information Using HY-1C/D CZI Satellite Data. *Mar. Pollut. Bull* **2025**, *215*, 117813. [CrossRef]
18. Wachmann, A.; Starko, S.; Neufeld, C.J.; Costa, M. Validating Landsat Analysis Ready Data for Nearshore Sea Surface Temperature Monitoring in the Northeast Pacific. *Remote Sens.* **2024**, *16*, 920. [CrossRef]
19. Sun, C.; Li, J.; Liu, Y.; Pan, T.; Shi, K.; Cai, X. Synthesizing Landsat Images Using Time Series Model-Fitting Methods for China's Coastal Areas against Sparse and Irregular Observations. *GIsci. Remote Sens.* **2024**, *61*, 2421574. [CrossRef]
20. Shen, A.; Shao, J. Spectral Harmonization Landsat-8 and Sentinel-2A: The Matching Bands Adjustment before the Missing Bands Prediction Method. *IEEE Trans. Geosci. Remote Sens.* **2024**, *62*, 1–14. [CrossRef]
21. Yan, L.; Roy, D.P. Using Landsat 8 and 9 Operational Land Imager (OLI) Data to Characterize Geometric Distortion and Improve Geometric Correction of Landsat Multispectral Scanner (MSS) Imagery. *Remote Sens. Environ.* **2025**, *321*, 114679. [CrossRef]
22. St-Denis, F.; Tremblay, L.B.; Mahoney, A.R.; Takata-Glushkoff, K.P.L.M. Coastal Sea Ice Concentration Derived from Marine Radar Images: A Case Study from Utqiaġvik, Alaska. *Remote Sens.* **2024**, *16*, 3357. [CrossRef]
23. Xu, J.; Liu, Z.; Hong, G.; Cao, Y. A New Machine-Learning-Based Calibration Scheme for MODIS Thermal Infrared Water Vapor Product Using BPNN, GBDT, GRNN, KNN, MLPNN, RF, and XGBoost. *IEEE Trans. Geosci. Remote Sens.* **2024**, *62*, 1–12. [CrossRef]
24. Bondur, V.; Chernikova, V.; Chvertkova, O.; Zamshin, V. Spatiotemporal Variability of Anthropogenic Film Pollution in Avacha Gulf near the Kamchatka Peninsula Based on Synthetic-Aperture Radar Imagery. *J. Mar. Sci. Eng.* **2024**, *12*, 2357. [CrossRef]
25. Yi, Y.; Chen, K.; Xu, J.; Luo, Q. Tracking the Dynamics of Salt Marsh Including Mixed-Vegetation Zones Employing Sentinel-1 and Sentinel-2 Time-Series Images. *Remote Sens.* **2025**, *17*, 56. [CrossRef]
26. Welikanna, D.R.; Jin, S. A Data Driven Oil Spill Mapping Using GMM Clustering and Damping Ratio on X-Press Pearl Ship Disaster in the Indian Ocean. *Mar. Pollut. Bull.* **2024**, *203*, 116392. [CrossRef]
27. Oluwadare, T.S.; Chen, D.; Oluwafemi, O.; Babadi, M.; Hossain, M.; Ibukun, O. Reconstructing Snow-Free Sentinel-2 Satellite Imagery: A Generative Adversarial Network (GAN) Approach. *Remote Sens.* **2024**, *16*, 2352. [CrossRef]
28. Li, C.; Kim, D.-j.; Park, S.; Kim, J.; Song, J. A Self-Evolving Deep Learning Algorithm for Automatic Oil Spill Detection in Sentinel-1 SAR Images. *Remote Sens. Environ.* **2023**, *299*, 113872. [CrossRef]
29. Fan, Y.; Rui, X.; Zhang, G.; Yu, T.; Xu, X.; Poslad, S. Feature Merged Network for Oil Spill Detection Using Sar Images. *Remote Sens.* **2021**, *13*, 3174. [CrossRef]
30. Chen, F.; Balzter, H.; Ren, P.; Zhou, H. SRCNet: Seminal Image Representation Collaborative Network for Oil Spill Segmentation in SAR Imagery. *IEEE Trans. Geosci. Remote Sens.* **2024**, *62*, 1–18. [CrossRef]
31. Zhu, Q.; Zhang, Y.; Li, Z.; Yan, X.; Guan, Q.; Zhong, Y.; Zhang, L.; Li, D. Oil Spill Contextual and Boundary-Supervised Detection Network Based on Marine SAR Images. *IEEE Trans. Geosci. Remote Sens.* **2022**, *60*, 5213910. [CrossRef]
32. Cheng, L.; Li, Y.; Zhang, X.; Xie, M. An Analysis of the Optimal Features for Sentinel-1 Oil Spill Datasets Based on an Improved J–M/K-Means Algorithm. *Remote Sens.* **2022**, *14*, 4290. [CrossRef]
33. Song, D.; Huang, Q.; Gao, H.; Wang, B.; Zhang, J.; Chen, W. Adaptive Oil Spill Detection Network for Scene-Based PolSAR Data Using Dynamic Convolution and Boundary Constraints. *Int. J. Appl. Earth Obs. Geoinf.* **2024**, *130*, 103914. [CrossRef]
34. Ji, Y.; Wang, Y.; Sun, W.; Zhang, H.; Li, F.; Huang, W. Target Monitoring Capability Analysis for Shipborne HFSWR under Different Platform Motions. *IEEE Trans. Geosci. Remote Sens.* **2024**, *62*, 1–17. [CrossRef]

35. Lau, T.K.; Huang, K.H. A Timely and Accurate Approach to Nearshore Oil Spill Monitoring Using Deep Learning and GIS. *Sci. Total Environ.* **2024**, *912*, 169500. [CrossRef] [PubMed]
36. Chen, R.; Jia, B.; Ma, L.; Xu, J.; Li, B.; Wang, H. Marine Radar Oil Spill Extraction Based on Texture Features and BP Neural Network. *J. Mar. Sci. Eng.* **2022**, *10*, 1904. [CrossRef]
37. Liu, P.; Li, Y.; Liu, B.; Chen, P.; Xu, J. Semi-Automatic Oil Spill Detection on X-Band Marine Radar Images Using Texture Analysis, Machine Learning, and Adaptive Thresholding. *Remote Sens.* **2019**, *11*, 756. [CrossRef]
38. Martinetz, T.; Schulten, K. A "Neural-Gas" Network Learns Topologies. In *Artificial Neural Networks*; Elsevier: Amsterdam, The Netherlands, 1991; pp. 397–402.
39. Cottrell, M.; Hammer, B.; Hasenfuß, A.; Villmann, T. Batch and Median Neural Gas. *Neural Netw.* **2006**, *19*, 762–771. [CrossRef]
40. Dirgová Luptáková, I.; Šimon, M.; Huraj, L.; Pospíchal, J. Neural Gas Clustering Adapted for Given Size of Clusters. *Math. Probl. Eng.* **2016**, *2016*, 9324793. [CrossRef]
41. Qin, A.K.; Suganthan, P.N. Robust Growing Neural Gas Algorithm with Application in Cluster Analysis. *Neural Netw.* **2004**, *17*, 1135–1148. [CrossRef]
42. Qin, A.K.; Suganthan, P.N. Enhanced Neural Gas Network for Prototype-Based Clustering. *Pattern Recognit.* **2005**, *38*, 1275–1288. [CrossRef]
43. Wang, R.B.; Hu, R.B.; Geng, F.D.; Xu, L.; Chu, S.C.; Pan, J.S.; Meng, Z.Y.; Mirjalili, S. The Animated Oat Optimization Algorithm: A Nature-Inspired Metaheuristic for Engineering Optimization and a Case Study on Wireless Sensor Networks. *Knowl. Based Syst.* **2025**, *318*, 113589. [CrossRef]
44. Huang, L.K.; Wang, M.J.J. Image Thresholding by Minimizing the Measures of Fuzziness. *Pattern Recognit.* **1995**, *28*, 41–51. [CrossRef]
45. Goh, T.Y.; Basah, S.N.; Yazid, H.; Aziz Safar, M.J.; Ahmad Saad, F.S. Performance Analysis of Image Thresholding: Otsu Technique. *Measurement* **2018**, *114*, 298–307. [CrossRef]
46. Ma, G.; Yue, X. An Improved Whale Optimization Algorithm Based on Multilevel Threshold Image Segmentation Using the Otsu Method. *Eng. Appl. Artif. Intell.* **2022**, *113*, 104960. [CrossRef]
47. Tan, J.; Tang, Y.; Liu, B.; Zhao, G.; Mu, Y.; Sun, M.; Wang, B. A Self-Adaptive Thresholding Approach for Automatic Water Extraction Using Sentinel-1 SAR Imagery Based on OTSU Algorithm and Distance Block. *Remote Sens.* **2023**, *15*, 2690. [CrossRef]
48. Najafi, M.H.; Salehi, M.E. A Fast Fault-Tolerant Architecture for Sauvola Local Image Thresholding Algorithm Using Stochastic Computing. *IEEE Trans. Very Large Scale Integr. VLSI Syst.* **2016**, *24*, 808–812. [CrossRef]
49. Mohamed Razali, M.R.; Ahmad, N.S.; Mohd Zaki, Z.; Ismail, W. Region of Adaptive Threshold Segmentation between Mean, Median and Otsu Threshold for Dental Age Assessment. In Proceedings of the I4CT 2014—1st International Conference on Computer, Communications, and Control Technology, Langkawi, Malaysia, 2–4 September 2014; Institute of Electrical and Electronics Engineers Inc.: New York, NY, USA, 2014; pp. 353–356.
50. Sauvola, J.; Pietikäinen, M. Adaptive Document Image Binarization. *Pattern Recognit.* **2000**, *33*, 225–236. [CrossRef]
51. Xu, J.; Pan, X.; Jia, B.; Wu, X.; Liu, P.; Li, B. Oil Spill Detection Using LBP Feature and K-Means Clustering in Shipborne Radar Image. *J. Mar. Sci. Eng.* **2021**, *9*, 65. [CrossRef]
52. Li, B.; Xu, J.; Pan, X.; Ma, L.; Zhao, Z.; Chen, R.; Liu, Q.; Wang, H. Marine Oil Spill Detection with X-Band Shipborne Radar Using GLCM, SVM and FCM. *Remote Sens.* **2022**, *14*, 3715. [CrossRef]
53. Golubović, D.; Erić, M.; Vukmirović, N. High-Resolution Doppler and Azimuth Estimation and Target Detection in HFSWR: Experimental Study. *Sensors* **2022**, *22*, 3558. [CrossRef]
54. Golubovic, D.; Eric, M.; Vukmirovic, N. High-Resolution Method for Primary Signal Processing in HFSWR. In Proceedings of the 2022 30th European Signal Processing Conference (EUSIPCO), Belgrade, Serbia, 29 August–2 September 2022; Volume 2022, pp. 912–916.
55. Golubovic, D.; Eric, M.; Vukmirovic, N. Improved Detection of Targets on the High-Resolution Range-Doppler Map in HFSWRs. In Proceedings of the 2024 23rd International Symposium INFOTEH-JAHORINA (INFOTEH 2024), East Sarajevo, Bosnia and Herzegovina, 20–22 March 2024; Institute of Electrical and Electronics Engineers Inc.: New York, NY, USA, 2024.
56. Jiang, Y.; Wang, Y.H.; Li, Y.; Chen, X. Eigenvalue-Based Ground Target Detection in High-Resolution Range Profiles. *IET Radar Sonar Navig.* **2020**, *14*, 1747–1756. [CrossRef]

remote sensing

MDPI

Article

Detection of Oil Spills in the Northern South China Sea Using Landsat-8 OLI

Xiaorun Hong [1,†], Lusheng Chen [2,†], Shaojie Sun [2,3,4,5,*], Zhen Sun [6], Ying Chen [2], Qiang Mei [7] and Zhichao Chen [2]

1 School of Geography and Planning, Sun Yat-sen University, Guangzhou 510006, China
2 School of Marine Sciences, Sun Yat-sen University, Zhuhai 519082, China
3 Southern Marine Science and Engineering Guangdong Laboratory (Zhuhai), Zhuhai 519000, China
4 Guangdong Provincial Key Laboratory of Marine Resources and Coastal Engineering, Guangzhou 510275, China
5 Pearl River Estuary Marine Ecosystem Research Station, Ministry of Education, Zhuhai 519000, China
6 School of Marine Engineering and Technology, Sun Yat-sen University, Zhuhai 519082, China
7 Navigation College, Jimei University, Xiamen 361021, China
* Correspondence: sunshj7@mail.sysu.edu.cn; Tel.: +86-186-6697-2696
† These authors contributed equally to this work.

Abstract: Petroleum extraction, transportation, and consumption in the marine environment contribute to a large portion of anthropogenic oil spills into the ocean. While previous research focuses more on large oil spill accidents from oil tankers or offshore oil platforms, there are few systematic records on occasional regional oil spills. In this study, optical imagery from Landsat-8 OLI was used to detect oil slicks on the ocean surface through spatial analysis and spectral diagnosis in the northern South China Sea (NSCS). The source of the slicks was identified through datasets from traffic density and platform locations. A total of 632 oil slicks were detected in the NSCS from 2015 to 2019, where 57 were from platforms sources, and 490 were from ships. The average area of the detected slicks was 4.8 km^2, and half of the slicks had areas <1.7 km^2. Major oil spill hot spots included coastal Guangdong (ship origins), southeast and northeast Dongsha Island (ship origins), middle of south Beibu Gulf (ship and platform origins), and southeast Pearl River Estuary (platform origins). Through this study, we demonstrate the capability of medium-resolution optical imagery in monitoring regional oil spills. Such results and methods may help in near real-time oil spill monitoring and further environmental assessments.

Keywords: oil spills; remote sensing; spill origin; northern South China Sea

1. Introduction

Oil spills in the ocean may have an adverse impact on the marine environment. The 2010 Deepwater Horizon oil spill leaked more than 3 million barrels of oil into the Gulf of Mexico (GoM) and was demonstrated to have greatly affected the marine habitat and the local environment [1]. The Iranian oil tanker SANCHI, carrying 1 million barrels of condensate oil, collided with a grain freighter in 2018, causing fires and sinking and released carried oil into the East China Sea [2]. In 2020, the Japanese bulk carrier Wakashio hit Pt d'Esny Coral Reef in Mauritius, leaking oil into a lagoon composed of a variety of habitats including beaches, mangroves, swamps, seagrass beds, and coral reefs [3].

Such large oil spill accidents attract lots of public attention; however, small-sized oil spills happen more frequently and may leak more volume than large spill accidents. According to statistics from ITOPF [4], over 80% of oil spill incidents recorded were small spills (<7 tons) since 1970. Moreover, accident spills and operational discharges from non-tanker vessels alone contribute to >20% of the total spilled volume annually, though they usually would not lead to large spill accidents [5]. Sources of oil in the marine environment can be categorized into four types: oil extraction, transportation (including oil tankers, pipelines, etc.), consumption (including land runoff, non-tank vessel spill, and operational discharge), and natural seeps [5,6], which correspond to 3%, 12%, 38%, and 47% of total spilled volume worldwide according to statistics from 1990 to 1999 [5]. A recent study has found that anthropogenic oil spills are one order of magnitude larger than spills from natural seeps globally (94% vs. 6%) [7], according to data from 2014 to 2019.

Monitoring all the oil spills in the ocean is not trivial [8,9]. The U.S. National Response Center theoretically records leaked oil that can cause surface discoloration or deposited sludge/emulsion below the ocean surface in U.S. waters [10]. However, such records are reported mainly by responsible parties (polluters) or third parties (passers-by), and their reports on spill location and magnitude are not validated independently [11]. Some researchers statistically report regional oil spills in a certain period [12,13]. Though statistical data show oil spill area and frequency, we know little about spatial distributions of the oil spills. Moreover, such records are usually not up-to-date. In addition, since such records usually focus on large spill accidents, data of those small-sized spills are regularly blank.

With characteristics of wide coverage, and rapid, continuous, and periodic imaging, remote sensing nowadays plays an important role in oil spill detection on the sea surface [14–16]. Many remote sensing techniques serve this purpose, including Synthetic Aperture Radar (SAR), optical remote sensing, and thermal remote sensing [17–19]. The Sentinel-1 SAR and ENVISAT ASAR were used to interpret oil spills in the western Java Sea [20]. Oil spills in the Eastern Arabian Sea were extracted by Sentinel-1 SAR to study the year-to-year variability of oil spills [21]. Both the above studies were conducted through visual inspection and manual delineation/clustering of oil pixels based on the spatial contrast and shape of the oil spills. During the 2010 Deepwater Horizon oil spill, the released oil was effectively mapped by SAR through a Textural Classifier Neural Network Algorithm [22], with the resulting cumulative oil coverage in the GoM comparable to results from the optical MODIS and MERIS interpretation (within 16%) [23].

Oil spills can also be differentiated by optical remote sensing through spatial and spectral analyses. The optical MODIS images are used to detect natural oil slicks in the northwestern GoM through spatial contrast under sun glint, with the comparable slick area in same-day SAR detection [24]. Oil slick spatial contrasts and spectral characteristics have been interpreted in Landsat (TM, ETM+, OLI) and Sentinel-2 MSI images to manually delineate chronic spills from an abandoned platform in the GoM with areal relative error ~50% from same-day detection of SAR and other optical sensors [25]. In this study, the recurring, long time series Landsat OLI imagery was used to detect and statistically analyze oil spills (small- and large-sized) in the northern South China Sea (NSCS) from 2015 to 2019. The study has two objectives. First, we hope to demonstrate remote sensing capabilities in oil spill monitoring on a regional scale. Second, by accessing databases from oil production platforms and marine traffic, the sources of the detected oil slicks can be determined.

2. Data

2.1. Study Area

Figure 1 shows the study area of this research in the NSCS (~105°–121°E, ~16°–25°N). The South China Sea (SCS) is a semi-enclosed basin, located between the Pacific Ocean and the Indian Ocean, and is one of the world's four major oil and gas accumulation centers, with proven oil reserves of 6.4 billion tons and natural gas reserves of 980 billion m^3 [26]. Main oil and gas reserves of Pearl River Mouth Basin, Beibu Gulf Basin, Yinggehai Basin, and southeast Qiong basin are included in the study area [26]. Moreover, the SCS is an important hub of international strategic material transportation for East Asian countries, with particularly high ship traffic in the NSCS. According to Chang and Yang [27], oil transported through the SCS accounts for half of the world's total oil transportation volume each year.

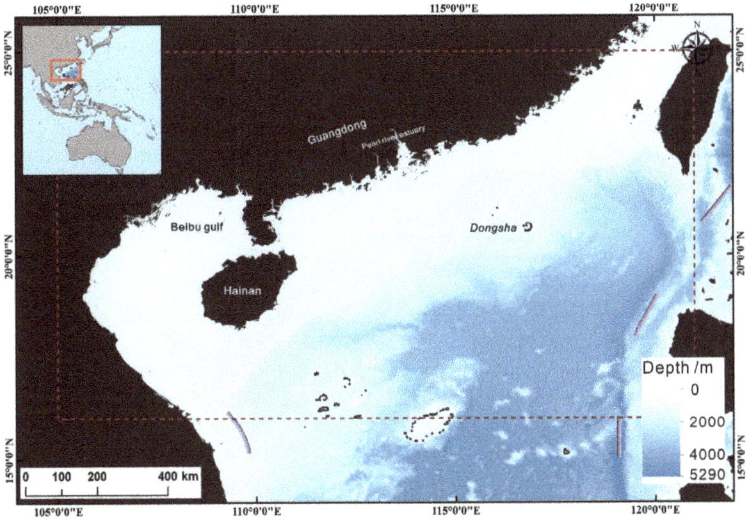

Figure 1. The study area (longitude ~105°–121°E, latitude ~16°–25°N).

2.2. Satellite Datasets

Satellite images used in this study include the Landsat-8 OLI from 2015 to 2019 and selected PlanetScope imagery in 2018 and 2019. Landsat-8 OLI images were downloaded from the United States Geological Survey (USGS), with a spatial resolution of 30 m for multispectral bands. OLI images were then processed to Rayleigh-corrected reflectance (Rrc, dimensionless) using the ACOLITE (version 20211124.0, https://github.com/acolite, accessed on 18 January 2022). ACOLITE is a generic processor developed at the Royal Belgian Institute of Natural Sciences for atmospheric correction and processing for aquatic applications of medium to high resolution optical imagery, including Landsat, Sentinel-2, PlanetScope, etc. [28]. PlanetScope 3-band Top of Atmospheric reflectance data (~3 m spatial resolution) were obtained from Planet labs [29]. It is noted that the standard PlanetScope data are usually not provided for regions >15 km from shoreline except for the 3-band data in red-green-blue channels. Same day PlanetScope images at the determined slick locations by OLI were downloaded from March to October 2018. The 3-band RGB images were then used to identify oil presence and delineate slick distribution. Table 1 listed the specifications of the satellite imagery used in this study.

Table 1. Specifications of the satellite image used in this study.

Sensor	Resolution	Revisit Time	Wavelength (nm)	Bands	Running Time	Source
OLI	30 m	16 days	430–2290	8	2013–now	USGS
PlanetScope	~3 m	1–2 days	455–875	3–8	2014–now	Planet labs

Of the many remote sensing techniques used for oil spill detection, SAR and optical remote sensing are most frequently used [14,19]. The optical OLI imagery was used in this study for oil detection because OLI has regular coverage both for near shoreline and offshore waters, and its dataset was publicly available over the study period. The PlanetScope imagery in this study was used for cross-checking purposes over the selected period.

2.3. Oil Slick Detection by OLI

In optical imagery, spatial contrasts between oil and clear water pixels are mainly caused by two factors: oil has its unique optical properties, which feature strong absorption in the blue and exponentially decay toward longer wavelengths; the coverage of an oil slick can dampen sea surface capillary and short gravity waves, enhancing the contrast between oil and water under sun glint [30]. The latter also applies for oil slick detection by SAR [14]. Because of the high absorption, thick crude oil would usually display lower than background water reflectance in the visible wavelengths, with the largest contrast in the blue–green bands depending on background water optical properties [30]. The oil–water mixing in the oil emulsions enables high scattering and thus elevated reflectance in the red, Near InfraRed (NIR), and ShortWave InfraRed (SWIR) bands. Under sun glint, oil would show either positive or negative contrast with nearby uncontaminated water, which demonstrates higher or lower than water reflectance over the spectral bands [30,31]. Moreover, oil slicks usually have elongated shapes and exhibit sharp edges in a sun glint image because of oil's dampening effect [32].

The above principles were used to discriminate oil from uncontaminated seawater in this study. Firstly, enhanced RGB true color images were visually inspected to check spatial anomalies; the anomalies were then diagnosed spatially and spectrally to eliminate oil look-alikes. After spatial/spectral diagnoses, the identified oil slicks were finally delineated in software ArcMap (version 10.6, Environment System Research Institute, Redlands, CA, USA). The flowchart in Figure 2 summarizes the oil slick extraction procedures.

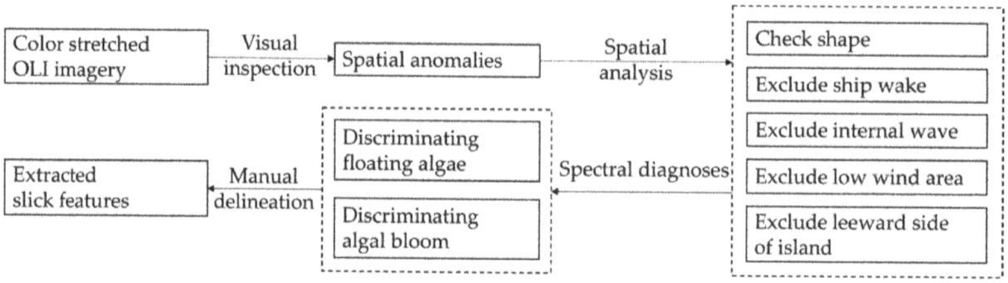

Figure 2. Flowchart of oil slick feature extraction procedures.

As mentioned, oil displays either positive or negative contrast or partly positive and partly negative contrast with water [24,32]. In this study, the bright slick or slick part (positive oil–water contrast) of the anomaly feature was spectrally diagnosed. The corresponding dark slick or slick part under sun glint (negative oil–water contrast), mostly resampled water spectra but with a lower reflectance, was usually diagnosed through the feature's morphology (e.g., typical shapes from ship wake, internal waves, etc.). Figure 3a shows Rrc spectra from oil slicks of the dark and bright parts and seawater under sun glint. Generally, bright slicks' reflectance was higher than water's, while dark slick reflectance was lower than water's (Figure 3b), which is mainly attributed to the sun glint effect caused by oil's wave dampening.

Oil look-alikes were ruled out through a combination of spectral diagnosis and shape analysis. Such oil-resembled features in optical imagery include floating algae, ship wake, internal waves, etc. Ship wake is a kind of trace left on the water surface by a moving ship, which is usually presented by one or two straight lines and could extend over several kilometers. Once formed, the ship wake would last from minutes to several hours, with the shape of the slicks changing with time because of the current movement. Ship wakes generally appear as a V-shape with a dark region between two wake arms in remote sensing imagery [33,34]. Figure 4 shows a transect of the ship wake perpendicular to the ship's driving direction, presenting two bright lines at the edge of the V-shape and a dark region within the V-shape following the ship. As shown in Figure 4, the reflectance elevation at the edge and depression between the two lines were clearly demonstrated from the reflectance of the 561-nm and 865-nm bands. Similar brightness reversal happens in images with internal waves. Figure 5 shows internal waves that happened 120 km west of the Dongsha Island in May 2016, which featured alternating bands of light and dark strips in the OLI image. Such a reflectance anomaly was caused by variations of roughness on the surface (Figure 5b), which can be effectively differentiated from oil slicks in OLI imagery.

In addition to ship wakes and internal waves, the leeward side of an island and low wind areas can also be differentiated because of direct attachment to islands and extensive dark areas with scattered dark patches at the borders, respectively.

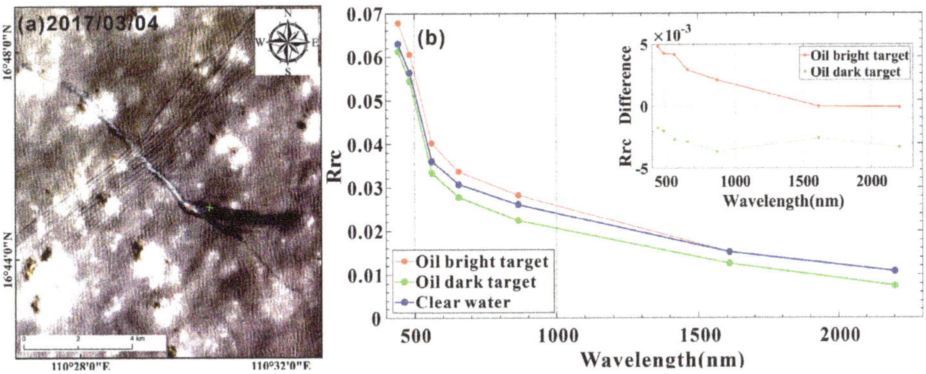

Figure 3. (**a**) Outlined oil slicks in the NSCS: (**b**) Rrc reflectance spectra from the marked points in panel a. The inset shows the Rrc difference (oil–water) between oil and clear water.

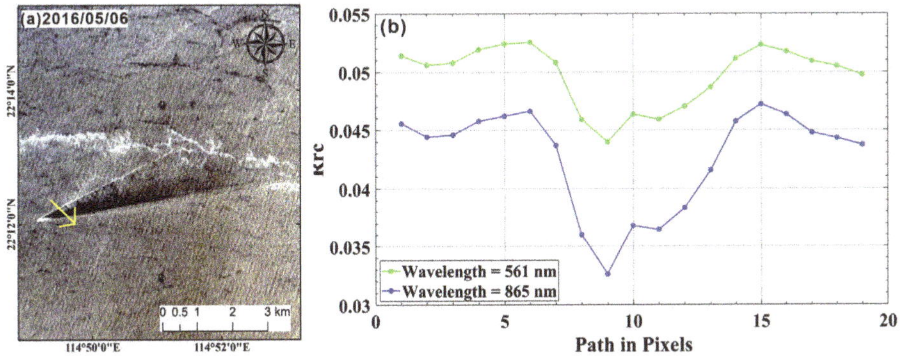

Figure 4. (**a**) Ship wakes on the ocean surface; (**b**) Rrc spectra at 561-nm and 865-nm bands along the transect marked in panel a (yellow arrow).

Floating algae of various forms could resample oil slicks in optical imagery [35]. tides are widely distributed along the coast of Guangdong and Guangxi Province, China, with red tide blooms happening mostly from March to June each year [36,37]. Figure 6 shows typical red tide events that usually happen near the Guangdong coast, presenting dark patches in the RGB image. The Rrc spectra show decreased reflectance in the blue to green bands (Figure 6b). The dark color and the decreased reflectance in blue–green bands resemble thick crude oil in optical imagery [30]. However, such surface algal blooms are usually sparsely distributed, while oil slicks typically demonstrate sharp edges (drastic reflectance change across the edge). Considering the distribution pattern and the bloom season, such a red tide could still be discriminated from those small-sized spills from ships or platforms. Figure 7 demonstrates the appearance of blooms from Red *Noctiluca Scintillans*, which feature high absorption in the blue bands and elevated reflectance in the near-infrared bands [38]. With these above characteristics, the spatial contrast caused by algal blooms or floating algae can be excluded through spatial and spectral analysis.

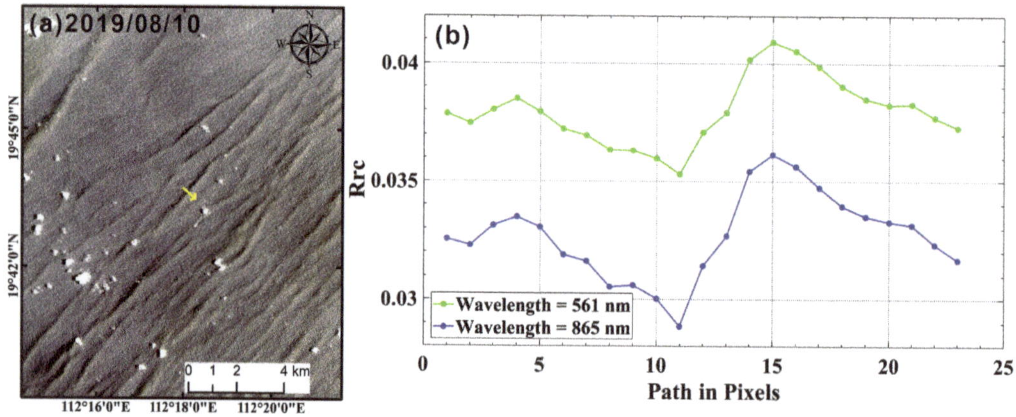

Figure 5. (**a**) Internal wave presented as alternating bands of light and dark strips; (**b**) Rrc reflectance at 561-nm and 865-nm bands along the transect marked in panel a (yellow arrow).

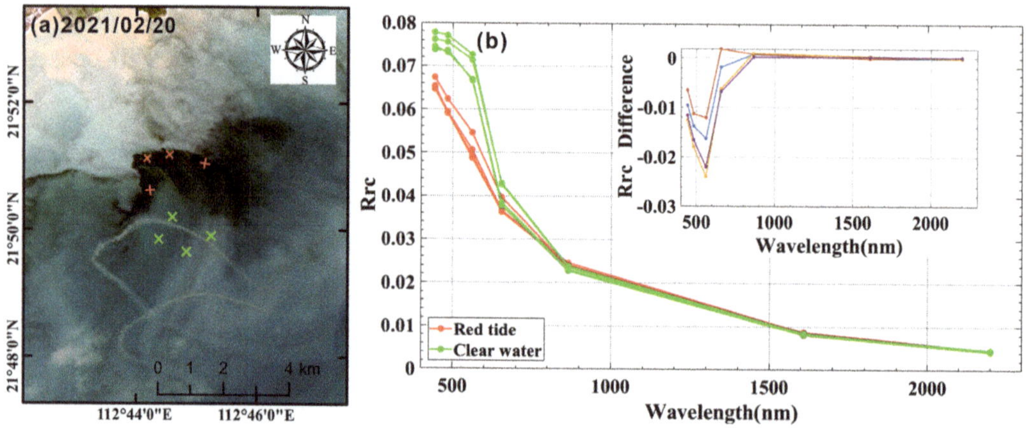

Figure 6. (**a**) Typical red tide blooms near Yangjiang; (**b**) spectral curves of selected red tide blooms and nearby water pixels marked as red and green crosses in panel a, respectively; The inset shows the Rrc difference between red tide and nearby water.

Slick features partially blocked by small cloud (scattered cloud) or other objects would be connected through visual interpretation in the delineation process (e.g., in Figure 2). If the slick feature was under complete cloud cover, it would not be detected through the OLI image. Or, if a large portion of the slicks was under cloud cover, such slicks would not be delineated for areal analysis. To be conservative, all delineated slicks with an area <0.2 km^2 (corresponding to slicks ~1 km in length and ~200 m in width, not including the detached ones from a major slick, where they were defined as one slick in this study) were excluded from the dataset.

2.4. Offshore Platform Database and Traffic Densities

An offshore platform database in the study area was overlaid to determine if the detected oil slicks originated from oil and gas platforms. The offshore platform database was retrieved through a time series remote sensing approach [39], which includes information on platform location, its status (physically removed or not), the installation date, and the removal date (if removed) of the platforms. In this study, if a platform was directly associated with one end of the oil slick, then the detected slick was determined to be originating from the corresponding platform.

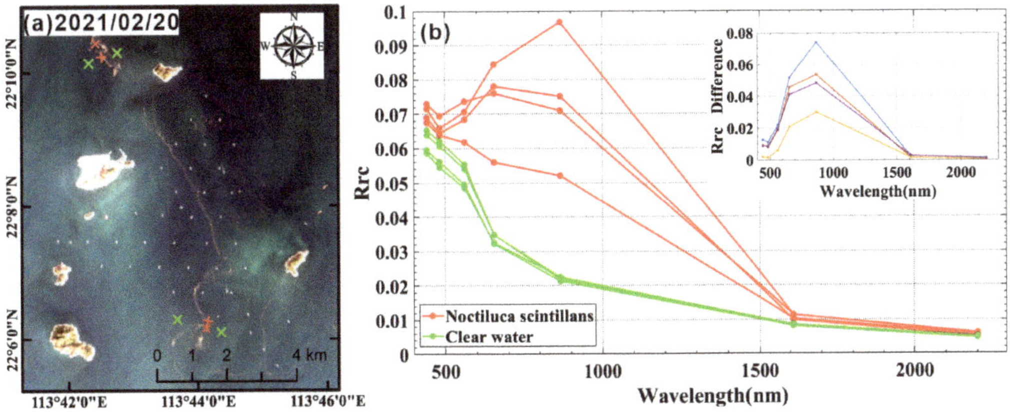

Figure 7. (**a**) Red *Noctiluca Scintillans* bloom in the Pearl River Estuary; (**b**) spectral curves of selected Red *Noctiluca Scintillans* and nearby water pixels marked as red and green crosses in panel a, respectively. The inset shows the Rrc difference between Red *Noctiluca Scintillans* and nearby water.

Traffic density data were used in this study to determine oil slicks from ship origins. Here a global shipping traffic density dataset was used, which was created using hourly AIS positions received between January 2015 and February 2021 from the International Monetary Fund (IMF). The value of the traffic density data represents the total number of AIS positions that have been reported by ships in each grid cell (0.005° spatial resolution). The dataset was then down-sampled to 0.05° spatial resolution and divided by the total months in the above period to obtain the monthly average number of ships reported within the grid (0.05°, ~5.5 km at the equator).

After the detected slicks were determined not from oil platforms, a check of connected ships from the slick was executed. If a ship was directly associated with one end of the oil slick, the slick or nearby slicks would be classified as a vessel-originated slick. Otherwise, a simple traffic density threshold would be used to determine if the slicks originate from ships. The median traffic density not including the value of zero in the study area were chosen as the threshold. Here such a median value is 991, which means an average of 991 ships are reported to be within the extent of the grid in one month's time. Detected slicks with traffic densities surpass the threshold would be designated as vessel oil spills. If below the threshold, the detected slicks would be designated as from an unknown source.

3. Results

3.1. Delineated Oil Slicks

After inspecting and interpreting ~4800 Landsat 8 OLI images following the procedures described in Section 2.3, we identified 632 oil slicks in the NSCS from 2015 to 2019, among which 490 records were identified from ships, 57 from platforms, and 85 from unknown sources, accounting for 78%, 9%, and 13% of the total number of oil spills, respectively. Sources of oil slicks from ships mostly appeared as long strips (Figure 8), while some slicks showed irregular shapes. The shape of the slicks is usually determined by source types, discharged time, current and wind force. For example, the long narrow slick in Figure 8d was freshly released from a moving vessel (arrow pointed), and the earlier released part became wider because of the longer spreading time. The lower left of Figure 8d displays ship wakes from a moving vessel, which showed a typical V-shape with a dark region left behind. Figure 9 presents oil slicks determined from platform sources. As can be seen, all the slicks were directly connected to one or more of the platforms, usually with elongated shapes to one side of the platforms because of current and wind forces. Compared with ship-originated slicks, the shape of slicks from platform sources was often irregular (Figure 8 vs. Figure 9).

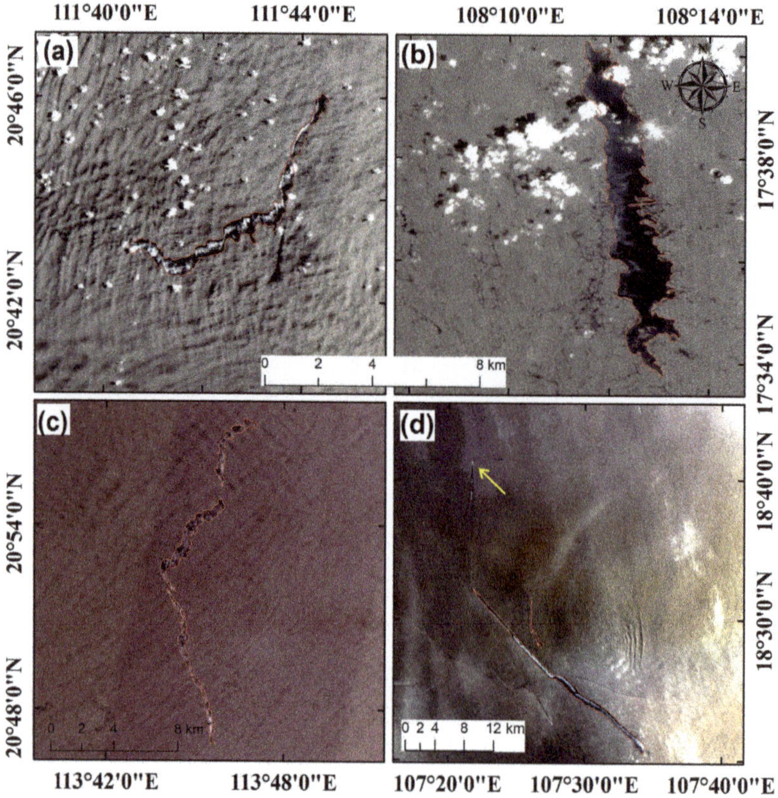

Figure 8. RGB imagery of oil slicks determined from ship sources in cases (**a**–**d**). Yellow arrow in panel d refers to the associated ship with the slicks.

3.2. Distribution of Oil Spills

The derived oil slick distribution map demonstrated large variability spatially and temporally (Figure 10). Year 2017 witnessed the highest oil-covered area (1208 km^2), 2018 the lowest (383 km^2), while there were fewer variations with other years (498–653 km^2). After combining all detected slicks, a cumulative oil pollution area of 3242 km^2 was determined in the NSCS from 2015 to 2019 (Figure 10). For individual slicks, the maximum area of detected slicks was 327 km^2 (35.5 after removing extremely large slicks). The average area of detected slicks was 4.8 km^2, with a median area of 1.7 km^2. The median area was significantly smaller than the mean because of extremely large spills pulling up the mean value. Several of the below regions showed a high incidence of oil spills: coastal areas from Shenzhen to Shantou, offshore regions from Wenchang to Yangjiang, the northeastern Beibu Gulf, and the middle of the south Beibu Gulf (Figure 10b). Overall, slicks were more frequently observed in the coastal region than in offshore waters. In reality, 50% of the oil spills were found within 33 km of the shoreline over the NSCS, and 90% of the oil spills were within 112 km of the shoreline. Particularly in the coastal Guangdong, the derived oil slicks were mostly in elongated shape and parallel to shorelines. In the Beibu Gulf, however, slicks were found to be distributed mostly offshore, where detected slicks tended to be in irregular shapes.

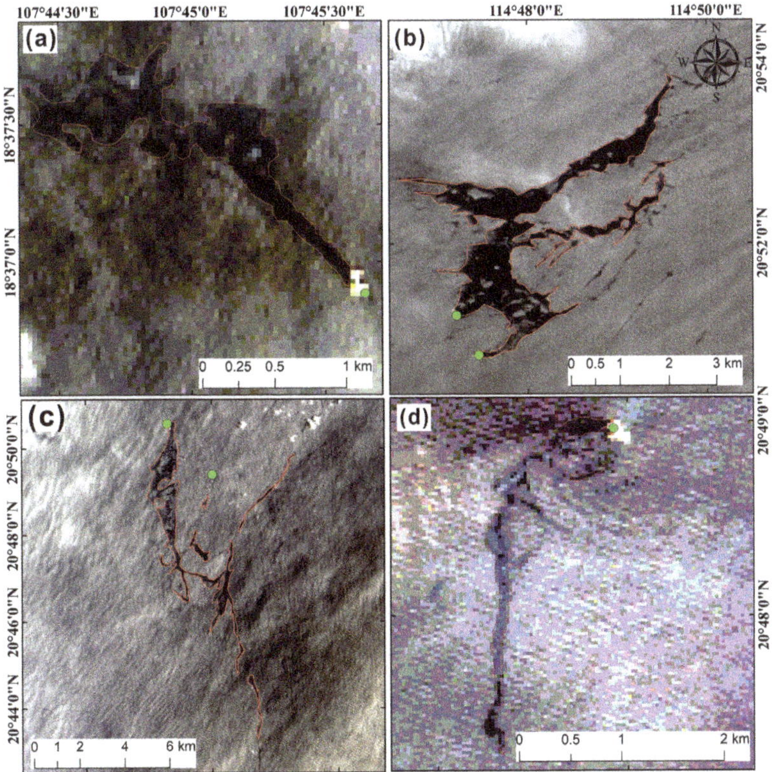

Figure 9. RGB imagery of oil slicks determined from platform sources in cases (**a**–**d**). The green dots denote oil platform locations.

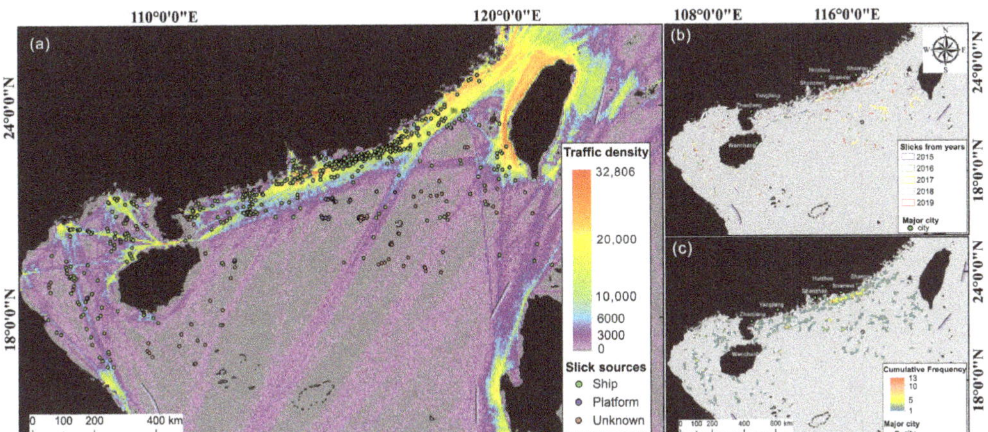

Figure 10. (**a**) Traffic density and the determined slick origins in the study area; (**b**) distribution of oil spills in the NSCS; (**c**) oil slick cumulative distribution (grid: 10×10 km^2).

The extracted oil slicks were further binned into 10×10 km^2 grids (Figure 10c), where the value of the grid denoted the number of times the detected slicks covered this region during 2015–2019. Oil slicks were most frequently observed south of Huizhou city, where a total of 13 oil spills were detected. The coastal area from Shenzhen to Shantou represented a high incidence of oil spill regions in the NSCS (Figure 10c). Eight oil slicks were detected in the offshore area east of Hainan Island. However, the above hot spot was within one grid, where surrounding waters scarcely detected oil presence.

We further select two sub-regions to present detailed information on oil slick distribution spatially and temporally: the Guangdong–Hong Kong–Macao Greater Bay Area (GBA, Figure 11a) and the northeastern Beibu Gulf (Figure 11b). GBA includes three of the world's busiest ports (Guangzhou, HongKong, and Shenzhen) and is an important marine traffic hub in Southeast Asia. The Beibu Gulf Basin is a major oil and gas reserve located in the northeastern Beibu Gulf as stated in Section 2. Oil slicks in GBA were most frequently observed off the coast in the east side of the Pearl River Estuary (PRE), most of which were from ships. Oil spill cases in this region ranged from 30 to 50 yearly in this region, with average oil spills of ~40 cases each year (Table 2). The years 2016 and 2019 witnessed the largest slicks that exceeded 30 km^2, with the rest years at ~23 km^2. The years 2015, 2016, and 2019 witnessed the largest total polluted area (~160 km^2). The median slick area was ~1.6 km^2 each year. Seasonally, there was a clear pattern that spills were more frequently observed in spring and summer than in fall and winter in the GBA (Table 2).

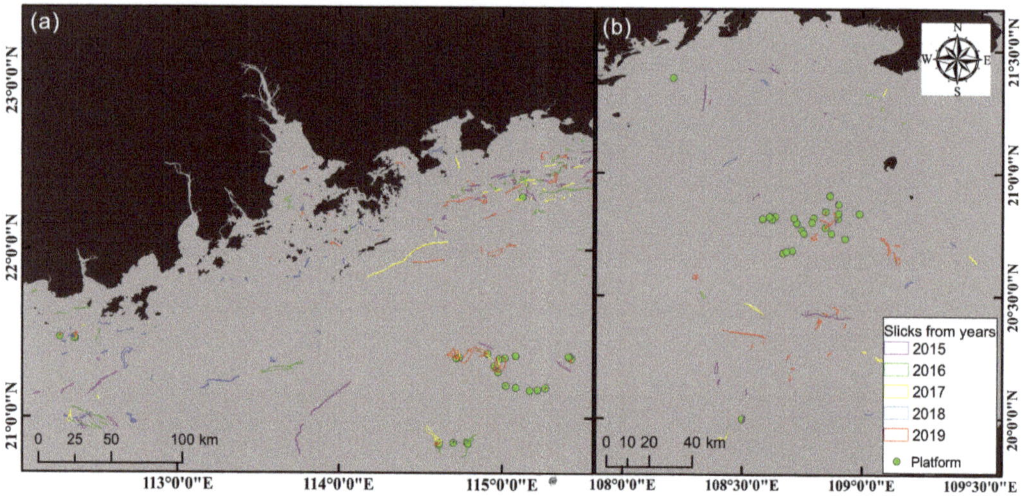

Figure 11. Distribution of oil slicks in: (**a**) the Guangdong–Hong Kong–Macao Greater Bay Area; and (**b**) northeastern Beibu Gulf. Black points denote oil platforms location.

Table 2. Statistical data for delineated oil slicks in the Guangdong–Hong Kong–Macao Greater Bay Area.

Year	Spring (Times)	Summer (Times)	Fall (Times)	Winter (Times)	Total (Times)	Maximum Area (km^2)	Minimum Area (km^2)	Average Area (km^2)	Median Area (km^2)	Total Area (km^2)
2015	27	8	7	0	42	21.1	0.24	3.7	2.0	154.4
2016	17	21	6	0	44	35.5	0.21	3.8	2.0	165.0
2017	3	22	0	2	27	22.7	0.30	3.5	1.1	94.3
2018	25	21	0	0	46	26.9	0.21	2.7	1.0	124.8
2019	22	0	13	4	39	33.8	0.21	4.4	2.1	170.0

In the northeastern Beibu Gulf, a similar seasonal trend can be observed in 2015, 2016, and 2018 (Table 3). Only occasional oil spills (seven cases) were detected to originate from platforms in this region. Other than that, most of the detected slicks (39 cases) were from ships. The overall percentage of oil spills from platforms was only 14% in the number of slicks and 7% in the area, respectively. The same percentage became 15% in number of slicks and 21% in area in the GBA. From 2016 to 2018, both the number of oil slicks (~6 vs. ~15) and total oil polluted area (<10 vs. >20 km^2) were relatively small compared to those of slicks detected in 2015 and 2019 in this region. The average yearly slick area was almost one magnitude smaller (~15 vs. 140 km^2) than areas in the GBA, partly because of the smaller size of this sub-region.

Table 3. Statistical data for delineated oil slicks in the Beibu Gulf.

Year	Spring (Times)	Summer (Times)	Fall (Times)	Winter (Times)	Total (Times)	Maximum Area (km^2)	Minimum Area (km^2)	Average Area (km^2)	Median Area (km^2)	Total Area (km^2)
2015	9	0	0	5	14	9.1	0.25	1.6	0.8	23.3
2016	6	0	0	1	7	2.9	0.36	1.1	0.5	7.8
2017	0	3	0	3	6	3.6	0.34	1.5	1.2	9.3
2018	2	4	0	0	6	2.4	0.24	1.0	0.6	6.1
2019	6	1	10	0	17	4.7	0.22	1.4	0.6	24.0

3.3. Cross-Check

Because of the ocean surface dynamics and oil weathering processes, it is rather difficult to validate remote sensing interpretation with field measurements [30,40]. In this study, we use imagery from high-resolution PlanetScope to cross-check the OLI detections. A total of 26 same-day PlanetScope/OLI matching pairs were found from March to October 2018. All of the matched same-day PlanetScope images detected oil slicks in the same region. The corresponding PlanetScope images were then delineated for area comparison. The comparison of oil slick delineated results from the two sensors are listed in Table 4. Here the Unbiased Mean Relative Error (UMRE) was used to estimate the relative detection area error between the two sensors (Equation (1)), with UMRE calculated to be 17% [25]. Not all of the slick pairs were compared for area because of incomplete spill size by cloud cover or by image swath limit. Figure 12 shows oil slicks by the OLI/PlanetScope image pairs. Because both OLI and PlanetScope are optical sensors, their imaging time of the same-day is usually within a few hours or even a few minutes. The oil slick in OLI imagery was similar to the oil slick in PlanetScope in shape and area (Figure 12a–c, UMRE < 47%), with the largest oil area difference in Figure 12d (UMRE = 133%).

$$\text{UMRE} = \frac{1}{n}\sum_{i=1}^{n}\left|\frac{x_i - y_i}{0.5x_i + 0.5y_i}\right| \quad (1)$$

where x_i, y_i are the area of the delineated oil slicks from the two images, respectively.

Table 4. Comparison of oil slick delineated results from the two sensors.

Slick	Imaging Time	Location	Time Difference (Min)	Slick Area from OLI (m^2)	Slick Area from PlanetScope (m^2)	UMRE
1	14 October 2018	20°42′N, 109°24′E	19	1,388,000	1,390,000	0.002
2	20 September 2018	18°54′N, 107°18′E	13	11,160,000	11,590,000	0.038
3	20 September 2018	20°42′N, 109°24′E	22	25,400,000	26,940,000	0.059
4	21 August 2018	20°00′N, 109°00′E	3	287,600	211,100	0.307
5	7 August 2018	18°54′N, 112°42′E	18	6,473,000	9,454,000	0.374
6	7 August 2018	21°30′N, 112°42′E	3	1,286,000	1,224,000	0.049
7	2 August 2018	23°18′N, 117°24′E	13	4,359,000	2,865,000	0.414
8	1 July 2018	24°42′N, 119°00′E	24	1,248,000	784,800	0.456
9	18 June 2018	17°00′N, 108°30′E	11	7,358,000	6,761,000	0.085
10	9 June 2018	21°00′N, 108°30′E	4	825,000	723,900	0.130
11	23 May 2018	22°36′N, 119°48′E	24	3,826,000	4,458,000	0.153
12	23 May 2018	22°36′N, 119°48′E	24	1,392,000	1,389,000	0.003

Table 4. *Cont.*

Slick	Imaging Time	Location	Time Difference (Min)	Slick Area from OLI (m^2)	Slick Area from PlanetScope (m^2)	UMRE
13	21 May 2018	22°36′N, 116°36′E	22	1,276,000	1,251,000	0.020
14	21 May 2018	22°24′N, 116°30′E	22	3,438,000	3,887,000	0.123
15	19 May 2018	21°18′N, 112°36′E	21	24,150,000	26,220,000	0.082
16	17 May 2018	20°42′N, 109°24′E	10	278,300	355,100	0.243
17	14 May 2018	23°24′N, 118°18′E	23	19,640,000	15,270,000	0.250
18	1 May 2018	18°12′N, 108°30′E	20	706,820	628,937	0.117
19	1 April 2018	21°54′N, 114°00′E	27	1,317,000	2,061,000	0.441
20	29 March 2018	20°42′N, 109°24′E	26	7,935,000	9,129,000	0.140

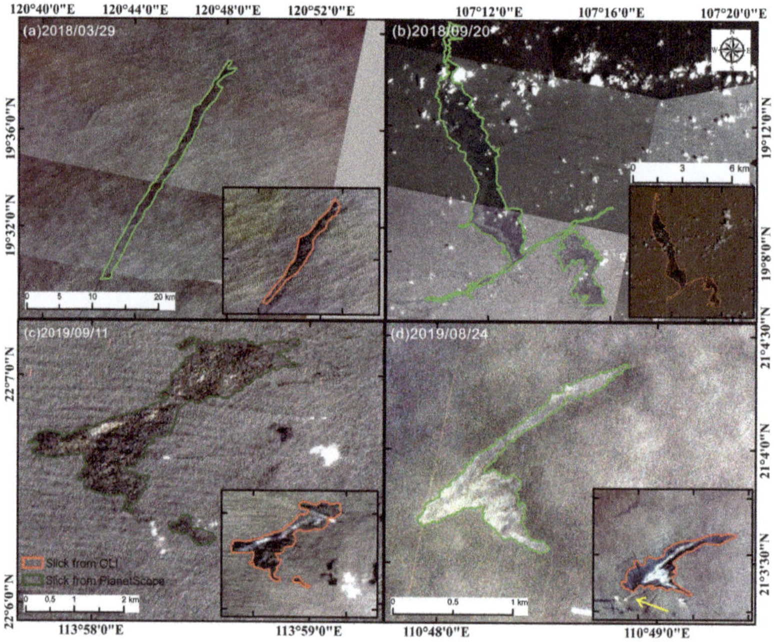

Figure 12. Same day image pairs from PlanetScope and OLI (insets), with a time difference of: (**a**) 26 min; (**b**) 13 min; (**c**) 13 min; and (**d**) 18 min, respectively. The yellow arrow from the panel d inset is pointed to moving ship directly connected to the slicks, which was determined to be not released from the pointed ship.

4. Discussion

4.1. Slick Distribution from Different Sources

Origins of the delineated slicks are shown in Figure 10a. Spatial cluster analysis was performed using Getis-Ord Gi * in GeoDa (Version:1.18, The University of Chicago, Chicago, U.S.) to analyze the clustering effect of oil spill distribution from different sources and areas [41,42]. Figure 13a shows regions of significant slick spatial clustering from offshore platforms (red color) and ships (blue) ($p < 0.05$), which indicated oil slick clusters from offshore platforms in the southeast of the PRE, east the Hainan Island, the middle of the south Beibu Gulf, and oil slick clusters from ship sources east of the PRE along the Guangdong coast, northeast of Dongsha Island, east of Zhanjiang city, and in the middle of the south Beibu Gulf. Figure 13b shows results of spatial clustering from slick areas. Significant large slicks were mainly distributed southwest and northeast of Dongsha Island, while significant small slicks were distributed slightly east of the PRE, south of Yangjiang, and scattered in the northeastern Beibu Gulf. As can be seen, smaller areas of slicks were mainly clustered along the coast (half of the slicks were within 54 km of the shoreline), while the significant large slicks were clustered in offshore waters (all the slicks were >147 km from the shoreline). Moreover, none of the significant large slick clusters was associated with oil platforms, indicating that platform spills tended to have smaller oil spill sizes in the study area. The smaller size of platform slicks can also be summarized from slick area histograms, which clearly presented larger area distribution from ship or unknown sources than from platform sources (Figure 14a).

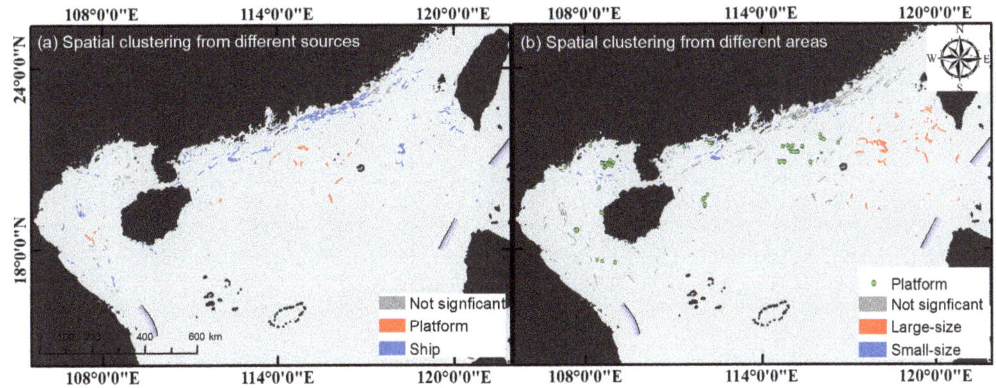

Figure 13. The spatial cluster of oil slicks: (**a**) from different sources: and (**b**) from different areas determined by the Gi * method.

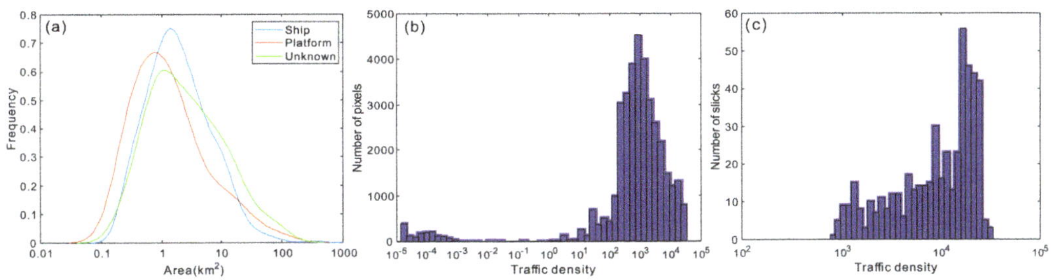

Figure 14. (**a**) The frequency distribution of slick areas from different sources; the frequency distribution histogram of traffic density: (**b**) in the study area; and (**c**) in the slick regions of ship origins.

The cumulative oil distribution map demonstrated wide slick cover over the joint fishing zone between China and Vietnam in the Beibu Gulf (Figure 10c, [43]). Moreover, a noticeable disparity of oil distributions can be found between the east and west side of the PRE (Figure 10b), and a significant oil slick cluster from ship sources is found in the east side of the PRE mouth. Such disparity coincides with the several major maritime traffic belts in this region, which travels from east of Hainan Island to the GBA and then to the east and north along the Guangdong coast, as also demonstrated in the traffic density map (Figure 10a) [44,45]. Similar results have been found from statistical analysis of crude oil tanker accidents, which reveal a positive correlation between traffic density and crude oil tanker accidents [46].

4.2. Uncertainties

Same-day image pairs were used to cross-check the OLI result. The OLI/Sentinel-1 SAR image pairs were searched, with sensing time differences ranging from 5 to 19 h.

Part of the slicks was detected in both SAR and OLI, with small variations in shape or area (Figure S1). However, some of the OLI-determined slicks were not detected by the same-day SAR imagery, or the shapes of the SAR-detected slick were quite different from OLI. A hindcast model from natural slicks (with slick lengths ranging from 1 to 50 km) in the GoM suggests a life span of 6.4 ($\pm$5.7) h of the detected slicks [47]. Slicks from natural seeps in the Lower Congo Basin are found to have residence time between 1 and 7 h, with an average of 3 h 15 min [48]. With the above time difference between SAR and OLI imaging, the surface slick or part of the surface slicks could disappear because of the oil weathering processes. Therefore, only results from the optical PlanetScope with sensing time a few minutes apart from OLI were compared for areal statistical analysis (Table 4). Figure 12d shows oil slicks detected by OLI and PlanetScope at a time difference of 2 min, but with an area difference of 133%. Clearly, little slick variation was expected between such a small time difference. When we looked into the two images, we found that the bright part of the two images had similar shapes, with an area difference of 20%. The dark part of the slicks in OLI was not effectively detected by PlanetScope, which could be because of the weak sun glint intensity in the PlanetScope image or because of the relatively low signal-to-noise ratio of the PlanetScope image. The estimated relative error of the OLI-determined area from PlanetScope was 17%. Such a difference represents uncertainties from manual delineation, scattered cloud cover (scattered cloud moved during the time difference), and the oil detection abilities of different sensors.

Thick oil can be detected through the reflectance response in the visible-NIR-SWIR range, while thin oil slicks are detectable because of the sun glint in optical imagery [38,39]. In fact, both of the mechanisms are used in optical imagery to detect the slick-caused spatial contrast and to rule out oil look-alikes spectrally [30,40]. However, in the absence of sun glint ($L_{GN} < 10^{-6}$ sr^{-1} for MODIS for example, because of unfavorable solar and view angels, and wind conditions), thin oil is undetectable in optical imagery [40]. Such sun glint strength is relatively higher in April to September than during the rest of the year in the mid-latitude Northern Hemisphere [25,40]. In reality, 99.5% of the detected slicks were within optimal wind conditions for slick detection under sun glint (1–8 m/s, Figure S2) [31]. Moreover, optical imagery suffers from cloud cover, which is useless if the slick is fully covered by cloud. Therefore, the derived slick distribution by the optical OLI imagery in this study represented a subset of all slicks on the ocean surface during the time period.

Other than the traffic density data, we used a rule of ship-connection to justify slick origin from ships, which was not always true. As shown in the Figure 12d inset, there was a moving ship connected to the slick, but the slick was most possibly not released from the ship considering slick width (>2 km) (time required for spreading) as the ship was still moving at the scene time. Nevertheless, the slick was still identified to be from ships because of the traffic density in this region (11,800 > threshold 991), and it might have been released by another ship outside the scene at the imaging time.

5. Conclusions

In this study, the medium resolution optical Landsat-8 OLI imagery was used to delineate oil slicks in the NSCS. The source of the detected oil slicks was determined from an offshore oil and gas platform database and the traffic density data in the study area. A total of 632 oil slicks were identified, with cumulative oil pollution of 3242 km^2 in the NSCS from 2015 to 2019. Of all the slicks detected, 57 were determined to be from platform sources, 490 from ship sources, and the rest from unknown sources. The average area of detected slicks was 4.8 km^2, with half of the slick area <1.7 km^2. Our results show that 90% of the oil spills were found within 112 km of the shoreline, and half of the slicks were within 33 km from shorelines. Spatial clustering analysis indicated that slicks from ship sources were significantly clustered east of the PRE along the Guangdong coast, northeast of the Dongsha Island, east of Zhanjiang city, and in the middle of the south Beibu Gulf, and slicks from platform sources were significantly clustered southeast of the PRE, east Hainan Island, and in the middle of the south Beibu Gulf. Moreover, slicks from ship or unknown sources tended to have larger areas than slicks from platforms in the study area. Results from this study demonstrate the capability of medium-resolution OLI imagery in detecting and monitoring oil spills regionally, and the derived oil distribution map in the NSCS would hopefully provide useful information for future environmental assessments of the chronic oil spills.

Supplementary Materials: The following supporting information can be downloaded at: https://www.mdpi.com/article/10.3390/rs14163966/s1, Figure S1: Comparison of same-day image pairs from Sentinel-1 SAR and Landsat-8 OLI; Figure S2: Histogram of wind speed for all the delineated slicks.

Author Contributions: Conceptualization, S.S.; methodology, X.H., L.C. and S.S.; writing—original draft preparation, X.H. and L.C.; writing—review and editing, X.H., L.C., S.S., Z.S., Y.C., Q.M. and Z.C.; funding acquisition, S.S. and L.C. All authors have read and agreed to the published version of the manuscript.

Funding: This research was funded by the Guangdong Basic and Applied Basic Research Foundation, grant number 2020A1515110957, the National Natural Science Foundation of China, grant number 42106173, the Sun Yat-sen University Undergraduate Innovation Program, grant number 20201430, and the Dragon 5 Cooperation, grant number 59193.

Data Availability Statement: Landsat-8 data are openly accessible from https://earthexplorer.usgs.gov/, accessed on 11 July 2022, global shipping traffic density data are openly accessible from https://datacatalog.worldbank.org/search/dataset/0037580/Global-Shipping, accessed on 11 July 2022.

Acknowledgments: The authors would like to thank USGS for providing Landsat data, Planet Labs for providing PlanetScope data, and IMF for providing the shipping traffic density data.

Conflicts of Interest: The authors declare no conflict of interest.

References

1. DOINews. *DOINews: Scientific Teams Refine Estimates of Oil Flow from BP's Well Prior to Capping*; U.S. Department of the Interior: Washington, DC, USA, 2010; Volume 2022.
2. Sun, S.; Lu, Y.; Liu, Y.; Wang, M.; Hu, C. Tracking an oil tanker collision and spilled oils in the East China Sea using multisensor day and night satellite imagery. *Geophys. Res. Lett.* **2018**, *45*, 3212–3220. [CrossRef]
3. Hebbar, A.A.; Dharmasiri, I.G. Management of marine oil spills: A case study of the Wakashio oil spill in Mauritius using a lens-actor-focus conceptual framework. *Ocean Coast. Manag.* **2022**, *221*, 106103. [CrossRef]
4. International Tanker Owners Pollution Federation (ITOPF). Oil Tanker Spill Statistics 2021. Available online: https://www.itopf.org/knowledge-resources/data-statistics/statistics (accessed on 11 July 2022).
5. National Research Council (US). *Oil in the Sea III: Inputs, Fates, and Effects*; National Academies Press (US): Washington, DC, USA, 2003.
6. Fingas, M. *The Basic of Oil Spill Clean Up*, 3rd ed.; CRC Press: Boca Raton, FL, USA, 2012.
7. Dong, Y.; Liu, Y.; Hu, C.; MacDonald, I.R.; Lu, Y. Chronic oiling in global oceans. *Science* **2022**, *376*, 1300–1304. [CrossRef] [PubMed]
8. Laffon, B.; Pásaro, E.; Valdiglesias, V. Effects of exposure to oil spills on human health: Updated review. *J. Toxicol. Environ. Health Part B Crit. Rev.* **2016**, *19*, 105–128. [CrossRef] [PubMed]
9. Keramea, P.; Spanoudaki, K.; Zodiatis, G.; Gikas, G.; Sylaios, G. Oil spill modeling: A critical review on current trends, perspectives, and challenges. *J. Mar. Sci. Eng.* **2021**, *9*, 181. [CrossRef]
10. U.S. Environmental Protection Agency (EPA). Oil Discharge Reporting Requirements. Available online: https://www.epa.gov/sites/default/files/2014-06/documents/spccfactsheetspillreportingdec06-1.pdf (accessed on 11 July 2022).
11. Daneshgar Asl, S.; Amos, J.; Woods, P.; Garcia-Pineda, O.; MacDonald, I.R. Chronic, anthropogenic hydrocarbon discharges in the Gulf of Mexico. *Deep Sea Res. Part II Top. Stud. Oceanogr.* **2016**, *129*, 187–195. [CrossRef]
12. Chen, Q.; Hu, S. Study on offshore oil spill accidents in China. *Mar. Dev. Manag.* **2020**, *37*, 49–53.
13. Gong, Y.; Zhao, P.; Lan, D.; Zhu, R.; Yan, X.U.; Bao, C.; Chunyan, Y.U. Characteristics and trend analysis of marine oil spill accidents in China. *Ocean Dev. Manag.* **2018**, *35*, 42–45.
14. Brekke, C.; Solberg, A. Oil spill detection by satellite remote sensing. *Remote Sens. Environ.* **2005**, *95*, 1–13. [CrossRef]
15. Fingas, M.; Brown, C. Review of oil spill remote sensing. *Mar. Pollut. Bull.* **2014**, *83*, 9–23. [CrossRef]
16. Leifer, I.; Lehr, W.J.; Simecek-Beatty, D.; Bradley, E.; Clark, R.; Dennison, P.; Hu, Y.; Matheson, S.; Jones, C.E.; Holt, B.; et al. State of the art satellite and airborne marine oil spill remote sensing: Application to the BP Deepwater Horizon oil spill. *Remote Sens. Environ.* **2012**, *124*, 185–209. [CrossRef]

17. Kokaly, R.F.; Couvillion, B.R.; Holloway, J.; Roberts, D.A.; Ustin, S.L.; Peterson, S.H.; Khanna, S.; Piazza, S.C. Spectroscopic remote sensing of the distribution and persistence of oil from the Deepwater Horizon spill in Barataria Bay marshes. *Remote Sens. Environ.* **2013**, *129*, 210–230. [CrossRef]
18. Lu, Y.; Zhan, W.; Hu, C. Detecting and quantifying oil slick thickness by thermal remote sensing: A ground-based experiment. *Remote Sens. Environ.* **2016**, *181*, 207–217. [CrossRef]
19. Garcia-Pineda, O.; MacDonald, I.; Hu, C.; Svejkovsky, J.; Hess, M.; Dukhovskoy, D.; Morey, S.L. Detection of floating oil anomalies from the Deepwater Horizon oil spill with synthetic aperture radar. *Oceanography* **2013**, *26*, 124–137. [CrossRef]
20. Mohr, V.; Gade, M. Marine oil pollution in an area of high economic use: Statistical analyses of SAR data from the Western Java Sea. *Remote Sens.* **2022**, *14*, 880. [CrossRef]
21. Trinadha Rao, V.; Suneel, V.; Raajvanshi, I.; Alex, M.J.; Thomas, A.P. Year-to-year variability of oil pollution along the Eastern Arabian Sea: The impact of COVID-19 imposed lock-downs. *Mar. Pollut. Bull.* **2022**, *175*, 113356. [CrossRef]
22. Garcia-Pineda, O.; MacDonald, I.R.; Li, X.; Jackson, C.R.; Pichel, W.G. Oil spill mapping and measurement in the Gulf of Mexico with Textural Classifier Neural Network Algorithm (TCNNA). *IEEE J. Sel. Top. Appl. Earth Obs. Remote Sens.* **2013**, *6*, 2517–2525. [CrossRef]
23. Hu, C.; Weisberg, R.H.; Liu, Y.; Zheng, L.; Daly, K.L.; English, D.C.; Zhao, J.; Vargo, G.A. Did the northeastern Gulf of Mexico become greener after the Deepwater Horizon oil spill? *Geophys. Res. Lett.* **2011**, *38*, L09601. [CrossRef]
24. Hu, C.; Li, X.; Pichel, W.G.; Muller-Karger, F.E. Detection of natural oil slicks in the NW Gulf of Mexico using MODIS imagery. *Geophys. Res. Lett.* **2009**, *36*, L1604. [CrossRef]
25. Sun, S.; Hu, C.; Garcia-Pineda, O.; Kourafalou, V.; Le Hénaff, M.; Androulidakis, Y. Remote sensing assessment of oil spills near a damaged platform in the Gulf of Mexico. *Mar. Pollut. Bull.* **2018**, *136*, 141–151. [CrossRef]
26. Chen, T. Study on rapid prediction of oil spill drift in the east of the South China Sea. *China Pet. Chem. Stand. Qual.* **2020**, *5*, 168–172.
27. Chang, P.; Yang, Q. Necessity and feasibility of establishing a particularly sensitive sea area system in the South China Sea. *Water Transp. Manag.* **2018**, *15*, 6–8.
28. Vanhellemont, Q. Adaptation of the dark spectrum fitting atmospheric correction for aquatic applications of the Landsat and Sentinel-2 archives. *Remote Sens. Environ.* **2019**, *225*, 175–192. [CrossRef]
29. Planet Team. Planet Team. Planet Application Program Interface. In *Space for Life on Earth*; Planet Team: San Francisco, CA, USA, 2017. Available online: https://api.planet.com (accessed on 11 July 2022).
30. Sun, S.; Hu, C. The challenges of interpreting oil–water spatial and spectral contrasts for the estimation of oil thickness: Examples from satellite and airborne measurements of the Deepwater Horizon oil spill. *IEEE Trans. Geosci. Remote* **2019**, *57*, 2643–2658. [CrossRef]
31. Sun, S.; Hu, C. Sun glint requirement for the remote detection of surface oil films. *Geophys. Res. Lett.* **2016**, *43*, 309–316. [CrossRef]
32. Sun, S.; Hu, C.; Feng, L.; Swayze, G.A.; Holmes, J.; Graettinger, G.; Macdonald, I.; Garcia, O.; Leifer, I. Oil slick morphology derived from AVIRIS measurements of the Deepwater Horizon oil spill: Implications for spatial resolution requirements of remote sensors. *Mar. Pollut. Bull.* **2016**, *103*, 276–285. [CrossRef]
33. Liu, Y.; Zhao, J.; Qin, Y. A novel technique for ship wake detection from optical images. *Remote Sens. Environ.* **2021**, *258*, 112375. [CrossRef]
34. Liu, Y.; Deng, R. Ship wakes in optical images. *J. Atmos. Ocean. Technol.* **2018**, *35*, 1633–1648. [CrossRef]
35. Hu, C.; Feng, L.; Hardy, R.F.; Hochberg, E.J. Spectral and spatial requirements of remote measurements of pelagic Sargassum macroalgae. *Remote Sens. Environ.* **2015**, *167*, 229–246. [CrossRef]
36. Bin, Y.I.; Chen, K.; Zhou, J.; Yihua, L.U. Characteristics of red tide in coastal region of South China from 2009 to 2016. *Trans. Oceanol. Limnol.* **2018**, *2*, 23–31.
37. Tian, Y.; Li, T.; Hu, S.; Xie, X.; Liu, S. Temporal and spatial characteristics of harmful algal blooms in Guangdong coastal area. *Mar. Environ. Sci.* **2020**, *39*, 1–8.
38. Qi, L.; Tsai, S.F.; Chen, Y.; Le, C.; Hu, C. In search of red Noctiluca scintillans blooms in the East China Sea. *Geophys. Res. Lett.* **2019**, *46*, 5997–6004. [CrossRef]
39. Liu, Y.; Hu, C.; Sun, C.; Zhan, W.; Sun, S.; Xu, B.; Dong, Y. Assessment of offshore oil/gas platform status in the northern Gulf of Mexico using multi-source satellite time-series images. *Remote Sens. Environ.* **2018**, *208*, 63–81. [CrossRef]
40. Hu, C.; Lu, Y.; Sun, S.; Liu, Y. Optical remote sensing of oil spills in the ocean: What is really possible? *J. Remote Sens.* **2021**, *2021*, 9141902. [CrossRef]
41. Getis, A.; Ord, J.K. The Analysis of Spatial Association by Use of Distance Statistics. *Geogr. Anal.* **1992**, *24*, 189–206. [CrossRef]
42. Mitchell, A. *The Esri Guide to GIS Analysis Volume 2: Spatial Measurements & Statistics*, 2nd ed.; Esri Press: Redlands, CA, USA, 2021.
43. Pan, P.; Luo, J.C.; Yi-Yun, H.U. Review of fishery cooperative agreement in Beibu Gulf between the People's Republic of China and the Socialist Repubulic of Vietnam. *Chin. Fish. Econ.* **2016**, *34*, 22–26.
44. Wu, L.; Xu, Y.; Wang, Q.; Wang, F.; Xu, Z. Mapping global shipping density from AIS data. *J. Navig.* **2017**, *70*, 67–81. [CrossRef]
45. Mei, Q.; Lin, W.U.; Peng, P. Typical spatial distribution of merchant vessels and trade flow in South China Sea. *J. Geo-Inf. Sci.* **2018**, *20*, 632–639.
46. Wang, J.; Zhou, Y.; Zhuang, L.; Shi, L.; Zhang, S. Study on the critical factors and hot spots of crude oil tanker accidents. *Ocean Coast. Manag.* **2022**, *217*, 106010. [CrossRef]

47. Daneshgar Asl, S.; Dukhovskoy, D.S.; Bourassa, M.; Macdonald, I.R. Hindcast modeling of oil slick persistence from natural seeps. *Remote Sens. Environ.* **2017**, *189*, 96–107. [CrossRef]
48. Jatiault, R.; Dhont, D.; Loncke, L.; Dubucq, D. Monitoring of natural oil seepage in the Lower Congo Basin using SAR observations. *Remote Sens. Environ.* **2017**, *191*, 258–272. [CrossRef]

remote sensing

MDPI

Article

Study of the Response of Environmental Factors of the Coastal Area in Zhoushan Fishery to Typhoon In-fa Based on Remote Sensing

Rong Tang [1,2], Lina Cai [1,2,*], Xiaojun Yan [1], Xiaomin Ye [2], Yuzhu Xu [2] and Jie Yin [1]

1 Marine Science and Technology College, Zhejiang Ocean University, Zhoushan 316004, China
2 Key Laboratory of Space Ocean Remote Sensing and Application, Ministry of Natural Resources, Beijing 100081, China
* Correspondence: clnown@zjou.edu.cn

Abstract: The response of typical environmental factors in Zhoushan Fishery, including sea surface temperature (SST), sea surface salinity (SSS), and chlorophyll a (Chl-*a*), before and after Typhoon In-fa was analyzed using satellite data and reanalysis data in this study. Additionally, this study simultaneously elucidated the mechanism by which the typhoon affected these factors. The results showed that: (1) the strong vertical mixing caused by In-fa provoked a decrease in SST, while the asymmetric typhoon wind stress and vertical difference in temperature structure before the typhoon caused a more robust cooling of SST on the right side of the In-fa track; (2) despite the strong mixing and inflow of hypersaline seawater increasing SSS, the combined effect of intense rainfall and diluted water inflow caused an overall decrease in SSS after In-fa's landing; (3) In-fa caused the Chl-*a* concentration to decrease first and then increase. The high cloudiness and low Chl-*a* seawater inflow inhibited the phytoplankton growth during the typhoon, while the abundant light, rich surface nutrients under the upwelling effect, and transport of rich land-based substances induced rapid phytoplankton reproduction after the typhoon; and (4) the change in Chl-*a* concentration, current, temperature, and salinity induced by a typhoon are essential factors that affect fish behavior and community composition in fisheries. This study provides a point of reference to reveal the response of environmental factors to typhoons and their effects on fishery resources in fisheries located on nearshore estuarine shallow waters with intensive islands.

Keywords: typhoon; Zhoushan Fishery; environmental factors; fishery resource

1. Introduction

Typhoons (tropical cyclones or storms) are a common oceanic phenomenon. Among the numerous typhoons, about 6.5 of them pass by the eastern coast of China [1,2], making it one of the most frequent natural disasters in this area. The transit of typhoons can cause dramatic disturbances to the marine environment, causing varying degrees of change to the marine environment, climate, and ecosystems [3,4]. A crucial concern has been the response of the ocean to environmental factors.

Temperature, salinity, and chlorophyll *a* (Chl-*a*) responses to typhoon events have been extensively studied in the northwest Pacific Ocean (NOP). Numerous studies have demonstrated that strong winds induce ocean cooling during typhoons through vertical entrainment processes, resulting in sea surface temperature (SST) decreases ranging from 1 to 10.8 °C [5–8]. Substantial rainfall from tropical cyclones can reduce sea surface salinity (SSS). At the same time, evaporation, mixing, and Ekman pumping can increase SSS. The different responses of SSS to typhoons may result from variations in the strength of the effects of the four processes [4,9].

The location, depth, and duration of the subsurface chlorophyll maximum layer (SCML) are controlled by processes such as intense mixing, upwelling, and entrainment induced by strong winds. These processes also significantly contribute to increased phytoplankton biomass in the upper ocean [10,11]. In coastal waters, abundant rainfall and runoff can cause a substantial influx of nutrients from terrestrial sources, and it is coupled with advection transport, which results in coastal phytoplankton blooms [12–14].

Estuaries and nearshore waters serve as concentration zones for land–sea interaction, where physical, chemical, biological, and geological processes are coupled and variable, contributing to a sensitive and fragile ecological environment [15]. Typhoons cause significant disruptions to the maritime environment, leading to increased estuary runoff, reduced dissolved oxygen, water clarity and temperature changes, and other factors that may impact fish communities [16,17]. These disturbances can directly alter aquatic habitats, potentially resulting in the escape and mortality of marine organisms [18,19]. Conversely, the enhanced primary productivity in the estuary can increase fishery production [20,21].

Due to global warming, typhoons in the NOP have become more frequent and intense [22,23]. Based on the critical relationship between spatial movement, community and density of fish, and changes in aquatic habitats, typhoons by no means have a negligible impact on habitat elements and marine organisms within the fishery. Until now, studies have examined changes in fish biomass density and species richness in the outer sea [24], nearshore waters (estuaries, bays, coral reef areas) [17,19,20], and their inland watersheds [25] after typhoons.

Zhoushan Fishery is adjacent to the Yangtze River estuary, Hangzhou Bay, the Yellow Sea, and the East China Sea, serving as the convergence point for various runoff currents (Yangtze diluted water (YDW), Qiantang River) and sea currents (Taiwan warm current (TWC), Subei coastal current, Zhemin coastal current (ZMCC)). There are special geographical conditions, and the fishery ecosystem contributes to the area being the most significant and excellent fishery in China [26,27]. However, the area is also more frequently affected by typhoons, posing a threat to the stability of the estuary and to the habitat of Zhoushan Fishery. Given that the environmental conditions of fisheries in each sea area and the characteristics of transiting typhoons are very different [28], there is a lack of research on the effects of typhoons on Zhoushan Fishery in the northwest part of the East China Sea, and this needs to be further explored.

Typhoon effects on the ocean and shelf marine environment have been primarily studied in the past using data from ocean buoys [29], remote sensing observations [30,31], numerical simulations [9,32], and combinations of in situ observations and numerical simulations [33,34]. However, the rapid changes in the path and intensity of typhoons make it challenging to achieve large-scale, simultaneous sampling due to the high cost of setting up buoys and conducting cruise measurements as well as the low temporal resolution [29]. Moreover, technical problems with single satellite data, such as missing values due to cloud cover [35] and the low temporal resolution of the dataset [36], can result in a lack of observational information during typhoon transit. To overcome these limitations, reanalysis products that combine in situ observation sites, multi-source satellite remote sensing observations, and numerical simulations provide an effective method [34,37] that offers more comprehensive and systematic spatial and temporal information for periods of severe conditions like typhoons.

In the study, based on the remote sensing data from the Moderate Resolution Imaging Spectroradiometer (MODIS) and Gaofen-4 (GF-4) satellite as well as reanalysis products, which was combined with a comparison of environmental factors in the fishery before and after a typhoon, we revealed the response characteristics and influence mechanisms of SST, SSS, and Chl-*a*, typical environmental factors in Zhoushan Fishery, to In-fa. Moreover, the possible changes in fishery resources after the typhoon were discussed. This study was conducted to provide support for a future comprehensive assessment of the influence of typhoons on the coastal estuarine ecosystem.

2. Materials and Methods

2.1. Satellite Data

The daily Chl-*a* concentration product was obtained from MODIS-Aqua ocean color remote sensing data (https://oceancolor.gsfc.nasa.gov/ (accessed on 17 September 2022)). This dataset is a sea surface remote sensing reflectance product that has undergone preprocessing steps, including bow-tie effect removal, geometric correction, and atmospheric correction. For this study, four satellite images taken on 17, 21, 31 July and 8 August 2021, before and after the landing of In-fa, were used. Given that the suspended sediment concentration (SSC) in Hangzhou Bay tended to be higher, reaching up to 1.103 g/L [38], we effectively extracted the Chl-*a* concentration from the area spanning 121°48′E–123°50′E and 29°21′N–31°30′N in order to decrease the impact of SSC on the spatial distribution of Chl-*a* concentration.

The GF-4 satellite, successfully launched on 29 December 2015 [39], operates in geosynchronous earth orbit at an orbital height of 36,000 km. It is equipped with a panchromatic multispectral sensor (PMS) with a resolution of 50 m, mid-wave infrared of 400 m, and width of over 400 m (Table 1). With high temporal and spatial resolution advantages, the satellite has excellent potential and vast space in meteorological weather monitoring, disaster detection, and coastal zone management [40]. This offers quick, dependable, and stable optical remote sensing data for typhoons and other natural disasters. Therefore, we selected four GF-4/PMS L1-level remote sensing images (taken on 20, 21, 31 July and 9 August 2021) from the study area with relatively complete coverage (https://data.cresda.cn/#/home (accessed on 16 November 2022)) to perform the inversion of SSC concentration in Zhoushan Fishery.

Table 1. Band information of Gaofen-4 (GF-4) satellite.

Type	Band No.	Spectral Range (μm)	Spatiotemporal Resolution (m)	Width (km)	Revisit Time (s)
Near-Infrared (VINR)	B1 (Pan)	0.45–0.90	50	400	20
	B2 (Blue)	0.45–0.52			
	B3 (Green)	0.52–0.60			
	B4 (Red)	0.63–0.69			
	B5 (NIR)	0.76–0.90			
Middle Infrared (MWIR)	B6 (MWIR)	3.5–4.1	400		

After preprocessing the satellite data using radiometric calibration, atmospheric correction, and orthorectification with ENVI 5.3 software, the changes in SSC concentration in Zhoushan Fishery before and after the typhoon were analyzed using models from previous studies [41]. The equations were as follows:

$$\rho = 314.435 \times B3 + 3805.982 \times B4 + 28.54 \quad (1)$$

where ρ is SSC concentration, and B3 and B4 are divided into green and red band reflectance in the GF-4 image, respectively.

2.2. Reanalysis and Observed Data

Due to the thick cloud cover during typhoons and the relatively frequent fishing activities in the East China Sea, meeting the continuous observation of the response of upper ocean variables after typhoons is challenging [29,35]. In this study, the surface ocean data, including SST, SSS, Chl-*a*, sea-surface wind, and current ocean data, were all obtained from reanalysis products, which provide more comprehensive and systematic spatial and temporal information (Table 2).

Table 2. Data sources and related information.

Field (Units)	Dataset	Data Source	Spatiotemporal Resolution	Recourses
SST (°C)	MUR-SST	NASA JPL PO. DAAC	2 km/1 d	https://registry.opendata.aws/mur (accessed on 29 October 2022).
SSS (psu)	HYCOM GOFS 3.0	HYCOM	0.08°/3 h	https://www.hycom.org/ (accessed on 8 November 2022)
Chl-*a* (μg/L)	OCEANCOLOUR_GLO_BGC_L4_MY_009_104	CMEMS	4 km/1 d	https://data.marine.copernicus.eu/ (accessed on 18 November 2022)
10 m *u*/*v* components of wind (m/s)	ERA5	ECMWF	0.25°/1 h	https://marine.copernicus.eu/ (accessed on 15 October 2022)
Total precipitation (m/d)	ERA5	ECMWF	0.25°/1 h	https://marine.copernicus.eu/ (accessed on 24 November 2022)
Ocean current (m/s)	HYCOM GOFS 3.0	HYCOM	0.08°/1 h	https://www.hycom.org/ (accessed on 28 November 22)
Topography (m)	ETOPO1	NOAA/NGDC	1 arc-min	http://www.ngdc.noaa.gov/mgg/global/global.html (accessed on 12 September 2022)
Typhoon (lat/lon)	DWRZP	Typhoon track	--	https://typhoon.slt.zj.gov.cn/default.aspx (accessed on 31 August 2022)

MUR-SST: Multi-scale Ultra-high Resolution Sea Surface Temperature; HYCOM: Hybrid Coordinate Ocean Model; GOFS: global ocean forecasting system; OCEANCOLOUR_GLO_BGC_L4_MY_009_104: global ocean colour (Copernicus-GlobColour), biogeochemical, L4 (monthly and interpolated) from satellite observations; NASA: National Aeronautics and Space Administration; JPL: Jet Propulsion Laboratory; PO-DAAC: Physical Oceanography Distributed Active Archive Center; CMEMS: Copernicus Marine Environment Monitoring Service; ECMWF: European Centre for Medium-Range Weather Forecasts; NOAA/NGDC: National Oceanic and Atmospheric Administration/National Geophysical Data Center; DWRZP: Department of Water Resources of Zhejiang Province.

SST was derived from the global, gapless, gridded, daily Multi-scale Ultra-high Resolution Sea Surface Temperature (MUR-SST) dataset provided by the National Aeronautics and Space Administration, with a spatial resolution of up to 2 km. This dataset has a spatial resolution that is orders of magnitude more accurate than the majority of current reanalysis products [42]. The MUR-SST dataset integrates multiple data sources, including satellite, buoy, and ship observations; it is combined with data assimilation techniques to achieve a high-accuracy simulation and prediction of SST. It has been utilized in various fields, such as marine environment monitoring, climate change research, and fisheries resource management, providing valuable information services.

SSS and current data were obtained from the Hybrid Coordinate Ocean Model (HYCOM) ocean circulation model. Unlike the traditional ocean models, HYCOM employs a vertical hybrid coordinate model that combines isopycnic coordinates, vertical z coordinates, and terrain-following coordinates. This approach enhances the accuracy of simulation results for global or regional oceans, stratified or non-stratified oceans, oceans, or nearshore areas [43]. With its higher vertical resolution in the surface and shallower coastal regions, the model can represent the physical properties of the nearshore upper ocean well [44,45], which provides reliable data support for the observation of SSS and currents before and after typhoons in the coastal area of Zhoushan Fishery.

The Chl-*a* reanalysis product was derived from global ocean satellite observations at a 4 km spatial resolution as provided by the Copernicus Marine Environment Monitoring Service (CMEMS). This product includes daily and monthly averaged data on global ocean biogeochemical parameters. To assess the suitability of the Chl-*a* reanalysis product within the coastal area of Zhoushan Fishery, we compared it with the surficial in situ Chl-*a* data gathered on 23–24 October 2019 [46] and the CMEMS Chl-*a* data (Figure 1). We calculated the spatial average for the 4 km region surrounding each in situ location. The findings indicated that there was a good correlation between the in situ Chl-*a* and the CMEMS Chl-*a* (R^2 = 0.625, RMSE = 0.7662 μg/L), indicating that the CMEMS Chl-*a* was applicable within the coastal area of Zhoushan Fishery.

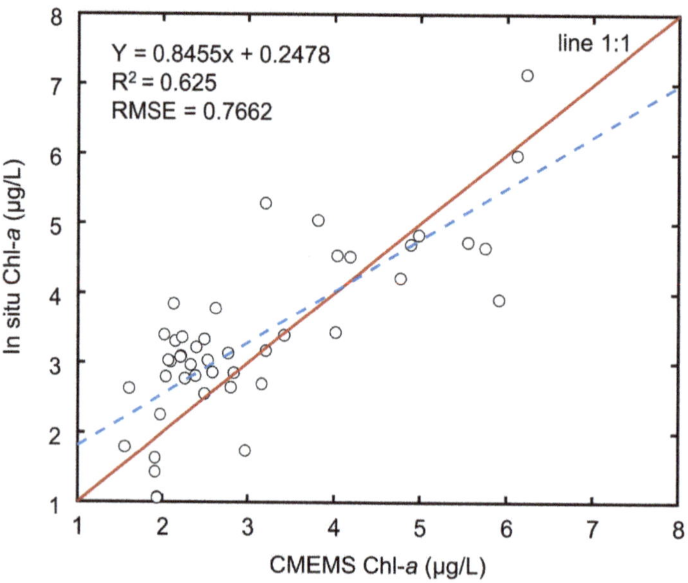

Figure 1. Analysis of the correlation between the in situ and CMEMS Chl-*a* (μg/L).

The National Oceanic and Atmospheric Administration (NOAA) (https://www.ncei.noaa.gov/maps/hourly/ (accessed on 11 December 2022)) provides hourly observations from weather sites around the world and has recorded temperature, pressure, wind direction, wind speed, and rainfall for the period from August 1956 to the present; it was used to obtain the daily average rainfall data before and after the landfall of Typhoon In-fa. This study selected the hourly rainfall data from the Dinghai and Baoying hydrological stations and calculated the daily cumulative rainfall for July 2021.

2.3. *Typhoon In-fa and Study Area*

The path of Typhoon In-fa is shown in Figure 2. In-fa was generated on 17 July 2021 at 18:00 UTC in the NPO, and it moved westward after it was generated. It intensified into a strong typhoon on the 21st of July at 3:00 UTC, reaching a maximum sustained wind speed of 42 m/s on the 22nd at 0:00 UTC. The minimum surface pressure recorded was 955 hPa, and it exhibited slow movement with a translational speed of 1.4 m/s (Figure 3). It made landfall in the Dinghai District of Zhoushan City, Zhejiang Province, at around 9:00 UTC on 25 July, then moved into Hangzhou Bay on the evening of 25 July, moving northwestward (Figure 2b). Figure 2c displays specific information regarding the study area (121°24′E–123°30′E, 29°21′N–31°30′N).

In-fa has the characteristics of strong wind, a slow-moving speed, long influence time, a wide rainfall range, and ample accumulated rainfall. When it made landfall in the coastal area of Zhoushan Fishery (Figure 3), the maximum sustained wind speed was 38 m/s, the minimum surface pressure was 965 hPa, and the translational speed was 2.8 m/s. The duration of rainfall impact was three times the average impact duration of typhoons, with a total precipitation amount of approximately 12.74 billion m^3, breaking the record for the largest typhoon landfall in Zhejiang Province (China Water Resources Annual Report 2021, http://www.mwr.gov.cn/ (accessed on 25 November 2022)). This profoundly impacted the marine environment and ecosystem of the estuary and coastal waters.

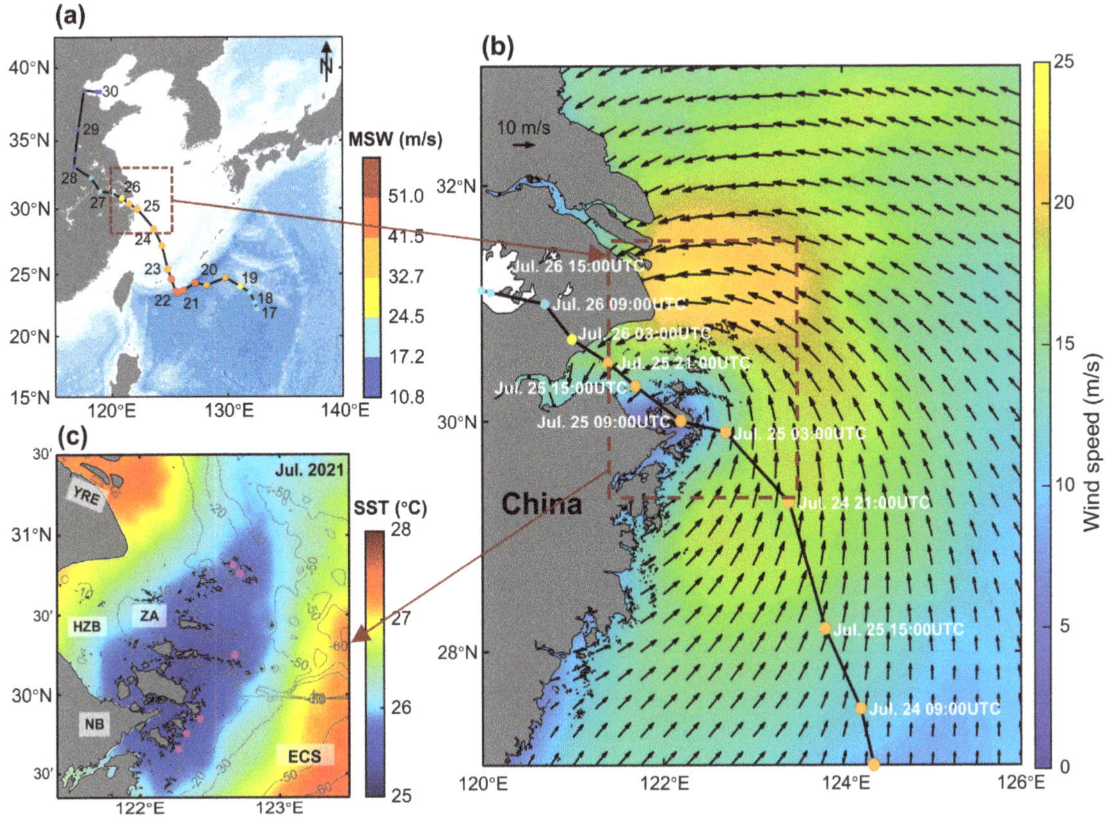

Figure 2. Typhoon In-fa path and study area. (**a**) The track and intensity of In-fa (MSW: the maximum sustained wind); (**b**) the wind field vector over East China Sea (ECS) at 09:00 UTC on 21 July 2021, where the color indicates the speed and the black arrow is the wind speed. (**c**) Sea surface temperature (SST) and its isobath within the research region on July 2021 (the magenta dots: the locations of national marine ranching demonstration areas). YRE: Yangtze River estuary; HZB: Hangzhou Bay; ZA: Zhoushan Archipelago; NB: Ning Bo.

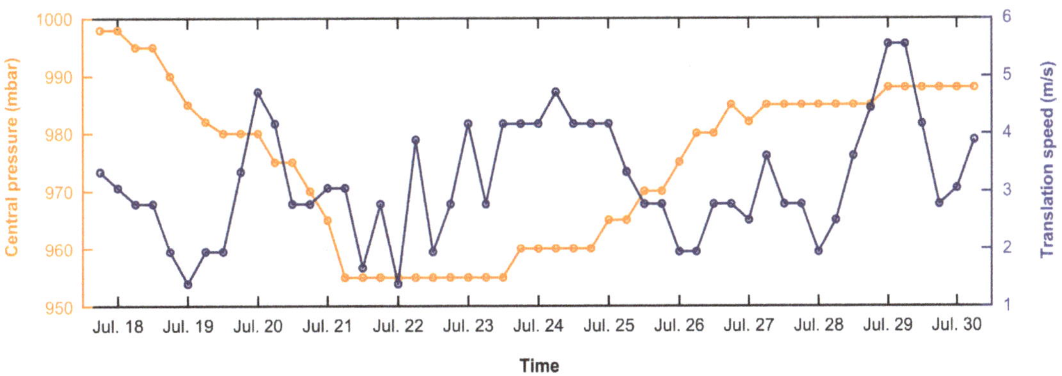

Figure 3. The translational speed and central stress of typhoon In-fa (17–30 July 2021).

2.4. Estimation of Typhoon Impact Factors

To evaluate the possible mechanisms of SST, SSS, and Chl-*a* concentration on In-fa response, we estimated the wind stress power (P_w), Ekman pumping velocity (EPV; $\times 10^{-5}$ m/s), coastal upwelling velocity, and Ekman current based on the 10 m u/v components of wind data provided by ERA5. The specific calculations for these four factors are as follows.

Wind stress at the sea surface (τ_0) is calculated as:

$$\tau_0 = \rho_{\text{air}} C_D U_{10}^2 \tag{2}$$

where U_{10} represents wind speed at a height of 10 m above mean sea level, as obtained from the ERA5 dataset. ρ_{air} denotes air density ($\approx$1.17 kg/m^3 at 30 °C), while C_D refers to the drag coefficient.

The P_w value in the upper layer of the ocean is a consequence of the wind-induced kinetic energy flux (W/m^2) [31];

$$P_w = \tau_0 U_{10} \tag{3}$$

The EPV is calculated as [47]:

$$EPV = \text{Curl}\left(\frac{\tau_0}{f \times \rho_w}\right) \tag{4}$$

where f denotes the Coriolis parameter, and the value of ρ_w corresponds to the density of seawater, which is 1025 kg/m^3.

The surface Ekman current (V_0) is calculated as [48]:

$$V_0 = \frac{0.0127}{\sqrt{\sin|\varphi|}} U_{10} \tag{5}$$

where φ is the latitude. In the Northern Hemisphere, the surface Ekman current flows at a direction 45° to the right of the wind direction.

The velocity of coastal upwelling (W_C) is calculated as [49];

$$W_C = \frac{\tau_a}{\rho_w f L} \tag{6}$$

where τ_a is the alongshore wind stress and L is the horizontal length scale of upwelling [50]. When W_C is positive, it represents the velocity of the water moving vertically towards the surface due to the upwelling-favorable alongshore wind.

3. Results

3.1. SST Responses before, during, and after In-fa

One of the significant oceanic responses to typhoons is SST [51] (Figure 4). In summer, the nearshore waters of the Zhoushan Islands exhibit an upwelling phenomenon, resulting in SST values that are 0–3 °C lower than the surrounding temperature (Figure 4(a1)). SST within shoreline waters was significantly affected by the landfall of In-fa on 25 July. SST in the research region dropped to about 26 °C with the phenomenon of seawater cooling (Figure 4(a2)), which was about 1–2 °C lower compared with that of before the typhoon (17 July). In addition, the decline in SST in the research region exhibited a distinctive saddle shape, as depicted in Figure 4(b2). Notably, SST obviously decreased on the right side (the sea area northwestern and southeastern of Gouqi Island, bounded by the red triangle), and the decrease was nearly 1–2 °C. Meanwhile, SST in the typhoon center (Zhoushan Islands) experienced a decrease of around 0–1 °C. In contrast, some sea areas in the northeast showed a slight warming indication.

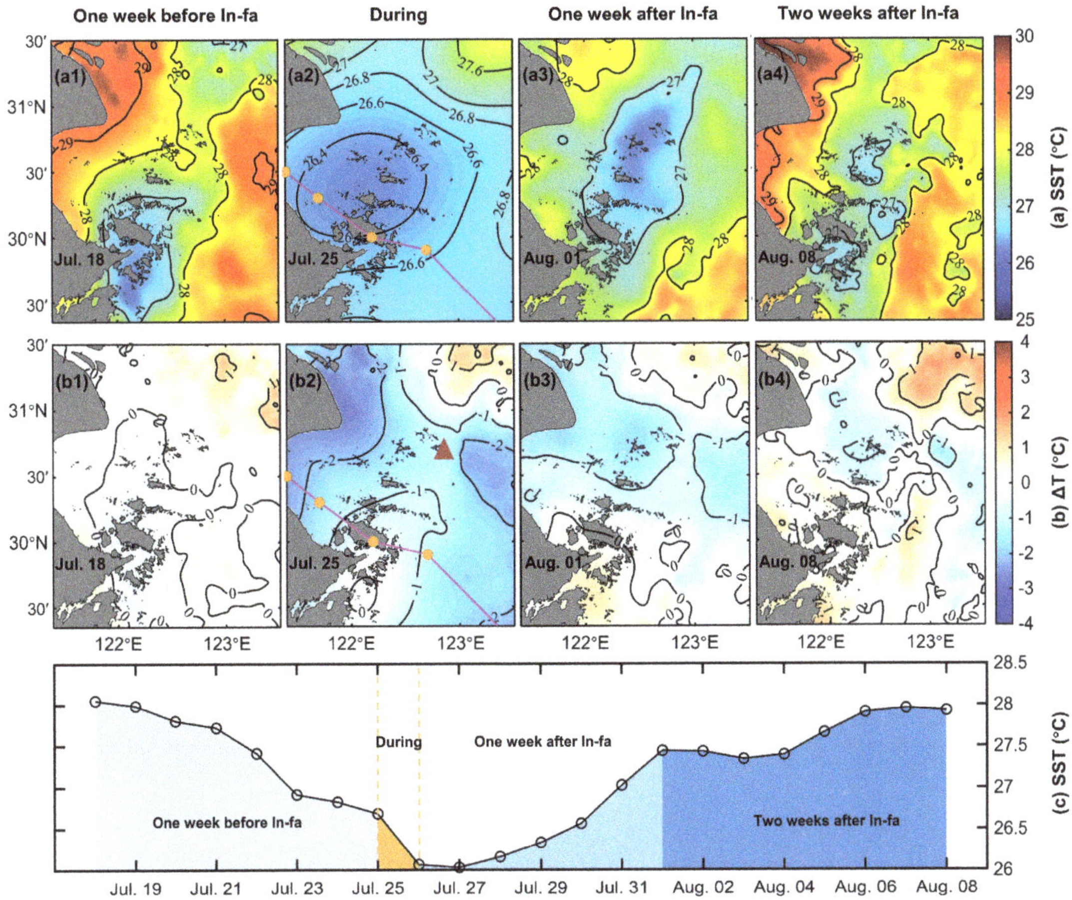

Figure 4. SST distribution before, during, and after the typhoon. (**a1–a4**): daily average of SST within the research region (**b1–b4**): ΔSST is a subtraction operation based on the SST of 17 July 2021; (**c**): time series of SST spatially averaged from 18 July to 8 August 2021. Red triangle: Gouqi Island.

One week after the typhoon (1 August), the SST in the area where the typhoon made landfall began to gradually recover, reaching 27–28 °C. However, low-temperature water persisted in the coastal waters and extended towards the northeast (Figure 4(a3,b3)). Two weeks after the typhoon (8 August), SST in the Yangtze River estuary had mostly returned to its pre-typhoon level, but the low-temperature seawater region had expanded further (Figure 4(a4,b4)).

Based on Figure 4c, the SST continuously decreased from 18 to 27 July as Typhoon In-fa moved northwestward, reaching its lowest value on 27 July with a 2 °C drop. Two weeks after the typhoon, the SST gradually warmed up and returned to its peak on 8 August. Thus, it is evident that the SST in Zhoushan Fishery showed a response to Typhoon In-fa.

3.2. SSS Responses before, during, and after In-fa

Figure 5 illustrates the variation trend of SSS in the research region before and after the period of typhoon transit. In summer, the SSS of the Zhoushan Islands sea area is about 24–33 psu. The SSS isopleth shows a distribution trend of lower western and higher eastern. The sea area to the west of 122°50′E exhibits a more densely distributed salinity gradient, with a flushing freshwater tongue extending eastward in the Yangtze River estuary. In contrast, the salinity isopleths in the ocean east of 122°50′E are sparse, and SSS is high (Figure 5(a1)).

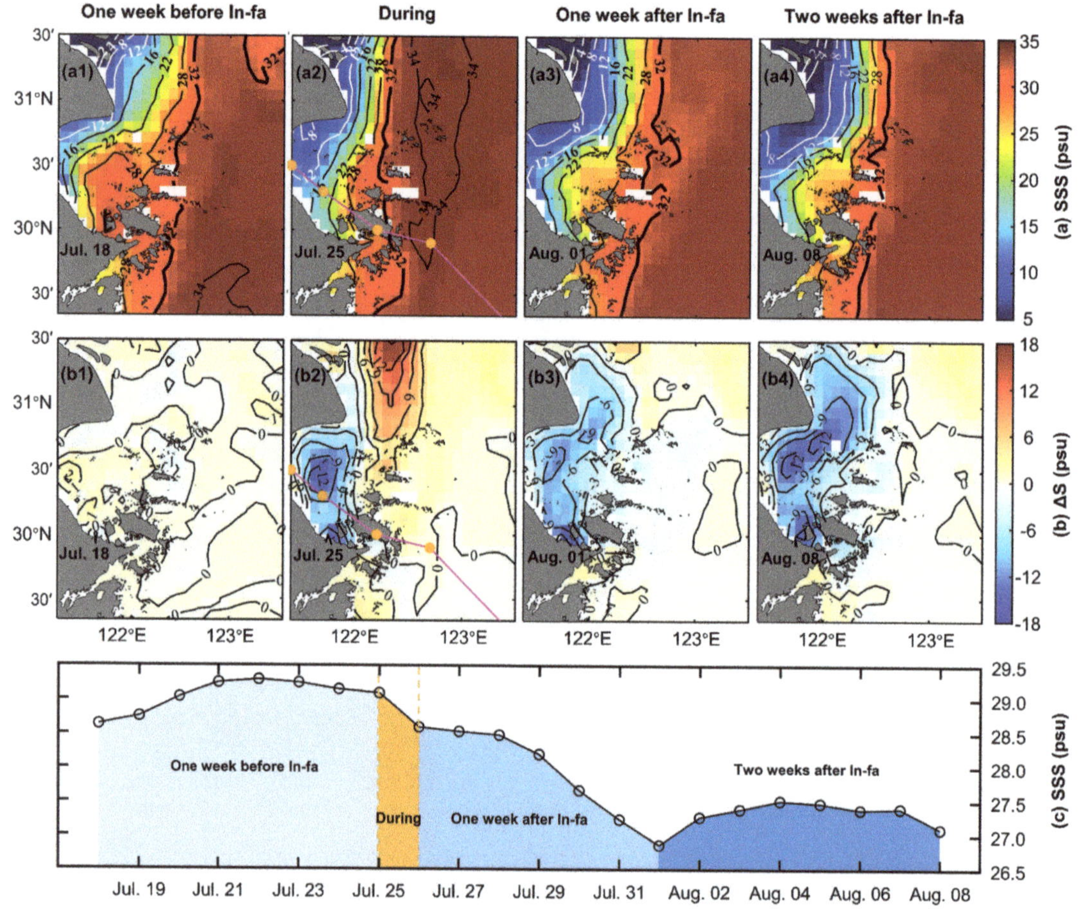

Figure 5. Sea surface salinity (SSS) distribution before, during, and after the typhoon. (**a1–a4**): daily average of SSS within the research region (**b1–b4**): ΔSST is a subtraction operation based on the SST of 17 July 2021; (**c**): time series of SSS spatially averaged from 18 July to 8 August 2021.

The variation in SSS during the landfall of Typhoon In-fa (25 July) differed within the impacted area of the typhoon (Figure 5(a2,b2)). High-salinity seawater advanced westward near the estuary (>32 psu). The SSS (of the Yangtze River estuary) significantly increased by about 0–15 psu (compared with the salinity value on 17 July), while the salinity in the sea around the typhoon center (the Hangzhou Bay estuary) decreased by 0–12 psu. During the two weeks following the typhoon (1 and 8 August) (Figure 5(a3,a4,b3,b4)), SSS at the inlet exhibited more noticeable changes, and the low-salinity seawater continued to spread northeastward, becoming more widely distributed. SSS at the Yangtze River Estuary (121°24′E–122°40′E, 30°45′N–31°30′N) transitioned from an increasing trend to a decreasing trend, decreasing by about 0–9 psu.

According to Figure 5c, the average daily SSS within the research region exhibited an overall increasing trend during the week preceding the typhoon (18–24 July). However, after the typhoon made landfall (25 July–1 August), SSS began to decline from 29.09 to 26.86 psu (down 7.6%). Two weeks after the typhoon's landfall (2–8 August), SSS rebounded, but it remained approximately 1–2 psu lower than the pre-typhoon SSS peak. These results suggest that typhoons have a delayed and persistent impact on SSS within the research region. To further understand the factors contributing to the distribution of SSS variations, the combined effects of changes in vertical mixing, rainfall, and currents on SSS will be analyzed in Section 4.1.2.

3.3. Chl-a Responses before, during, and after In-fa

During Typhoon In-fa, the Chl-*a* data may be significantly affected by cloud cover occlusion and the limited observation frequency of MODIS products. The reanalysis data can effectively supplement the missing data. Therefore, it is necessary to combine MODIS satellite data (Figure 6a) and CMEMS reanalysis data (Figure 6b) for a comprehensive observation of surface Chl-*a* concentration pre- and post-typhoon. Because the typhoon significantly altered the spatial distribution of Chl-*a* concentration within the research region, a differential analysis of Chl-*a* concentration was not performed in this section.

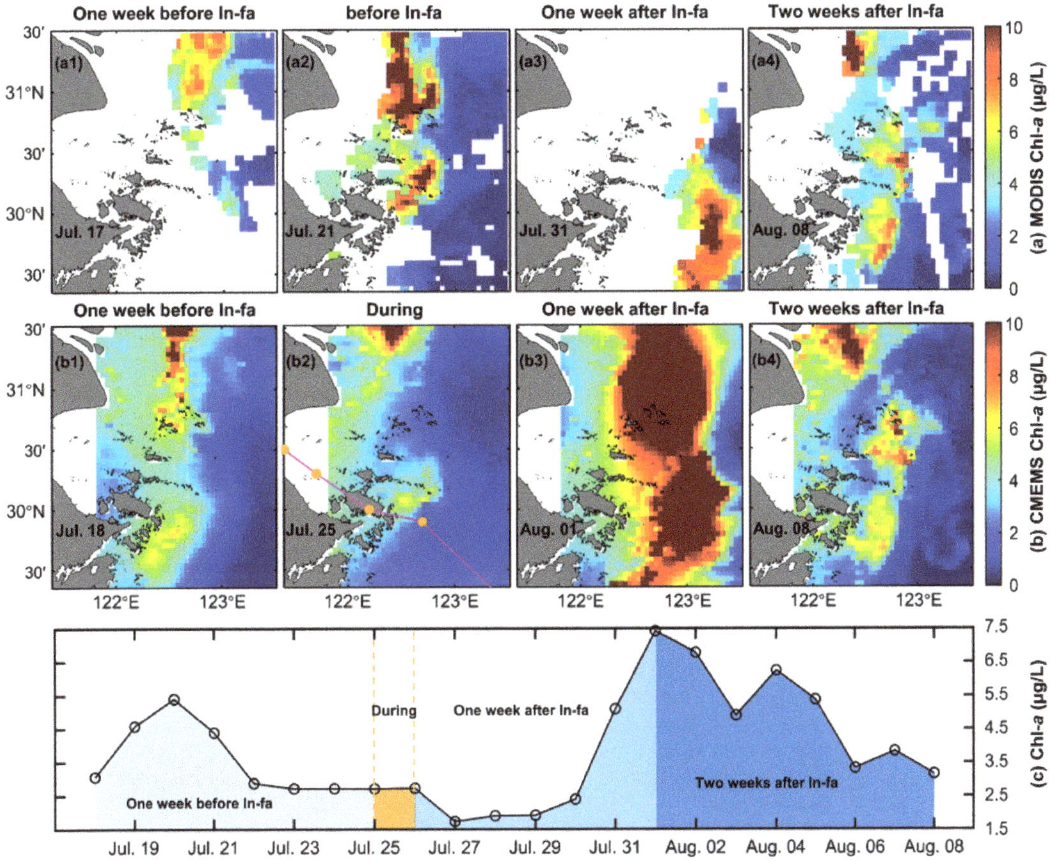

Figure 6. The change in sea surface Chl-*a* concentration before, during, and after the typhoon. (**a1–a4**,**b1–b4**): the daily average of Chl-*a* (**c**): time series of Chl-*a* concentration in terms of spatial means from 18 July to 8 August 2021.

During the pre-typhoon period (17 and 18 July), higher Chl-*a* concentration values of approximately 3–9 μg/L were observed in the island-dense areas of Zhoushan Islands, particularly in the waters west of 123°E (Figure 6(a1,a2,b1)). Chl-*a* concentration in the east of longitude 123°E was low, around 0.5–3 μg/L. During the landfall of the typhoon (25 July), waters with high Chl-*a* shifted towards the west. The concentration of Chl-*a* in the island-dense waters significantly decreased to about 1–4 μg/L. Meanwhile, the concentration of Chl-*a* in Hangzhou Bay and the Yangtze River estuary remained high, at approximately 3–9 μg/L (Figure 6(b2)).

One week after the typhoon (31 July and 2 August), there was a significant increase in Chl-*a* concentration within the research region, as shown in Figure 6(a3,b3). The waters with high Chl-*a* concentration shifted eastward, spanning from 123°E to 123°30′E, with concentrations ranging from about 5 to 10 μg/L. Large-scale phytoplankton blooms were observed within the research region. Two weeks after the typhoon (8 August) (Figure 6(a4,b4)), the spatial distribution of Chl-*a* concentration within the research region generally remained the same as before the typhoon.

The average daily Chl-*a* (Figure 6c) exhibited a decrease from 5.40 to 2.72 μg/L (49.7% decrease) in the four days before the typhoon's passage (21–24 July). Two days after the typhoon (27 July), it reached the lowest value of about 1.74 μg/L, followed by a continuous increase in Chl-*a* during the first week after the typhoon, reaching a maximum value of 7.42 μg/L on 1 August. This indicates that typhoons not only promote the flourishing of Chl-*a* but also have delayed effects, which is consistent with previous studies [52]. By two weeks after the typhoon (1–8 August), the Chl-*a* concentration had generally returned to the pre-typhoon level.

4. Discussion

4.1. The Contribution of the Typhoon to the Change in Typical Environmental Factors

To investigate the response mechanisms of typical oceanic environmental factors (SST, SSS, and Chl-*a*) to In-fa, we estimated the P_w, EPV, coastal upwelling velocity, and sea surface Ekman current using Equations (3)–(6). Additionally, we examined variations in sea surface currents, rainfall, and SSC concentration within the research region before and after In-fa to gain a comprehensive understanding of these mechanisms.

4.1.1. The Mechanism of Typhoon Effect on SST

Typhoons can cause significant disturbances in the upper ocean when they pass over it. One notable change in the upper ocean within the research region during the arrival of Typhoon In-fa was the cooling of the sea surface, especially on the right side of the typhoon (Figure 4). This response is related to the vertical mixing caused by the typhoon and the temperature structure in the vertical direction before the storm [7,53].

During the typhoon's landfall, the EPV within the research region was significantly enhanced (Figure 7a). This intense Ekman pumping resulted in the intensification of vertical seawater mixing. Furthermore, the relatively shallow water depth near the Zhoushan Islands (5–60 m) (Figure 2c) led to a rapid increase in upwelling velocity, which reached 0.957×10^{-4} m/s (Figure 7b). It has been observed that the coastal area of Zhoushan Fishery experiences significant thermal stratification during the summer, with a temperature difference of around 1–8 °C between the surface and bottom waters [54]. As a result of the strong vertical exchange and mixing between the cold bottom water and surface water, along with the increased net heat flow [55], the SST in the study area notably decreased by about 2 °C (Figure 4(a2)).

After the typhoon landed, there was a significant increase in upwelling strength within the research region, although the distribution of upwelling energy across space was uneven. The spatial distribution of P_w and EPV along the In-fa track (Figure 8) revealed that the intense P_w impacted the ocean surrounding the typhoon track except for its center. Notably, the right side of the typhoon track exhibited a particularly strong P_w (Figure 8a). Meanwhile, the physical values of EPV (red for upwelling, blue for downwelling) (Figure 8b) indicated that the vertical water exchange occurred along the center of the typhoon track, with the majority of the upwelling energy being centered on the right side of the typhoon (Figure 8(b3–b6)). The above phenomenon resulted in a stronger current response on the right side of the typhoon track within the research region, leading to increased mixing and a more pronounced weakening of SST (Figure 4(b2)). It is worth noting that the decrease in SST and strength of the Ekman effect are well-correlated [56–58].

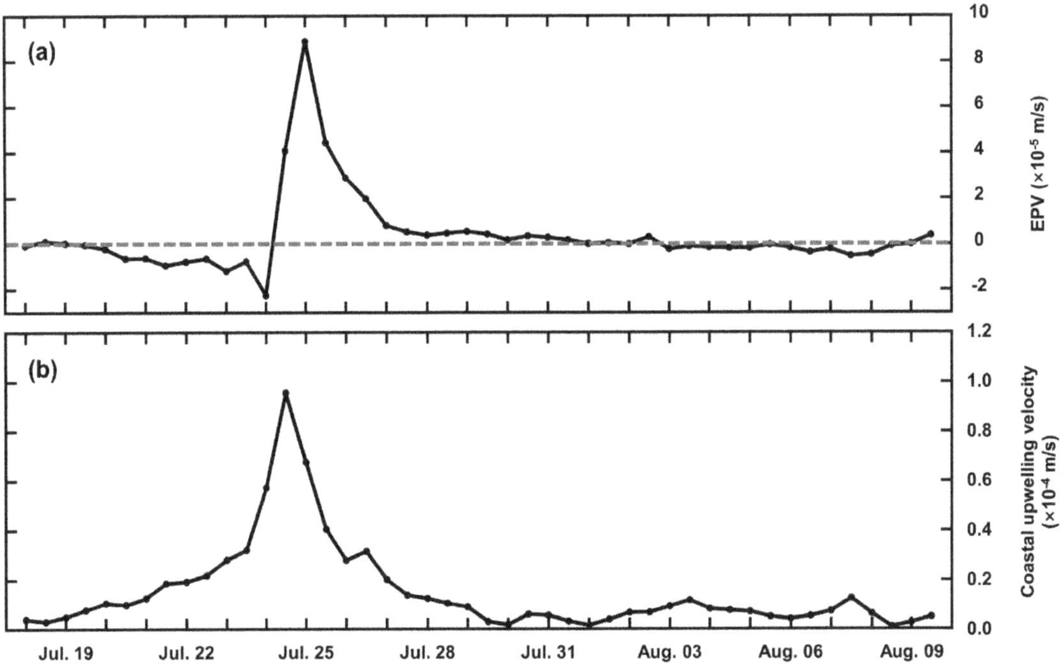

Figure 7. The average Ekman pumping velocity (*EPV*, positive upward) and coastal upwelling velocity in the coastal area of Zhoushan Fishery during the typhoon. (**a**): The average Ekman pumping velocity (**b**): coastal upwelling velocity.

Moreover, the variability in the spatial distribution of SST cooling is influenced by the pre-typhoon temperature structure in the vertical direction of the sea. SST cooling is typically observed in regions with a shallow mixed layer and steep vertical temperature gradient [6,58]. Prior to the typhoon's arrival, the coastal area of Zhejiang experienced an upwelling phenomenon in the sea surrounding the Zhoushan Islands due to the influence of the southwest monsoon and topography [59]. This resulted in a 1–3 °C reduction of sea surface temperature compared with neighboring regions (Figure 4(a1)). Moreover, the upwelling-affected area exhibited a smaller vertical gradient of temperature and deeper mixed layer depth compared with the Yangtze River estuary [54], contributing to the more prominent cooling of SST in the latter region (Figure 4(b2)).

During the passage of Typhoon In-fa, intense spiral updrafts were generated at the center of the typhoon. As the airflow reached a certain height and diverged, it mixed with the surrounding air and then sunk to the lower levels, often leading to the formation of high-temperature weather in the periphery of the typhoon. The study found signs of SST warming on the northeast side of the research region (Figure 4(b2)), which were related to the reduced upwelling effect due to the downwelling of seawater in the region during the landfall of In-fa (Figure 8(b4)), as well as the high-temperature climatic conditions.

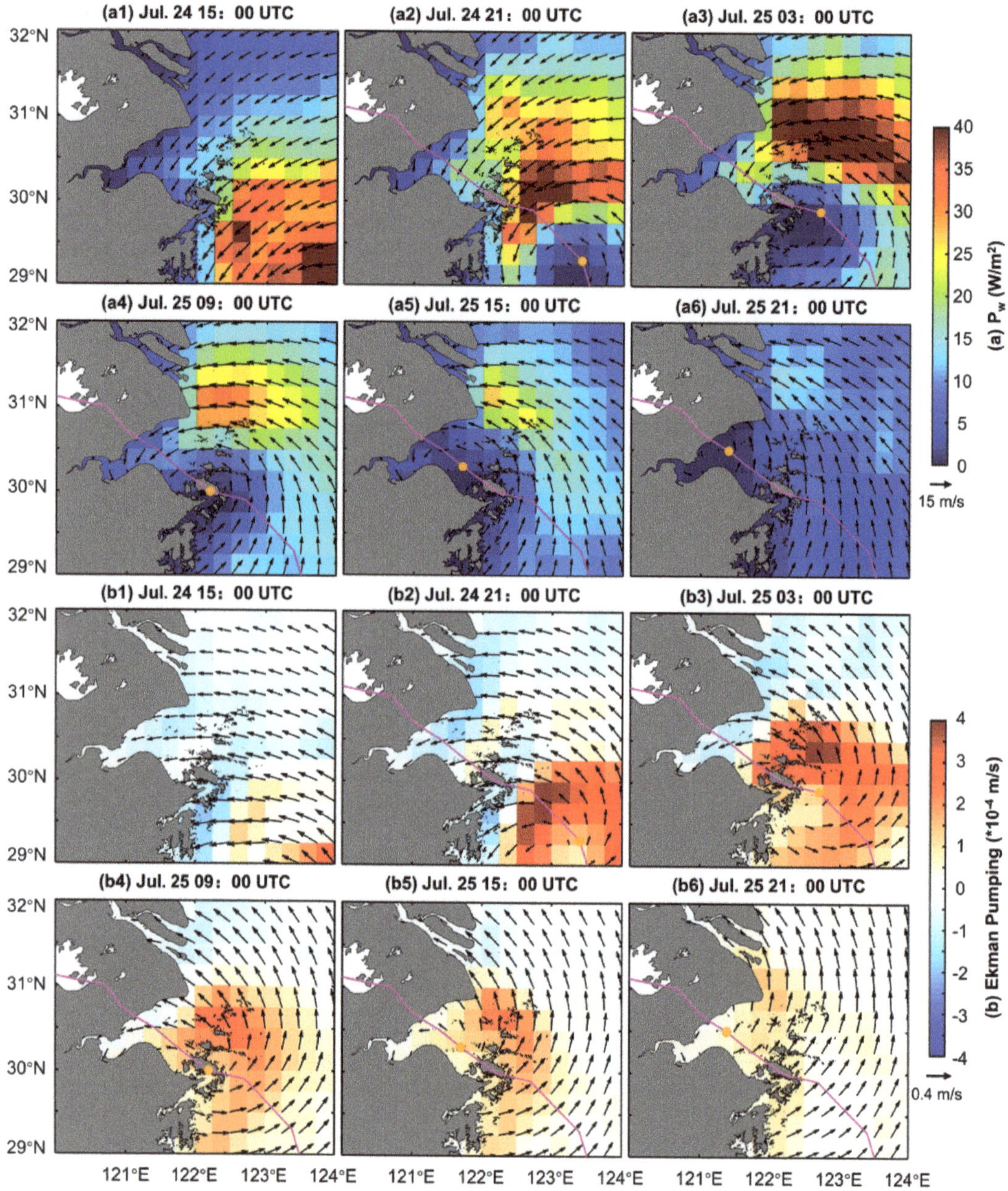

Figure 8. Spatial distribution of wind stress power (P_w) and *EPV*. (**a1–a6**): P_w and sea surface wind (arrows represent wind speeds and directions). (**b1–b6**): *EPV* and sea surface Ekman current (arrows represent current speeds and directions); the track of the typhoon and its intensity (magenta line; orange dots).

4.1.2. The Mechanism of Typhoon Effect on SSS

After Typhoon In-fa passed through, the SSS of the study area showed a decreasing trend and was spatially varied (Figure 5), which was related to the changes in the vertical mixing of seawater, rainfall, and sea surface currents brought about by the typhoon [32].

The coastal area of Zhoushan Fishery is located at the confluence of highly saline offshore waters and coastal waters, and the salinity of seawater is particularly stratified in summer [53]. The TWC, YDW, and ZMCC converging currents in the study area were deflected parallel northward and eastward before In-fa landfall. However, the surface currents were enhanced and significantly shifted after In-fa's landfall (Figure 9). Some of the TWC tributaries flowed into the northern part of the Zhoushan Islands and then into the Yangtze River estuary (Figure 9b), which resulted in the influx of high-salinity outer seawater towards the nearshore of the Yangtze River estuary (Figure 5(a2)). Meanwhile, the stronger P_w and EPV on the right side of the typhoon track (Figure 8) induced intense vertical mixing, resulting in high-salinity bottom water being brought to the sea surface. This increased SSS at the Yangtze River Estuary (Figure 5(b2)).

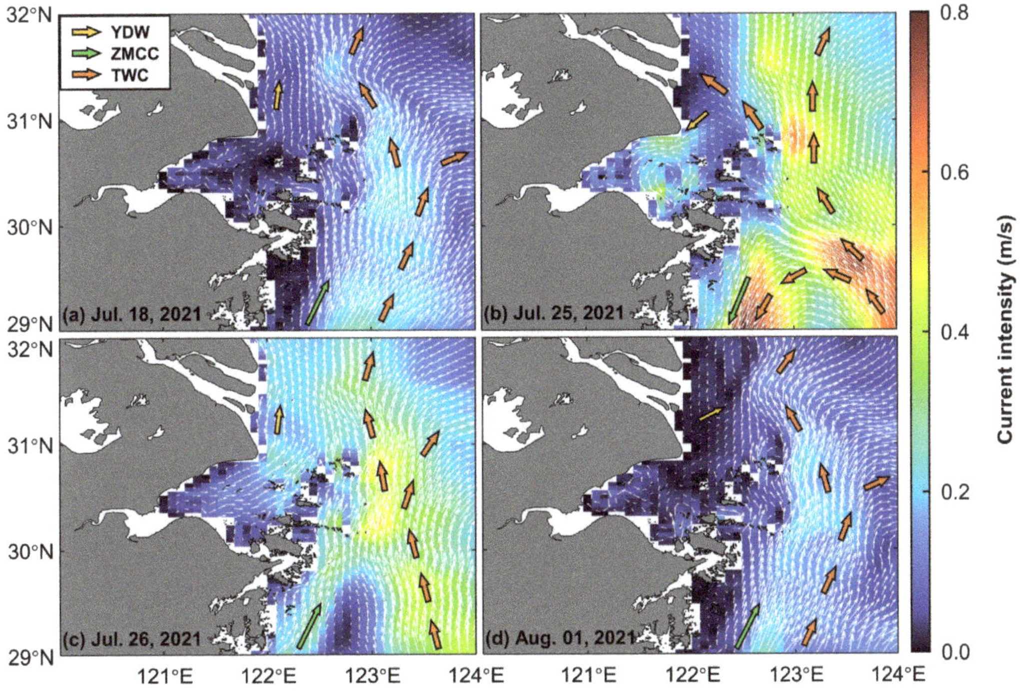

Figure 9. Sea surface currents distribution before, during, and after the typhoon. YDW: Yangtze diluted water; ZMCC: Zhemin coastal current; TWC: Taiwan warm current.

The study found that rainfall occurred two days before In-fa landfall in the waters of Zhoushan Islands (Figure 10(a1,a2)). On 24 July, the average daily rainfall exceeded 200 mm/d (Figure 10b). As In-fa transited, intense rainfall covered the entire Yangtze River estuary and Hangzhou Bay area (Figure 10(a3,a4)). On 26 July, the rainfall was as high as 355 mm/d, which was approximately five times the rainfall before In-fa made landfall (Figure 10b).

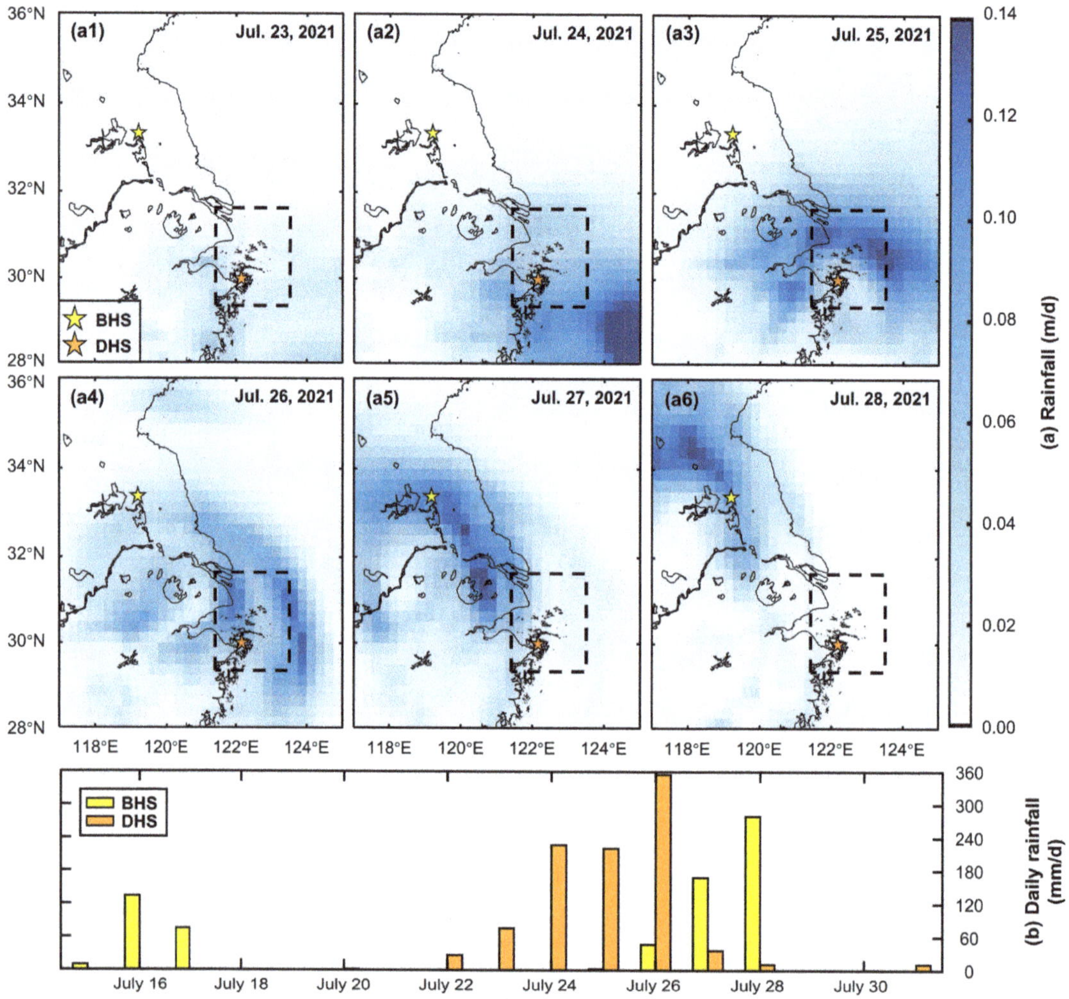

Figure 10. Spatial distribution and daily variation of rainfall within the research region before and after the passage of In-fa. (**a1**–**a6**): Spatial distribution of rainfall; (**b**): daily rainfall at hydrological stations from 15 to 31 July 2021. DHS: Dinghai hydrographic station; BHS: Baoying hydrographic station; black box: study area.

Furthermore, the typhoon's strong winds caused the ZMCC to deflect to the southwest, reducing the inflow of high-salinity seawater into the Hangzhou Bay from nearshore, while the YDW flushed into the bay in the southwest direction (Figure 9b). Under the combined effect of these factors, SSS in Hangzhou Bay rapidly decreased (Figure 5(b2)). The integrated impact of intense Ekman and vortex upwelling at this time had little impact on SSS in the center of the typhoon track compared with intense rainfall [60].

Although the rainfall in the research region decreased after the transit of In-fa, rainstorms continued to erode the runoff in the upper reaches of the Hangzhou Bay and the Yangtze River estuary (Figure 10(a5,a6,b)), which resulted in a surge in runoff volume and floods in multiple locations. This increased the possibility of a massive influx of diluted water into the coastal estuary after the typhoon. Meanwhile, lower salinity seawater from Hangzhou Bay flowed into the Yangtze River estuary in the northeast (Figure 9c,d). As a result, the SSS in the Yangtze River estuary changed from an increase to a decrease after the typhoon, and the lower salinity seawater continued to spread to the northeast (Figure 5(a3,a4,b3,b4)).

In-fa induced Ekman pumping significantly contributed to the rise in SSS, while rainstorms and changes in currents efficiently decreased SSS; this is a phenomenon that has also been observed in other nearshore seas [33,58,61].

4.1.3. The Mechanism of Typhoon Effect on Chl-*a*

Previous studies [46,62,63] have shown that Chl-*a* concentrations are higher in the coastal area of Zhoushan Fishery in summer (Figure 6(a1,a2)), which is probably due to the rich nutrients carried by the Yangtze diluted water, coastal upwelling, and sufficient sunlight. The maximum Ch1-*a* concentration usually occurs at the surface or subsurface layers, and the concentration decreases with depth [62].

Chl-*a* concentrations in the coastal area of Zhoushan Fishery exhibited a relatively swift response to the typhoon (Figure 6). In the four days leading up to the typhoon's landfall (20–24 July), the study area experienced downwelling (negative *EPV* values indicate the presence of wind-induced downwelling) (Figure 7a). This downwelling caused the transport of nutrients rich in surface seawater to the bottom layer, and the nutrients required for phytoplankton growth were not immediately replenished. Consequently, there was a rapid decline in Chl-*a* concentration in the upper layer of coastal seawater (Figure 6c).

Chl-*a* concentrations within the research region decreased during the five days after the typhoon (Figure 6c), and this phenomenon has also been reported in other coastal waters [12,48,64]. On the one hand, the seawater with high salinity and low Chl-*a* concentration continued to invade the nearshore under the northwestward currents (Figures 6(a2) and 9b). On the other hand, the high cloudiness during the typhoon impact could not provide sufficient light for good phytoplankton growth, which limited the growth of Chl-*a* concentration to some extent [65].

The study found that the SSC concentration corresponded well with the high and low Chl-*a* concentration during the week before and after the typhoon (Figure 11(a1,a2,a3,b1,b2,b3)). Before Typhoon In-fa, the estuary (the sea west of 121°48′E) exhibited a high suspended matter concentration. Subsequently, the scouring effect of In-fa, combined with rainstorms in mountainous and river entrances, as well as the vertical mixing of seawater, resulted in an increased SSC concentration in nearshore waters after the rainstorm, which then spread to the outer sea (Figure 11(b3)). This process caused a large influx of nutrients from land-based sources into the coastal waters, while upwelling caused nutrient resuspension [12], creating favorable conditions for phytoplankton blooms in the week after the typhoon (Figure 6(a3,b3)).

Moreover, the lag time in nutrient uptake by phytoplankton growth (average turnaround time of 2–7 days) [52], combined with the transport of offshore currents (Figure 9d), contributed to extensive phytoplankton blooms and the peak Chl-*a* concentration as of the seventh day following typhoon landfall (Figure 6(b3,c)).

After the passage of a typhoon, seawater was stable regarding vertical mixing and upwelling. During a certain period, the nutrient concentration on the sea surface presented a comparatively steady state and could not promote unlimited chlorophyll growth [66]. In addition, the high concentration of SSC continues to persist in nearshore waters following typhoons, leading to a decrease in seawater transparency (Figure 11(b4)), while insufficient light can also limit the growth of phytoplankton in high-turbidity areas [67]. Therefore, during the second week after typhoon landfall, Chl-*a* concentration gradually decreased and returned to pre-typhoon landfall levels (Figure 6(a4,b4)).

In summary, the intense rainfall brought about by In-fa, the upwelling of low-temperature and high-salt bottom water resulting from enhanced vertical mixing, and the influx of diluted water have collectively contributed to an overall decrease in SST and SSS after the typhoon's transit. Meanwhile, this mixing and transport of land-based substances have introduced abundant nutrients to the surface layer of the sea, promoting the growth of phytoplankton. Consequently, Chl-*a* concentration increased and peaked a week after the typhoon's landfall.

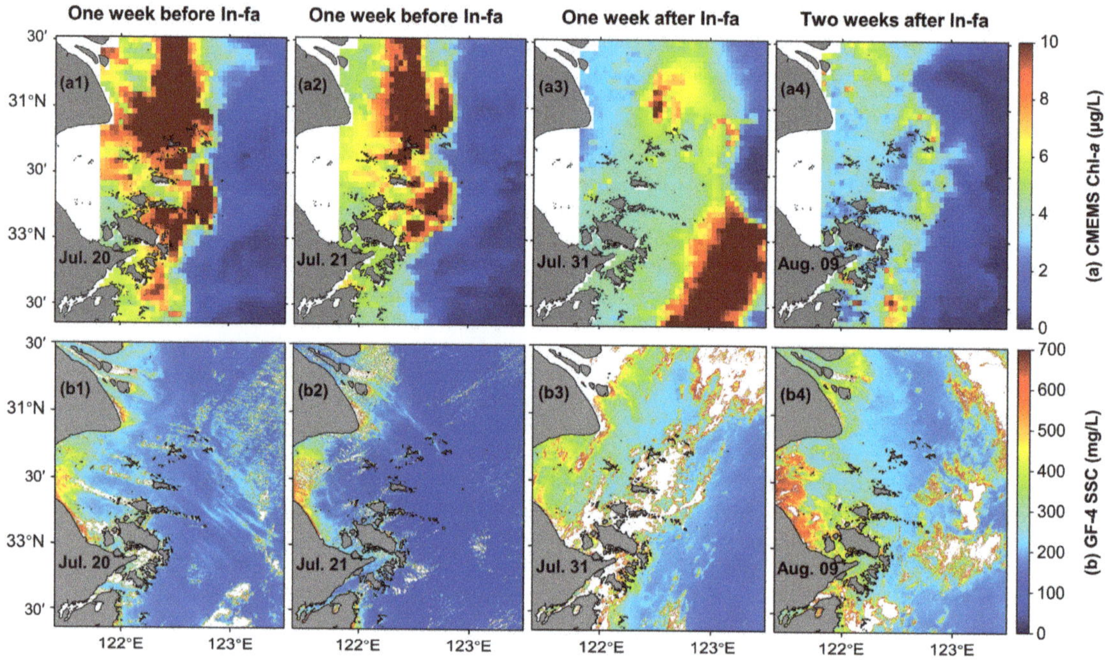

Figure 11. Variations in SSC and Chl-*a* concentration before and after the passage of In-fa. (**a1–a4**): Spatial distribution of the CMEMS Chl-*a* concentration; (**b1–b4**): spatial distribution of the SSC concentration.

4.2. The Contribution of Typhoon to Fishery Resources

The strong winds, decreased salinity, increased runoff, increased intensity of currents, upwelling phenomena, and local phytoplankton bloom production brought about by typhoons usually result in positive or negative changes in local fish behavior and community composition [17,24,68]. Summer is a crucial period for breeding the main economic fish and for the growth of juvenile fish in Zhoushan Fishery; at the same time, the landing of Typhoon In-fa directly or indirectly affects the biological processes in the six national marine ranching demonstration areas around the Zhoushan Islands.

Previous studies have shown a strong correlation between changes in the biomass density of fish eggs and larvae and variations in Chl-*a* concentration in the water [46], indicating that Chl-*a* is a reliable indicator of phytoplankton biomass and the prior status of the aquatic food chain [69]. Additionally, Chl-*a* levels can influence the fertility of waters and help evaluate the potential productivity of fishing grounds [70,71]. Chl-*a* concentrations at the sea surface within the research region were significantly increased and more widely distributed by In-fa (Figure 6(b3)). This was closely related to the sizeable terrestrial nutrient discharge in the estuary after the typhoon [13] and the enhanced upwelling effect [17], which increased nutrients. Fish actively migrate through feeding activities to nutrient-rich waters [17,18,72], making it an excellent place to catch fish. The increase in nutrient salinity after typhoons induces an enhancement in Chl-*a* concentration, which has a steady positive effect on fish.

Currents play an essential role in fish productivity [20,73]. The surface currents within the research region were significantly enhanced by In-fa (Figure 9b,c), and some weakfish will enter the nearshore waters due to strong currents [17]. In the week after the typhoon, the TWC formed some small flow sets, cyclones, and anticyclonic eddies in the waters near the Zhoushan Islands during its northward movement. The sea had upwelling and downwelling currents (Figure 9d). This process allows for sufficient seawater exchange, creates an ideal environment for nutrient salt operation, and creates different current conditions for fish selection, which positively impacts the increase in fishery resources [74,75]. In addition, typhoons lead to violent disturbances of seawater, which will scatter the density of fish and cause them to flee to calmer deep water or the leeward side of the island with less wind and waves, which is an excellent place to catch fish [76].

Species that can quickly adapt to environmental changes can survive in the original waters or enter the changing waters and vice versa, and they will flee from the original survival waters [17]. Several studies have indicated that small- and medium-sized nearshore fish dominate the record of increases after typhoons, while the proportion of small fish at relatively low trophic levels decreases [17,77]. The salinity and temperature in the fishery dramatically changed after the landing of In-fa (Figures 4 and 5). Rainstorms brought about by In-fa decreased SSS in the culture area, which cause fish that are adapted to higher-salinity to swim to higher-salinity waters offshore. In contrast, the freshwater fish population has increased [78]. The decrease in SST will reduce the number of fish that prefer a warm water environment, while the cool water-loving fish living at the bottom will swim to the sub-surface layer where the temperature decreases [77]. In addition, the strong vertical mixing of seawater caused by typhoons can weaken or eliminate the thermocline, causing fish to disperse and be challenging to catch. Lower pressure will cause bottom-dwelling fish to float, which affects bottom trawl production [76].

In summary, the mechanisms underlying fish resource changes in nearshore fisheries after typhoons are complex and can be influenced by various environmental factors, including temperature, salinity, Chl-*a*, and currents. Different species of fish have different degrees of adaptation to these environmental factors. This information helps fishermen obtain a better grasp of the relationship pattern between typhoons and fishing, provides a reference for the distribution of fish stocks, and predicts the direction of fishery movement.

5. Conclusions

Grounded in Zhoushan Fishery, this study used GF-4, MODIS satellite, and reanalysis data to investigate the responses of environmental factors, including SST, SSS, and Chl-*a*, in Zhoushan Fishery around the landfall of Typhoon In-fa, where we analyzed the mechanism.

Typhoon In-fa induced SST, SSS, and Chl-*a* changes. After the landing of In-fa, the intense vertical mixing and enhanced upwelling caused SST to decrease by about 2 °C. In the Northern Hemisphere, the asymmetrical typhoon wind pressure and vertical differences in the temperature structure in front of the typhoon (mixed layer depth and vertical temperature gradient) led to a more intense cooling of SST in the right In-fa track. SST returned to its pre-typhoon peak after two weeks. SSS decreased overall after In-fa landing, and the variation in SSS was spatially variable. SSS was significantly lower in Hangzhou Bay at the eye of the typhoon due to heavy rainfall and runoff; in comparison, the SSS was significantly higher at the Yangtze River estuary on the right side of the typhoon track, which is attributed to the strong EPV. MODIS satellite and reanalysis data revealed that the change process of Chl-*a* concentration at the sea surface manifested in a decrease and then an increase. On the seventh day following the typhoon, large-scale phytoplankton blooms appeared, and Chl-*a* concentration was higher than pre-typhoon levels. The spatial and temporal variation in Chl-*a* concentration was intimately correlated with the changes in light intensity, nutrients, current, and SSC after In-fa.

Furthermore, the alterations of physical and biotic variables within the fishery caused by In-fa can affect fishery fish behavior and community composition. The mechanisms of fish resource changes in the nearshore fishery after typhoons are complex and can lead to changes in temperature, salinity, Chl-*a*, and currents, among other environmental factors. This study provides some reference to reveal the response of environmental factors to typhoons and their effects on fishery resources in the fishery with intensive islands in nearshore estuarine shallow waters.

Author Contributions: Conceptualization, R.T. and L.C.; methodology, L.C.; software, R.T.; validation, R.T. and L.C.; formal analysis, R.T., L.C. and X.Y. (Xiaomin Ye); investigation, R.T., L.C., X.Y. (Xiaojun Yan), X.Y. (Xiaomin Ye), Y.X. and J.Y.; data curation, R.T. and L.C.; writing—review and editing, R.T.; writing—review and editing, L.C. and X.Y. (Xiaojun Yan); visualization, R.T.; supervision, L.C.; project administration, L.C.; funding acquisition, L.C. All authors have read and agreed to the published version of the manuscript.

Funding: This work was supported by the following research projects: the Science Foundation of Donghai Laboratory (DH-2022KF01010); the National Natural Science Foundation of China Key international (regional) cooperative research project (42020104009); and the Basic Public Welfare Research Program of Zhejiang Province (LGF21D010004).

Data Availability Statement: Publicly available datasets were analyzed in this study. GF-4/PMS satellite data are available at https://data.cresda.cn/#/home (accessed on 7 November 2022).

Acknowledgments: The authors would like to thank the China Centre for Resources Satellite Data and Application for providing the GF-4/PMS satellite data free of charge.

Conflicts of Interest: The authors declare no conflict of interest.

References

1. Chan, J.C.L. Interannual and interdecadal variations of tropical cyclone activity over the western North Pacific. *Meteorol. Atmos. Phys.* **2005**, *89*, 143–152. [CrossRef]
2. Ding, P.; Hu, K.; Kong, Y.; Hu, D. Numerical Simulation of Storm-induced Erosion/Deposition in Yangtze Estuary—A Case Study of Typhoon Jelawat. *J. Sediment Res.* **2003**, *6*, 18–24. [CrossRef]
3. Guo, J.; Zhang, T.; Xu, C.; Xie, Q. Upper ocean response to typhoon Kujira (2015) in the South China Sea by multiple means of observation. *J. Oceanol. Limnol.* **2019**, *38*, 314–333. [CrossRef]
4. Pudov, V.D.; Petrichenko, S.A. Trail of a typhoon in the salinity field of the ocean upper layer. *Izv. Atmos. Ocean Phys.* **2000**, *36*, 645–650.
5. Price, J.F. Upper Ocean Response to a Hurricane. *J. Phys. Oceanogr.* **1981**, *11*, 153–175. [CrossRef]
6. Price, J.F.; Morzel, J.; Niiler, P.P. Warming of SST in the cool wake of a moving hurricane. *J. Geophys. Res. Ocean.* **2008**, *113*, C07010. [CrossRef]
7. Chiang, T.L.; Wu, C.R.; Oey, L.Y. Typhoon Kai-Tak: An Ocean's Perfect Storm. *J. Phys. Oceanogr.* **2011**, *41*, 221–233. [CrossRef]
8. Jiang, X.; Zhong, Z.; Liu, C. The effect of typhoon-induced SST cooling on typhoon intensity: The case of Typhoon Chanchu (2006). *Adv. Atmos. Sci.* **2008**, *25*, 1062–1072. [CrossRef]
9. Wang, J.; Kuang, C.; Chen, K.; Fan, D.; Qin, R.; Han, X. Wave current interaction by Typhoon Fongwong on saline water intrusion and vertical stratification in the Yangtze River Estuary. *Estuar. Coast. Shelf Sci.* **2022**, *279*, 108138. [CrossRef]
10. Wang, Z.; Goodman, L. The evolution of a thin phytoplankton layer in strong turbulence. *Cont. Shelf Res.* **2010**, *30*, 104–118. [CrossRef]
11. Huisman, J.; Pham Thi, N.N.; Karl, D.M.; Sommeijer, B. Reduced mixing generates oscillations and chaos in the oceanic deep chlorophyll maximum. *Nature* **2006**, *439*, 322–325. [CrossRef]
12. Zheng, G.; Tang, D. Offshore and nearshore chlorophyll increases induced by typhoon winds and subsequent terrestrial rainwater runoff. *Mar. Ecol. Prog. Ser.* **2007**, *333*, 61–74. [CrossRef]
13. Zhao, H.; Tang, D.; Wang, D. Phytoplankton blooms near the Pearl River Estuary induced by Typhoon Nuri. *J. Geophys. Res. Ocean.* **2009**, *114*, 1–9. [CrossRef]
14. Zhao, H.; Han, G.; Zhang, S.; Wang, D. Two phytoplankton blooms near Luzon Strait generated by lingering Typhoon Parma. *J. Geophys. Res. Biogeosci.* **2013**, *118*, 412–421. [CrossRef]
15. Chen, J.; Chen, S. Estuarine and Coastal Challenges in China. *Mar. Geol. Front.* **2002**, *18*, 1–5. [CrossRef]
16. Guo, W.; Yang, Y.; Gong, Z.; Hu, M. Variation of salitity, PH, dissolved oxygen and COD in Xiamen Bay during typhoon and rainstorm process. *Mar. Sci.* **2000**, *25*, 1–5. (In Chinese)
17. Yu, J.; Tang, D.; Li, Y.; Huang, Z.; Chen, G. Increase in fish abundance during two typhoons in the South China Sea. *Adv. Space Res.* **2013**, *51*, 1734–1749. [CrossRef]
18. Kawabata, Y.; Okuyama, J.; Asami, K.; Okuzawa, K.; Yoseda, K.; Arai, N. Effects of a tropical cyclone on the distribution of hatchery-reared black-spot *tuskfish Choerodon schoenleinii* determined by acoustic telemetry. *J. Fish Biol.* **2010**, *77*, 627–642. [CrossRef]
19. Lassig, B.R. The effects of a cyclonic storm on coral reef fish assemblages. *Environ. Biol. Fishes* **1983**, *9*, 55–63. [CrossRef]
20. Houde, E.D.; Bichy, J.; Jung, S. Effects of hurricane Isabel on fish populations and communities in Chesapeake Bay. In Proceedings of the Chesapeake Research Consortum, Edgewater, MD, USA, 1 January 2005; pp. 193–199.
21. Qiu, Y.; Lin, Z.; Wang, Y. Responses of fish production to fishing and climate variability in the northern South China Sea. *Prog. Oceanogr.* **2010**, *85*, 197–212. [CrossRef]

22. Webster, P.J.; Curry, J.A.; Liu, J.; Holland, G.J. Response to Comment on "Changes in Tropical Cyclone Number, Duration, and Intensity in a Warming Environment". *Science* **2006**, *311*, 1713. [CrossRef]
23. Mendelsohn, R.; Emanuel, K.; Chonabayashi, S.; Bakkensen, L. The impact of climate change on global tropical cyclone damage. *Nature Clim. Chang.* **2012**, *2*, 205–209. [CrossRef]
24. Yu, J.; Tang, D.; Chen, G.; Li, Y.; Huang, Z.; Wang, S. The positive effects of typhoons on the fish CPUE in the South China Sea. *Cont. Shelf Res.* **2014**, *84*, 1–12. [CrossRef]
25. Chuang, L.; Shieh, B.S.; Liu, C.; Lin, Y.; Liang, S. Effects of Typhoon Disturbance on the Abundances of Two Mid-Water Fish Species in a Mountain Stream of Northern Taiwan. *Zool. Res.* **2008**, *47*, 564–573. Available online: https://zoolstud.sinica.edu.tw/Journals/47.5/564 (accessed on 14 February 2023).
26. Chang, N.; Shiao, J.; Gong, G. Diversity of demersal fish in the East China Sea: Implication of eutrophication and fishery. *Cont. Shelf Res.* **2012**, *47*, 42–54. [CrossRef]
27. Liang, J.; Wang, W.; Xu, H.; Zhou, Y.; Xu, K.; Zhang, H.; Lu, K. Diel and seasonal variation in fish communities in the Zhongjieshan marine island reef reserve. *Fish. Res.* **2020**, *227*, 105549. [CrossRef]
28. Schade, L.; Emanuel, K. The Ocean's Effect on the Intensity of Tropical Cyclones: Results from a Simple Coupled Atmosphere–Ocean Model. *J. Atmos. Sci.* **1999**, *56*, 642–651. [CrossRef]
29. Liu, Z.; Xu, J.; Sun, C.; Wu, X. An upper ocean response to Typhoon Bolaven analyzed with Argo profiling floats. *Acta Oceanol. Sin.* **2014**, *33*, 90–101. [CrossRef]
30. Tang, R.; Shen, F.; Ge, J.; Yang, S.; Gao, W. Investigating typhoon impact on SSC through hourly satellite and real-time field observations: A case study of the Yangtze Estuary. *Cont. Shelf Res.* **2021**, *224*, 104475. [CrossRef]
31. Pan, J.; Huang, L.; Devlin, A.T.; Lin, H. Quantification of Typhoon-Induced Phytoplankton Blooms Using Satellite Multi-Sensor Data. *Remote Sens.* **2018**, *10*, 318. [CrossRef]
32. Chen, J.; Jiang, C.; Wu, Z.; Long, Y.; Deng, B.; Liu, X. Numerical Investigation of Fresh and Salt Water Distribution in the Pearl River Estuary during a Typhoon Using a Fully Coupled Atmosphere-Wave-Ocean Model. *Water* **2019**, *11*, 646. [CrossRef]
33. Jeon, J.; Tomita, T. Investigating the Effects of Super Typhoon HAGIBIS in the Northwest Pacific Ocean Using Multiple Observational Data. *Remote Sens.* **2022**, *14*, 5667. [CrossRef]
34. Ma, C.; Zhao, J.; Bin, A.; Sun, S.; Zhang, G.; Huang, W.; Wang, G. Assessing responses of phytoplankton to consecutive typhoons by combining Argo, remote sensing and numerical simulation data. *Sci. Total Environ.* **2021**, *790*, 148086. [CrossRef] [PubMed]
35. Groom, S.; Sathyendranath, S.; Ban, Y.; Bernard, S.; Brewin, R.; Brotas, V.; Brockmann, C.; Chauhan, P.; Choi, J. Satellite Ocean Colour: Current Status and Future Perspective. *Front. Mar. Sci.* **2019**, *6*, 485. [CrossRef] [PubMed]
36. Banzon, V.; Smith, T.M.; Chin, T.M.; Liu, C.Y.; Hankins, W. A long-term record of blended satellite and in situ sea-surface temperature for climate monitoring, modeling and environmental studies. *Earth Syst. Sci. Data* **2016**, *8*, 165–176. [CrossRef]
37. Le Traon, P.Y.; Reppucci, A.; Fanjul, E.A.; Aouf, L.; Behrens, A.; Belmonte, M.; Bentamy, A.; Bertino, L.; Brando, V.E.; Kreiner, M.B.; et al. From Observation to Information and Users: The Copernicus Marine Service Perspective. *Front. Mar. Sci.* **2019**, *6*, 234. [CrossRef]
38. Cai, L.; Zhou, M.; Liu, J.; Tang, D.; Zuo, J. HY-1C observations of the impacts of islands on suspended sediment distribution in Zhoushan Coastal Waters, China. *Remote Sens.* **2020**, *12*, 1766. [CrossRef]
39. Liu, X.; Sun, L.; Yang, Y.; Zhou, X.; Chen, T. Cloud and Cloud Shadow Detection Algorithm for Gaofen-4 Satellite Data. *Acta Optica Sin.* **2019**, *39*, 438–449. [CrossRef]
40. Prakash, G.; Deng, L.; Nie, J. Effect of Image Fusion on Vegetation Index Quality—A Comparative Study from Gaofen-1, Gaofen-2, Gaofen-4, Landsat-8 OLI and MODIS Imagery. *Remote Sens. Environ.* **2020**, *12*, 1550. [CrossRef]
41. Cai, L.; Tang, D.; Levy, G.; Liu, D. Remote sensing of the impacts of construction in coastal waters on suspended particulate matter concentration—The case of the Yangtze River delta, China. *Int. J. Remote Sens.* **2016**, *37*, 2132–2147. [CrossRef]
42. Chin, T.M.; Vazquez-Cuervo, J.; Armstrong, E.M. A multi-scale high-resolution analysis of global sea surface temperature. *Remote Sens. Environ.* **2017**, *200*, 154–169. [CrossRef]
43. Wang, S.; Zhao, Y.; Yin, X.; Qiao, F. Current Status of Global Ocean Reanalysis Datasets. *Adv. Earth Res.* **2018**, *33*, 794–807. (In Chinese) [CrossRef]
44. Wang, H.; You, Z.; Guo, H.; Zhang, W.; Xu, P.; Ren, K. Quality Assessment of Sea Surface Salinity from Multiple Ocean Reanalysis Products. *J. Mar. Sci. Eng.* **2023**, *11*, 54. [CrossRef]
45. Wang, T.; Liu, G.; Gao, L.; Zhu, L.; Li, D. Biological responses to nine powerful typhoons in the East China Sea. *Reg. Environ. Chang.* **2017**, *17*, 465–476. [CrossRef]
46. Cai, L.; Tang, R.; Yan, X.; Zhou, Y.; Jiang, J.; Yu, M. The spatial-temporal consistency of chlorophyll-a and fishery resources in the water of the Zhoushan archipelago revealed by high resolution remote sensing. *Front. Mar. Sci.* **2022**, *9*, 2053. [CrossRef]
47. Price, J.F.; Sanford, T.B.; Forristall, G.Z. Forced Stage Response to a Moving Hurricane. *J. Phys. Oceanogr.* **1994**, *24*, 233–260. [CrossRef]
48. Zhao, H.; Shao, J.; Han, G.; Yang, D.; Lv, J. Influence of Typhoon Matsa on Phytoplankton Chlorophyll-a in the Northwest Pacific Ocean Offshore and Alongshore. *PLoS ONE* **2015**, *10*, e0137863. [CrossRef]

49. Han, G.; Ma, Z.; Chen, N. Hurricane Igor impacts on the stratification and phytoplankton bloom over the Grand Banks. *J. Mar. Syst.* **2012**, *100*, 19–25. [CrossRef]
50. Yin, W.; Ma, Y.; Wang, D.; He, S.; Huang, D. Surface Upwelling off the Zhoushan Islands, East China Sea, from Himawari-8 AHI Data. *Remote Sens.* **2022**, *14*, 3261. [CrossRef]
51. Yang, X.; Tang, D. Location of sea surface temperature cooling induced by typhoon in the South China Sea. *J. Trop. Oceanogr.* **2010**, *29*, 26–31. [CrossRef]
52. Behrenfeld, M.J.; Falkowski, P.G. Photosynthetic rates derived from satellite-based chlorophyll concentration. *Limnol. Oceanogr.* **2003**, *42*, 1–20. [CrossRef]
53. Guo, X.; Miyazawa, Y.; Yamagata, T. The Kuroshio Onshore Intrusion along the Shelf Break of the East China Sea: The Origin of the Tsushima Warm Current. *J. Phys. Oceanogr.* **2006**, *36*, 2205–2231. [CrossRef]
54. Lin, J.; Yan, Q.; Zhu, J.; Gong, F. Analysis of thermocline and hypoxia off the Changjiang Estuary in late summer. *J. Fish. Chn.* **2014**, *38*, 1747–1757. (In Chinese)
55. Yuan, J.; Liu, C. Improved scheme of axisymmetric typhoon bogus model its impact on numerical simulation of No. 0425 Typhoon. *J. Trop. Meteorol.* **2007**, *13*, 181–184.
56. Yang, Y.; Xian, T.; Sun, L.; Fu, Y.; Xun, S. Impacts of sequential typhoons on sea surface temperature and sea surface height in September 2008. *Acta Oceanol. Sin.* **2012**, *34*, 71–78. (In Chinese)
57. Sheng, J.; Zhai, X.; Greatbatch, R. Numerical study of the storm-induced circulation on the Scotian Shelf during Hurricane Juan using a nested-grid ocean model. *Prog. Oceanogr.* **2006**, *70*, 233–254. [CrossRef]
58. Wu, R.; Li, C. Upper ocean response to the passage of two sequential typhoons. *Deep Sea Res. Part I Oceanogr. Res. Pap.* **2018**, *132*, 68–79. [CrossRef]
59. Ding, Z. Influences of wind on vertical structures of temperature and salinity, and upwelling off Zhejiang coast in summer. *Chin. J. Oceanol. Limnol.* **1985**, *3*, 109–117. [CrossRef]
60. Zhang, H.; Liu, X.; Wu, R.; Chen, D.; Zhang, D.; Shang, X.; Wang, Y.; Song, X.; Jin, W.; Yu, L.; et al. Sea surface current response patterns to tropical cyclones. *J. Mar. Syst.* **2020**, *208*, 103345. [CrossRef]
61. Sun, L.; Li, Y.; Yang, Y.; Wu, Q.; Chen, X.; Li, Q.; Li, Y.; Xian, T. Effects of super typhoons on cyclonic ocean eddies in the western North Pacific: A satellite data-based evaluation between 2000 and 2008. *J. Geophys. Res. Atmos.* **2014**, *119*, 5585–5598. [CrossRef]
62. Gong, X.; Shi, J.; Gao, H.; Yao, X. Steady-state solutions for subsurface chlorophyll maximum in stratified water columns with a bell-shaped vertical profile of chlorophyll. *Biogeosciences* **2015**, *12*, 905–919. [CrossRef]
63. Oliveira, P.B.; Nolasco, R.; Dubert, J.; Moita, T.; Peliz, Á. Surface temperature, chlorophyll and advection patterns during a summer upwelling event off central Portugal. *Cont. Shelf Res.* **2009**, *29*, 759–774. [CrossRef]
64. Zhou, W.; Yin, K.; Harrison, P.J.; Lee, J.H.W. The influence of late summer typhoons and high river discharge on water quality in Hong Kong waters. *Estuar. Coast. Shelf Sci.* **2012**, *111*, 35–47. [CrossRef]
65. Pennock, J.R. Chlorophyll distributions in the Delaware estuary: Regulation by light-limitation. *Estuar. Coast. Shelf Sci.* **1985**, *21*, 711–725. [CrossRef]
66. Ritchie, J.C.; Cooper, C.M.; Schiebe, F.R. The relationship of MSS and TM digital data with suspended sediments, chlorophyll, and temperature in Moon Lake, Mississippi. *Remote Sens. Environ.* **1990**, *33*, 137–148. [CrossRef]
67. Díez-Minguito, M.; Swart, H.E. Relationships Between Chlorophyll-a and Suspended Sediment Concentration in a High-Nutrient Load Estuary: An Observational and Idealized Modeling Approach. *J. Geophys. Res. Ocean.* **2020**, *125*, e2019JC015188. [CrossRef]
68. Mukherjee, S.; Chaudhuri, A.; Sen, S.; Homechaudhuri, S. Effect of Cyclone Aila on estuarine fish assemblages in the Matla River of the Indian Sundarbans. *J. Trop. Ecol.* **2012**, *28*, 405–415. [CrossRef]
69. Bierman, P.; Lewis, M.; Ostendorf, B.; Tanner, J. A Review of Methods for Analysing Spatial and Temporal Patterns in Coastal Water Quality. *Ecol. Indic.* **2009**, *11*, 103–114. [CrossRef]
70. Huot, Y.; Babin, M.; Bruyant, F.; Grob, C.; Twardowski, M.S.; Claustre, H. Does chlorophyll a provide the best index of phytoplankton biomass for primary productivity studies? *Biogeosci. Discuss.* **2007**, *4*, 707–745. [CrossRef]
71. Bacher, C.; Grant, J.; Fang, J.; Zhu, M.; Besnard, M. Modeling the effect of food depletion on scallop growth in Sungo Bay (China). *Aquat. Living Resour.* **2003**, *16*, 10–24. [CrossRef]
72. Switzer, T.S.; Winner, B.L.; Dunham, N.M.; Whittington, J.A.; Thomas, M. Influence of sequential hurricanes on nekton communities in a Southeast Florida estuary: Short-term effects in the context of historical variations in freshwater inflow. *Estuaries Coasts* **2006**, *29*, 1011–1018. [CrossRef]
73. Lin, L.; Shen, J.; Zhang, J.; Wu, H. Storm surge of Typhoon Saomai and its effect on the stow net fishing. *Mar. Environ. Res.* **2007**, *105*, 537–540. (In Chinese)
74. Simon, N.S. Nitrogen cycling between sediment and the shallow-water column in the transition zone of the Potomac River and Estuary. II. The role of wind-driven resuspension and adsorbed ammonium. *Estuar. Coast. Shelf Sci.* **1989**, *26*, 483–497. [CrossRef]
75. Oyadomari, J.K.; Auer, N.A. Transport and growth of larval cisco (*Coregonus artedi*) in the Keweenaw Current region of Lake Superior. *Can. J. Fish. Aquat. Sci.* **2008**, *65*, 1447–1458. [CrossRef]
76. Zhang, Z. The relationship between typhoons and fisheries. *J. Aquac.* **1992**, *3*, 29–30. (In Chinese)

77. Yu, J.; Chen, B.; Chen, Z.; Huang, Z. Theimpacts of typhoon "Kai-tak" on fishery in west Guangdong fishing ground. *Mar. Environ. Res.* **2015**, *34*, 411–419. [CrossRef]
78. Greenwood, M.F.D.; Stevens, P.W.; Matheson, R.E. Effects of the 2004 hurricanes on the fish assemblages in two proximate southwest Florida estuaries: Change in the context of interannual variability. *Estuaries Coast.* **2006**, *29*, 985–996. [CrossRef]

remote sensing

MDPI

Article

Enhanced Oceanic Environmental Responses and Feedbacks to Super Typhoon Nida (2009) during the Sudden-Turning Stage

Jiagen Li [1], Yuanjian Yang [2,3], Guihua Wang [4], Hao Cheng [1] and Liang Sun [1,*]

1 School of Earth and Space Sciences, University of Science and Technology of China, Hefei 230000, China; ljg123@mail.ustc.edu.cn (J.L.); ch0127@mail.ustc.edu.cn (H.C.)
2 School of Atmospheric Physics, Nanjing University of Information Science and Technology, Nanjing 210000, China; yyj1985@nuist.edu.cn
3 Guangdong Key Laboratory of Ocean Remote Sensing, South China Sea Institute of Oceanology, Chinese Academy of Sciences, Guangzhou 510000, China
4 Department of Atmospheric and Oceanic Sciences, Fudan University, Shanghai 200000, China; wanggh@fudan.edu.cn
* Correspondence: sunl@ustc.edu.cn; Tel.: +86-0551-6360-6723

Abstract: The ocean surface and subsurface biophysical responses and their feedbacks to super typhoon Nida were comprehensively investigated based on a substantial dataset of multiple-satellite observations, Argo profiles, and reanalysis data. Nida experienced two Category 5 stages: a rapid intensification stage that was fast moving along a straight-line track, and a rapid weakening stage that was slowly moving along a sharp-left sudden-turning track. During the straight-line stage, Nida caused an average sea surface temperature (SST) cooling of 1.44 °C and a chlorophyll-*a* (chl-*a*) concentration increase of 0.03 mg m^{-3}. During the sudden-turning stage, *cyclonic* sudden-turning induced a strong cold cyclonic eddy (SSHA < −60 cm) by strong upwelling, which caused the maximum SST cooling of 6.68 °C and a long-lasting chl-*a* bloom of 0.6 mg m^{-3} on the left-hand side of the track, resulting in substantial impacts on the ocean ecological environment. Furthermore, the enhanced ocean cold wake and the longer air–sea interaction in turn decreased the average inner-core SST of 4 °C and the corresponding enthalpy flux of 780 W m^{-2}, which induced a notable negative feedback to the typhoon intensity by weakening it from Category 5 to Category 2. The left bias response and notable negative feedback are special due to sharp-left sudden-turning of typhoon. Comparing with the previously found slow translation speed (~4 m s^{-1}) of significant ocean response, the negative feedback requires even more restriction of translation speed (<2 m s^{-1}) and sharp sudden-turning could effectively relax restrictions by making equivalent translation speed lower and air-sea interaction time longer. Our findings point out that there are some unique features in ocean–typhoon interactions under sudden-turning and/or lingering tracks comparing with ordinary tracks.

Keywords: super typhoon Nida; air–sea interaction; oceanic environmental responses; left bias; negative feedbacks; cold wake; chlorophyll-*a* bloom

1. Introduction

Tropical cyclones are generated mainly near the tropics, known as typhoons in the northwestern Pacific (NWP) and hurricanes in the Atlantic, which can cause notable dynamical responses of the upper ocean [1]. The most significant one might be the cold wake often seen along the typhoon track [1] due to ocean surface enthalpy fluxes, vertical mixing, and upwelling [2–4]. Within the cold wake, there may also be marine phytoplankton blooms [5,6].

Since the cold wake caused by a typhoon can cool the sea surface temperature (SST) [7–9], it may have a negative feedback effect on the passing typhoon [10]. However, there is a long-term debate as to whether such a cold wake can immediately induce a significant negative feedback effect on the typhoon intensity itself [11]. For example, Chang and Anthes [12] concluded that the cold wake does not significantly affect the intensity of the typhoon itself under a typical moving speed, but Sutyrin and Khain [13] suggested that the cold wake could induce a significant negative feedback effect on the typhoon intensity.

In general, the ocean's negative feedback to typhoon intensity first requires that the typhoon can cause significant SST cooling. This is because a decrease in SST, especially an inner-core SST change, has a substantial impact on the observed typhoon intensity [14]. Emanuel [15] suggested that when the inner-core SST drops by 2.5 °C, the storm could terminate, and even a drop of 0.5 °C can have a substantial impact on the storm intensity. Second, negative feedback of the ocean to typhoon intensity also requires the condition that the typhoon has a long residence time over the SST cooling area. As there is a lag time in the response of a typhoon to the SST [12,16], it needs sufficient time to be affected by the cold wake. Therefore, the speed of the typhoon [17] and the forcing time [18] become important factors alongside other related factors including the SST, ocean mixed-layer depth (MLD), ocean stratification, typhoon size, and latitude [19].

Although these theoretical studies are instructive for understanding ocean–typhoon negative feedback, they are largely based on certain idealized conditions, such as a homogeneous spatial distribution of the background environment or the existence of idealized mesoscale eddies [20,21] and a straight-line typhoon track [12,13]. In the actual process of typhoon–ocean interaction, the ocean environment, typhoon track, and typhoon intensity are changeable. In particular, some typhoons turn sharply during their passage, which introduces new factors of change into the study of typhoon–ocean interaction. This type of typhoon–ocean interaction process caused by typhoon motion (nonuniform linear motion) is of great significance towards fully understanding typhoon–ocean interactions.

The NWP has the highest concentration of tropical cyclones in the world. According to the Joint Typhoon Warning Center (JTWC), super typhoon is defined as the typhoon with maximum sustained wind (MSW) exceeding 130 kts. Typhoon Nida was the twenty-sixth tropical storm and fourth super typhoon of the 2009 NWP typhoon season. Figure 1 shows the lifespan of typhoon Nida based on the best-track data from the JTWC. Nida generated on 21 November 2009 and achieved typhoon intensity with the MSW of 65 kts on 24 November. Thereafter, Nida kept fast moving northwest along a straight-line track with the moving speed of 6 m s^{-1}, and rapidly intensified to 155 kts on 25 November. It is surprised that Nida intensified by 90 kts within 24 h during the straight-line (SL) stage, which was three times greater than the definition of rapid intensification [22]. If Nida continued to move and develop in accordance with this trend, it was likely to hit China or Japan in the next few days and bring heavy disasters. However, Nida suddenly turned left on 28 November and weakened rapidly from 150 kts to 90 kts during the sudden-turning (ST) stage from 28 November to 30 November. After that, Nida weakened to a tropical storm on 2 December and disappeared on 3 December, and thus the populated areas on Northwest Pacific coast avoided a typhoon in the end.

In this paper, we focused on the interaction between ocean and Nida during the ST stage by comparing the ocean responses and feedback during the SL Stage. The scientific breakthroughs and new discoveries in this paper mainly include following three points. Firstly, the typical ocean SST response is right bias due to clockwise wind stress, as firstly pointed out by Price [1]. Some oceanic environments (e.g., previously existed cold eddy in left side) may shift this right bias (e.g., Zheng et al., [23]). This study showed a left bias response purely due to typhoon left-side sharp turning. To our best knowledge, this new phenomenon is has not been reported and explained before. Secondly, when a typhoon turns sharply, the equivalent moving speed of the typhoon will be slower, and the typhoon's forcing time on the ocean will be longer, resulting in more intense marine biological and physical responses. Thirdly and the most importantly, there was a long-term controversial theoretical problem "whether typhoon inducing cold wake can immediately induce a significant negative feedback effect on the typhoon intensity itself".

Comparing with the previously found slow translation speed (~4 m s^{-1}) of significant ocean response, we showed with observation that such negative feedback could only work under very restricted conditions with typical translation speed <2 m s^{-1} and we pointed out that sharp sudden-turning could effectively relax restrictions by making equivalent translation speed lower and air–sea interaction time longer. Comparing with speed, both the intensity and size play relatively minor roles. Our new findings provide positive evidence that enhanced oceanic environmental responses and feedbacks can occur under sudden-turning tracks, furtherly, we can improve our ability to forecast the variation of marine environment and rapidly change of typhoon intensity through typhoon track.

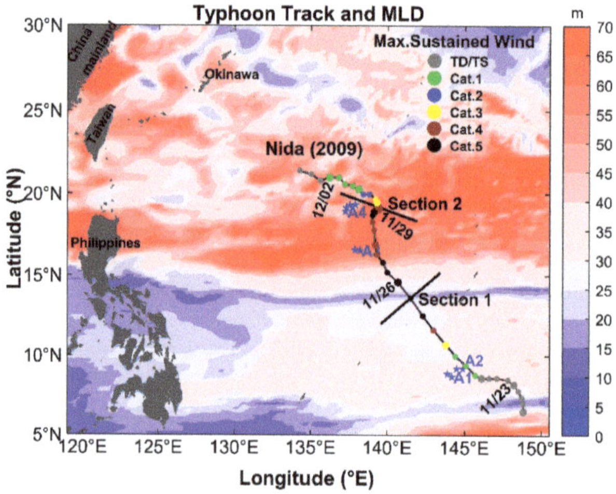

Figure 1. Best track of typhoon Nida (2009) from the JTWC. The blue stars denote the positions and trajectory of Argo floats (A1 to A4 are floats 5902121, 5901213, 5900976, and 2900401). The track line color indicates the MSW speed of the typhoon. The background pattern is the MLD field on 22 November before the passage of typhoon. The two black sections indicate the location of the strongest intensity (section 1) and maximum turning (section 2).

2. Materials and Methods

2.1. Typhoon Track Data

Each best-track file from JTWC contains the tropical cyclone center locations, the central pressure and the maximum sustained wind (MSW) speeds, the wind radius, and the radii of the specified winds (17, 25, 33, or 51 m s^{-1}) for four quadrants, at 6 h intervals. In this study, we defined the turning angle of a typhoon as the azimuth change of the path in the 12 h before and 12 h after a certain point (turn left, positive; turn right, negative).

Besides, we counted the translation speed and turning angle of typhoons between 1981 and 2018 based on JTWC dataset in order to count the proportion of turning typhoons and prove the importance of our research in the discussion.

2.2. Data

2.2.1. Sea Surface Temperature

We used daily SST products that were integrated with microwave and infrared optimally interpolated (MW_IR_OI) data from Remote Sensing Systems (RSS) (www.remss.com, (accessed on 1 April 2021)). The MW_IR_OI data products used included the Tropical Rainfall Measuring Mission (TRMM) Microwave Imager, the Advanced Microwave Scanning Radiometer, the WindSat Polarimetric Radiometer, and the Global Precipitation Measurement (GPM) microwave imager SST data. Also included were the SST data detected in the infrared band by the Moderate-Resolution Imaging Spectroradiometer onboard the Terra and Aqua satellites of the Earth Observation Satellite series and the Visible/Infrared Imaging Radiometer Suite onboard the Suomi National Polar-Orbiting Partnership. These instruments are suitable for analyzing the response of the upper ocean to the passage of typhoons because microwave imagers and radiometers are able to penetrate clouds and the infrared channel of radiometers onboard polar-orbiting satellites has a high spatial resolution [24]. In this study, we used daily data with a spatial resolution of 9×9 km.

2.2.2. Sea Surface Salinity

The sea surface salinity (SSS) data were from the three-dimensional global ocean reanalysis product produced by the Copernicus Marine and Environmental Monitoring Service Global Monitoring and Forecasting Centre (http://marine.copernicus.eu/services-portfolio/access-to-products/, (accessed on 1 April 2021)). The products assessed in this study are referenced as GLOBAL-REANALYSIS-PHY-001-031 for Global Reanalysis Multi-Model Ensemble Product. Global ocean reanalyses are homogeneous three-dimensional gridded descriptions of the physical state of the ocean spanning several decades, produced with a numerical ocean model constrained with data assimilation of satellite and in situ observations. The SSS data used in this study were the daily data with a spatial resolution of $0.25^\circ \times 0.25^\circ$.

2.2.3. Sea Surface Height Anomalies and Surface Geostrophic Velocity

The sea surface height anomaly (SSHA) and sea surface geostrophic velocity data, calculated based on geostrophic relationships, were satellite altimeter data from the Ssalto/Duacs multisensor gridded delay-time altimetry product provided by Archiving, Validation and Interpretation of Satellite Oceanographic Data (AVISO) (https://www.aviso.altimetry.fr/en/data/products/, (accessed on 1 April 2021)) and distributed by the Copernicus Marine and Environmental Monitoring Service. This product is processed from all satellite altimeter missions (Jason-3, Sentinel-3A, HY-2A, Saral/AltiKa, Cryosat-2, Jason-2, Jason-1, Topex/Poseidon, Environmental Satellite, Geosat Follow-On, European Remote Sensing satellites). The spatial resolution of the SSHA and geostrophic data is $0.25^\circ \times 0.25^\circ$ and the temporal resolution is 1 d.

2.2.4. Wind Vectors

The wind vector datasets were the wind at 10 m above the sea surface from the Cross-Calibrated Multi-Platform gridded surface vector winds provided by RSS. Cross-Calibrated Multi-Platform combines Version-7 RSS radiometer wind speeds, QuikSCAT and ASCAT scatterometer wind vectors, moored buoy wind data, and ERA-Interim model wind fields, and uses a variational analysis method to combine satellite observations and in situ wind measurements into gap-free wind fields, to produce four maps daily of 0.25° gridded vector winds. Therefore, the temporal resolution of the wind field data used is 6 h and the spatial resolution is $0.25^\circ \times 0.25^\circ$.

2.2.5. Precipitation

The GPM mission is an international network of satellites based on the TRMM. The GPM carries the first satellite-borne Ku/Ka-band Dual-frequency Precipitation Radar and a multichannel GPM Microwave Imager to improve the ability to measure light rain and provide global observations of rain and snow. This study used the final product from the Integrated Multi-satellite Retrievals for GPM, which combines data from all passive-microwave instruments in the GPM constellation and provides full coverage for latitudes 60°S–60°N. The National Aeronautics and Space Administration's Precipitation Processing System publicly provided this product in March 2014 (https://pmm.nasa.gov/data-access/downloads/gpm, (accessed on 1 April 2021)). The precipitation data used were daily data with a spatial resolution of $0.1^\circ \times 0.1^\circ$.

2.2.6. Chlorophyll-a

The ocean color merged datasets were obtained from GlobColour (http://hermes.acri.fr/index.php?class=archive, (accessed on 1 April 2021)). The GlobColour dataset provides daily, weekly, and monthly chlorophyll concentrations with a spatial resolution of 4, 25, and 100 km from 1997 to the present, and the sensors include single-sensor and merged products from SeaWiFS, MERIS, MODIS, VIIRS NPP, OLCI-A, VIIRS JPSS-1, and OLCI-B. Merged data products are coalesced from multiple mission observations into a single data product with better spatial and temporal coverage than the individual missions. In this study, we used daily and weekly ocean color data with a spatial resolution of 25×25 km. In addition, chl-a datasets obtained from the three-dimensional global ocean forecast product produced by the Copernican Marine Environmental Monitoring Service Center (http://marine.copernicus.eu/services-portfolio/access-to-products/, (accessed on 1 April 2021)) were used to show the sea surface two-dimensional fields of zooplankton. The two-dimensional fields of zooplankton data used were weekly data with a spatial resolution of $0.25° \times 0.25°$.

2.2.7. Argo Profiles

We extracted the Argo float profiles from the real-time quality-controlled Argo database of the China Argo Real-time Data Center (http://www.argo.org.cn, (accessed on 1 April 2021)). Argo is an array of 3887 active profiling floats, distributed almost uniformly across the global ocean, which provide temperature and salinity profiles from the surface to 2000 m depth. The array has approximately one float for every 3° of latitude and longitude and provides accurate and comprehensive measurements of changes to ocean temperature and salinity, enabling systematic assessments of the physical state of the upper ocean. The Argo data are collected and made freely available by the International Argo Project and the national programs that contribute to it. In this study, four Argo floats were used (shown in Figure 1): float A1, 5901213 (on 21 November), float A2, 5902121 (on 1 December), float A3, 5900976 (on 22 November and 10 December), and float A4, 2900401 (on 23 November and 28 November).

2.2.8. Mixed-Layer Depth

The mixed-layer depth (MLD) dataset were from the three-dimensional global ocean reanalysis product produced by the Copernicus Marine and Environmental Monitoring Service Global Monitoring and Forecasting Centre (http://marine.copernicus.eu/services-portfolio/access-to-products/, (accessed on 1 April 2021)). The products assessed in this study are referenced as GLOBAL-REANALYSIS-PHY-001-031 for Global Reanalysis Multi-Model Ensemble Product. Global ocean reanalyses are homogeneous three-dimensional gridded descriptions of the physical state of the ocean spanning several decades, produced with a numerical ocean model constrained with data assimilation of satellite and in situ observations. The MLD data used in this study were the daily data with a spatial resolution of $0.25° \times 0.25°$.

2.3. *Methods*

2.3.1. Ekman Pumping

The Ekman pumping velocity (EPV, unit: m s^{-1}) caused by Ekman pumping can be calculated as follows [1]:

$$\mathrm{EPV} = \nabla \times (\tau/\rho f), \tag{1}$$

where ∇ is a gradient operator (total differential in all directions of space), called the Hamilton operator, f is the Coriolis parameter, ρ is the density of seawater (unit: kg m^{-3}), and τ is the wind stress vector (unit: N),

$$\tau = \rho_a C_D |\mathbf{U}_{10}| \mathbf{U}_{10}, \tag{2}$$

$$C_D = \begin{cases} (4 - 0.6|\mathbf{U}_{10}|) \times 10^{-3} & \text{for } |\mathbf{U}_{10}| < 5\ m/s; \\ (0.737 + 0.0525|\mathbf{U}_{10}|) \times 10^{-3} & \text{for } 5m/s \le |\mathbf{U}_{10}| < 25\ m/s \\ 2.05 \times 10^{-3} & \text{for } |\mathbf{U}_{10}| \ge 25\ m/s \end{cases} \quad (3)$$

where ρ_a is the air density (unit: kg m^{-3}), C_D is the drag coefficient [25,26], and $\mathbf{U}_{10}$ is the 10 m wind vector (unit: m s^{-1}).

2.3.2. Enthalpy Flux and Energy Budget

The sensible heat flux Q_S (unit: W m^{-2}) and latent heat flux Q_L (unit: W m^{-2}) were calculated based on the bulk aerodynamic formula under the condition of tropical cyclone–ocean coupling [25,27] as follows:

$$Q_S = C_S V(T_s - T_a)\rho_a C_{pa}, \quad (4)$$

$$Q_L = C_L V(q_s - q_a)\rho_a L_{va}, \quad (5)$$

where C_L and C_S are the latent and sensible heat exchange coefficients, respectively. V is the wind speed derived from the modified Rankine vortex with n defined as in Mei et al. [28] as follows:

$$\begin{cases} V(r) = V_{max}\left(\frac{r}{R_{max}}\right) & \text{for } r < R_{max} \\ V(r) = V_{max}(r/R_{max})^{-n} & \text{for } r \ge R_{max} \end{cases}, \quad (6)$$

where n = 0.6. We used the model wind to calculate the heat flux because it is challenging for scatterometers to measure high and extreme winds [29], and the CCMP wind underestimated the wind speed by as much as 50% [30]. Ts and Ta are the during-typhoon SST and near-surface air temperature (unit: °C), and qs and qa are the surface and air specific humidity (unit: %), respectively. Ta and qa are from the National Centers for Environmental Prediction (NCEP) reanalysis dataset (http://apdrc.soest.hawaii.edu/erddap/search/index.html?page=1&itemsPerPage=1000&searchFor=NCEP+-ofes+-godas, (accessed on 1 April 2021)). C_{pa} is the specific heat of air at constant pressure, L_{va} is the latent heat of vaporization, ρ_a is the air density (unit: kg m^{-3}), and q_s is the saturated specific humidity (unit: %). The frictional heat D (unit: W m^{-2}) generated by the friction between the typhoon and ocean is calculated as follows:

$$D = \rho_a C_D V^3 \quad (7)$$

A typhoon is regarded as a Carnot heat engine, and the heat conversion

$$\text{efficiency} = (\text{SST} - \text{CTT})/\text{SST} \quad (8)$$

where CTT is the cloud top temperature (unit: °C). The ocean energy input into the typhoon W (unit: W m^{-2}) is:

$$W = (Q_S + Q_L + D) \times \text{efficiency} \quad (9)$$

The net energy input of the ocean to typhoon N (unit: W m^{-2}) is:

$$N = W - D \quad (10)$$

In accordance with previous studies [10,27], we did not consider the radiation term when calculating the change in typhoon energy, although the radiation term is an important variable in the energy budget.

2.3.3. Method to Evaluate Ocean Responses

The grid-based maximum response (GMR) method proposed by Li et al. [31] is a new method to accurately evaluate the SST response to a typhoon by using the asynchronous

SST fields. In the first step, the initial forcing time (Ti) and the terminal forcing time (Tt) of typhoon are recorded, and the SST response period is defined as the time from the start time (Ti − 2) to the end time (Tt + 5) in each grid. In the second step, the average SST of two days (Ti − 2) and one day (Ti − 1) before the initial forcing time are selected as the background SST, and the minimum SST between the initial forcing time (Ti) and the terminal time (Tt + 5) is selected as the responding SST. The SST cooling in each grid is evaluated by the difference between the responding SST and background SST. In the last step, the SST cooling response in each grid is displayed in the map as GMR SST cooling results. Schematic illustration of the GMR method can be found in Figure 2 in Li et al. [30]. Using daily data from satellite remote sensing, this method can accurately calculate the maximum amplitude of SST cooling and the location of SST cooling centers.

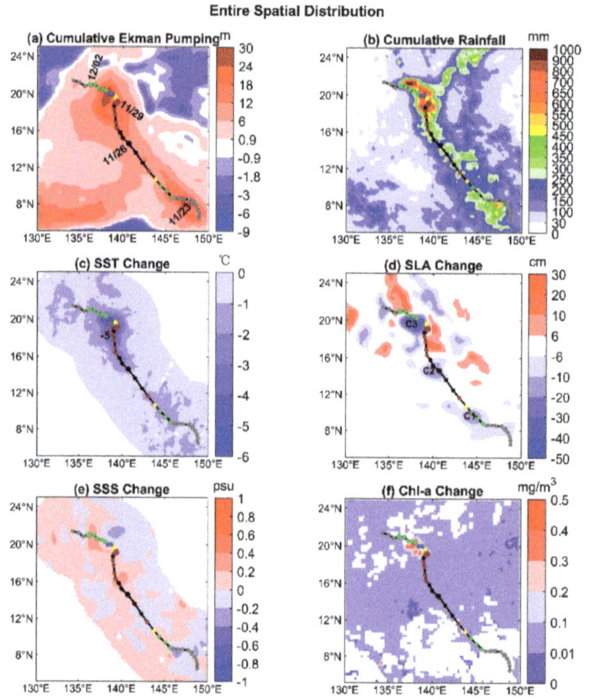

Figure 2. The entire spatial distribution of (**a**) cumulative Ekman pumping, (**b**) cumulative rainfall, (**c**) SST change (the red solid line shows the 5 °C contour), (**d**) SSHA change, (**e**) SSS change, and (**f**) chl-a change during the lifespans of Nida from 22 November 2009 to 3 December 2009. Ekman pumping and rainfall are the cumulative value during typhoon lifespans. Other graphs (such as SST change) are the maximum ocean responses caused by typhoon in its lifespans calculated by GMR method proposed by Li et al. [31].

2.3.4. Large-Scale Eddy Transport

Eddies were detected with the SSHA data [32]. Specifically, if there was a series of closed contours in the SSHA field, when the value of the outermost closed contour was greater than 6 cm, the region was defined as a warm eddy; when the outermost closed contour was less than −6 cm, the region was defined as a cold eddy (CE) [33,34].

A CE in NPWO would affect large-scale transport in several ways [33]. Previous studies stressed the negative transport by CE that is well known when CE dissipated in ocean interior [35]. However, if CE propagates westward to west boundary, it may have a positive transport for NPWO. This new mechanism was discovered recently [36].

In this study, we only stressed the transport by positive feedback mechanism, since the west propagating CE was very close to west boundary other than ocean interior. Zhang et al. [36] evaluated the eddy effect on Kuroshio transport based on the well-known turbulent Sverdrup balance as follows:

$$S_v = \frac{1}{\beta}\int |\mathbf{u_w}|\ \Delta Q dy\,, \tag{11}$$

where β is the beta effect, $\mathbf{u_w}$ denotes the eddies' westward propagation, and ΔQ denotes the potential vorticity anomaly. The formula is general for middle latitude ocean dynamics. Besides, the Typhoon Nida passing region is just within where Zhang et al. [36] studied (the Northwest Pacific and the Kuroshio area). According to the eddy structure model proposed by Wang et al., [37], a surface eddy consists of three parts and potential vorticity anomaly ΔQ is:

$$\Delta Q = \left(\frac{f+\zeta}{H1+h1+h2} - \frac{f}{H1}\right) \times (H1+h1+h2), \tag{12}$$

where h1 is the upper surface, h2 is the lower surface, H1 is the eddy body of height, ζ is the relative vorticity.

3. Results

3.1. Oceanic Responses in the Entire Area

Super typhoon Nida induced strong winds and severe rainfall around the track during its lifespan. Figure 2a shows the cumulative Ekman pumping caused by Nida during its passage. During the sudden turning dates (23 November and 30 November), Nida induced strong accumulation (>24 m) of Ekman pumping, which was in contrast to a very weak wind field and local Ekman pumping (<6 m) before Nida entered. During the passage of typhoon Nida, strong winds prevailed, with the upwelling velocity ranging from 0.5 to 4×10^{-4} m s^{-1} (Figure S1). Figure 2b shows the cumulative rainfall during the passage of the typhoon. The rainfall was mainly concentrated in the ST stage of the typhoon, during which the accumulated rainfall exceeded 1000 mm, and the rainfall induced by the typhoon was mainly located on the right-hand side of the typhoon track (for the daily evolution, see Figure S2).

Figure 2c shows the SST response field calculated using the GMR method. Before Nida entered, the SST in the study area was dominated by warm temperatures higher than 28 °C (Figure S3 in SM). In the SL stage, the cold wake was obviously lagging behind the typhoon center and the SST cooling showed an obvious right-hand bias. In contrast, the maximum SST cooling overlapped the typhoon center and was located on the left-hand side of the track in the ST stage. From Figure 2c, the SST cooling mainly occurred in the ST stage of the typhoon and almost all of the strong cooling area greater than 5 °C was located on the left-hand side of the typhoon track (with the maximum SST decrease of 6.68 °C about 80 km to the left of the typhoon track). This strong "left-bias" SST cooling induced by Nida is unusual, since most previous studies show a "right-bias" of the SST cooling owing to better wind-current resonance [1,38] and sub-mesoscale recirculation cells [39] on the right-hand side in the Northern Hemisphere. Therefore, the strong "left-bias" SST cooling induced by Nida should be paid particular attention and stressed in detail in Section 3.2.

The SSHA response to Nida calculated using the GMR method is shown in Figure 2d. The results suggest three main centers along the typhoon track where the SSHAs decreased, marked as C1, C2, and C3, respectively. C3 in the ST stage was strongest and the maximum SSHA reached −60 cm. This was because of the strong upwelling in the ST stage. In contrast, C1 and C2 in the SL stage were relatively weak. The evolution of the SSHA fields (Figure S4) show that the SSHAs in the study area were mostly higher than

-20 cm before the arrival of Nida, and then gradually decreased during the typhoon's passage.

Figure 2e shows the distribution of the SSS response field calculated using the GMR method. The results show that the area where the SSS increased by 0.6 psu corresponded to the strong Ekman pumping area, and that the area where the SSS decreased by 0.6 psu corresponded to the heavy rainfall area (for the daily evolution, see Figure S5). In summary, the change in SSS was mainly determined by a mechanism of competition between the asymmetric typhoon precipitation and asymmetric Ekman pumping.

Figure 2f shows the chl-*a* difference before and after Nida, indicating that the chl-*a* concentration increased by 0.5 mg m^{-3}, which was about five times that before the passage of Nida, and the chl-*a* bloom lasted for weeks after Nida passed. Compared with the ST stage, the chl-*a* concentration in the SL stage barely changed. Furthermore, the region with the highest chl-*a* increase (Figure S6c,g) corresponded to the region with the strongest SST cooling (Figure S3i,j) and SSHA decrease (Figure S4i,j).

Recent case studies have found that typhoons can modulate the large-scale flow, e.g., the Kuroshio Current, via cold eddy intensification [35,36,40]. In this study, we estimated the potential mass transport induced by the strong cold eddy caused by typhoon Nida in the sudden-turning stage. The height of the cold eddy body H1 was about 300 m from Argo measurements; the upper surface h1 was about -0.3 m from sea surface height anomaly field; and the lower surface h2 was about 200 times that of the eddy amplitude h1 according to Wang et al., [37]. These are typical values for the Northwest Pacific, which were obtained as a composite structure from Argo and other observations [41]. The westward speed of the eddy was about 0.27 m s^{-1}. Based on Equations (11) and (12), the strong cold eddy induced by Nida contributed about 0.25 Sv of mass transport to the large-scale mass transport (i.e., Kuroshio), which is about 0.6 times the total mass transport (0.4 Sv) in the South China Sea [35] and 0.4–0.6% of the total contribution to the mass transport (~40–60 Sv) by the Kuroshio. The temporal evolutions of the mean SSHA of the eddy and the transport induced by it are shown in Figure S7. Although the potential Kuroshio Current strengthening by of a single typhoon and eddy is not significant, there are 20–30 typhoons per year in this region, the integration of all typhoons and eddies may play a nonnegligible role for the Kuroshio Current (Zhang et al., [36] did this with composite analysis of typhoons between 1993 and 2014). The story of the relationship between typhoon and Kuroshio is a part of climate response of typhoon-ocean interaction, which is not very clear yet, so we simply add this part as an example or comparison for further studies.

3.2. Comparison of Oceanic Responses in Two Category 5 Stages

To summarize the regularity of the typhoon–ocean interaction under different typhoon motions, we quantitatively compared the ocean responses in the two Category 5 typhoon stages in detail. Two representative sections were selected for the two stages, as marked in Figure 1.

3.2.1. Direct Response during the Typhoon's Passage

Figure 3 shows the temporal evolution of the cross-track SSW, Ekman pumping, rainfall, SST, SSHAs, and SSS at section 1 during 23 November to 1 December 2009. The SSW field and Ekman pumping intensity at section 1 strengthened on 25 November and then weakened on 26 November. Meanwhile, the maximum SSW (14.2 m s^{-1}) and EPV (1.03×10^{-4} m s^{-1}) on the right-hand side were obviously stronger than on the left-hand side of the typhoon track in the SL stage. Figure 3c shows that the evolution of the rainfall corresponded to Ekman pumping in the time series, but the rainfall appeared with a relatively leftward bias. The SST began to decrease on 24 November and appeared with a rightward bias. Figure 3e shows that the maximum SST cooling center (1 °C) caused by Nida was located at about 100 km on the right-hand side of the typhoon path. Figure 3f,g show that the SSHA decreased by about 3 cm from 24 November to 1 December. The

SSS at section 1 changed slightly within the range from −0.02 psu to 0.02 psu (Figure 3h,i). In summary, at section 1, although the Ekman pumping velocity and rainfall were something notable due to the strongest intensity of 155 kts, the upper ocean responses (SST, SSHA, and SSS) were quite weak owing to the straight-line track and fast speed.

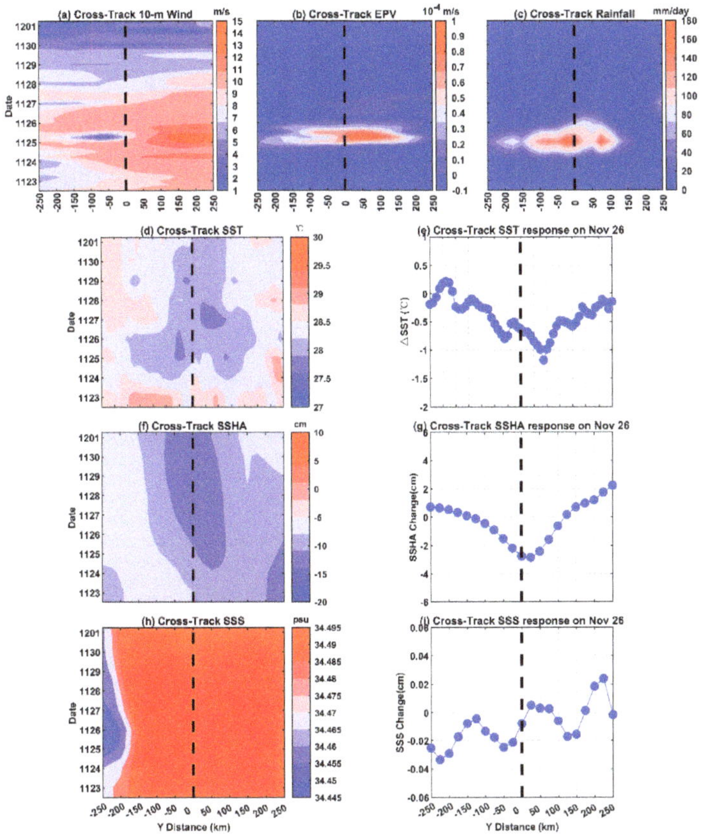

Figure 3. Temporal evolution of cross-track (**a**) SSW (shaded, m s^{-1}), (**b**) cumulative Ekman pumping (shaded, m), and (**c**) cumulative rainfall (shaded, mm) at location 1 during the passage of typhoon Nida (23 November–1 December 2009). (**d**,**e**) Temporal evolution of cross-track SST (shaded, °C) and the cross-track SST response on 26 November. (**f**,**g**) Temporal evolution of cross-track SSHA (shaded, cm) and the cross-track SSHA response on 26 November. (**h**,**i**) Temporal evolution of cross-track SSS (shaded, psu) and the cross-track SSS response on 26 November.

Figure 4 shows the temporal evolution of the cross-track SSW, Ekman pumping, rainfall, SST, SSHAs, and SSS at section 2 during 26 November to 4 December 2009. The SSW and Ekman pumping strengthened from 27 November to 30 November. It was exactly opposite to section 1, in that the maximum SSW (23.2 m s^{-1}) and EPV (3.61 × 10^{-4} m s^{-1}) on the left-hand side were obviously stronger than on the right-hand side SSW (15.9 m s^{-1}) and EPV (2.07 × 10^{-4} m s^{-1}) at section 2 (Figure S8). Figure 4c shows that the typhoon-induced rainfall was concentrated from 27 to 30 November and the strongest rainfall exceeded 320 mm d^{-1}. The cross-track SST at section 2 began to decrease on 27 November and appeared with an obvious leftward bias (Figure 4d). The maximum SST cooling center (>6 °C) was located 80 km to the left of the typhoon track, and was relatively larger than the SST cooling center (5.42 °C) on the right-hand side

(Figure 4e). Figure 4f,g show that the SSHAs decreased from 26 November to 4 December and the maximum SSHA decrease exceeded 30 cm on the left-hand side of track.

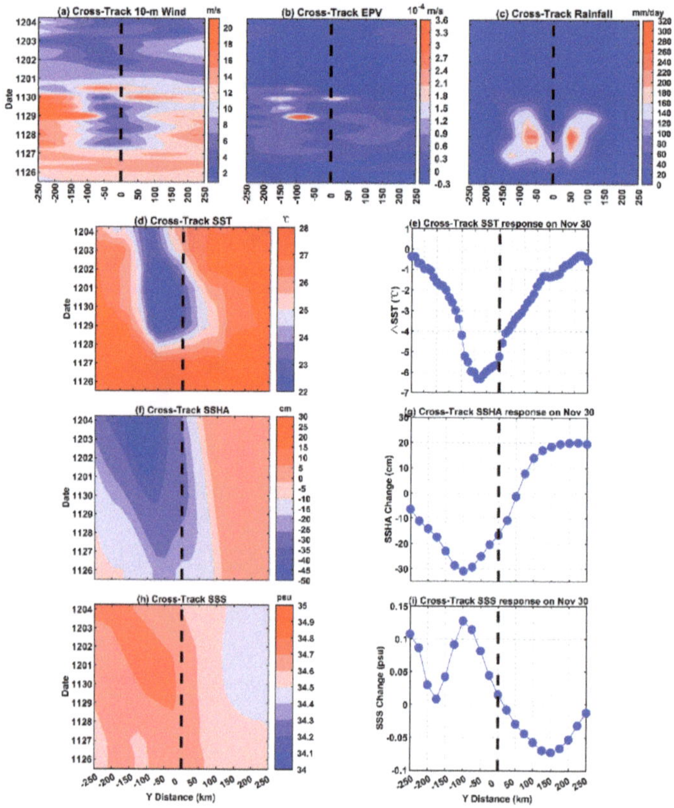

Figure 4. Temporal evolution of cross-track (**a**) SSW (shaded, m s^{-1}), (**b**) cumulative Ekman pumping (shaded, m), and (**c**) cumulative rainfall (shaded, mm) at location 2 during the passage of typhoon Nida (30 November–4 December 2009). (**d**,**e**) Temporal evolution of cross-track SST (shaded, °C) and the cross-track SST response on 30 November. (**f**,**g**) Temporal evolution of cross-track SSHAs (shaded, cm) and the cross-track SSHA response on 30 November. (**h**,**i**) Temporal evolution of cross-track SSS (shaded, psu) and the cross-track SSS response on 30 November.

Therefore, the left-bias SST cooling phenomenon during the ST stage could have resulted for the following reasons. First, the asymmetrical typhoon wind field was stronger on the left-hand side (Figure 4a,b) and the *cyclonic* left-turning made EPV stronger than on the right-hand side. Secondly, according to previous research [18], the forcing time of typhoon Nida was longer on the left-hand side due to the sudden left-turning of the track, both resulting in stronger cumulative Ekman pumping, as shown in Figure 1. The stronger cumulative Ekman Pumping consequently generated a cyclonic cold eddy, which can significantly enhance the SST cooling induced by the typhoon [3,42]. In summary, both the typhoon wind and the ocean condition were favorable to the left-bias SST cooling (Figure S8).

The SSS increased on the left-hand side of the track but decreased on the right-hand side, with the SSS response ranging from −0.05 to 0.05 psu at section 2 (Figure 4h). The asymmetric SSS response was influenced by asymmetric rainfall (Figure 4c). It can be clearly seen that the cross-track evolutions of SST (Figure 4d), SSHAs (Figure 4f) and SSS (Figure 4h) were highly consistent in the time series, due to the typhoon-induced Ekman pumping (Figure 4b). In summary, at section 2, both the ocean responses and rainfall were notable because of the long-term forcing of the strong EPV by cyclonic sudden-turning of this strong typhoon.

3.2.2. Long-Term Memory Response after the Typhoon's Passage

In addition to the direct response of the upper ocean, we studied the long-term impact of typhoon Nida on the marine environment by comparing the marine environment two weeks after the typhoon's passage. In the SL stage, the long-memory ocean responses one week and two weeks after Nida passed showed that the SSHAs, SST, and chl-a distribution indicated that the ocean responses recovered rapidly within a week (Figure 5).

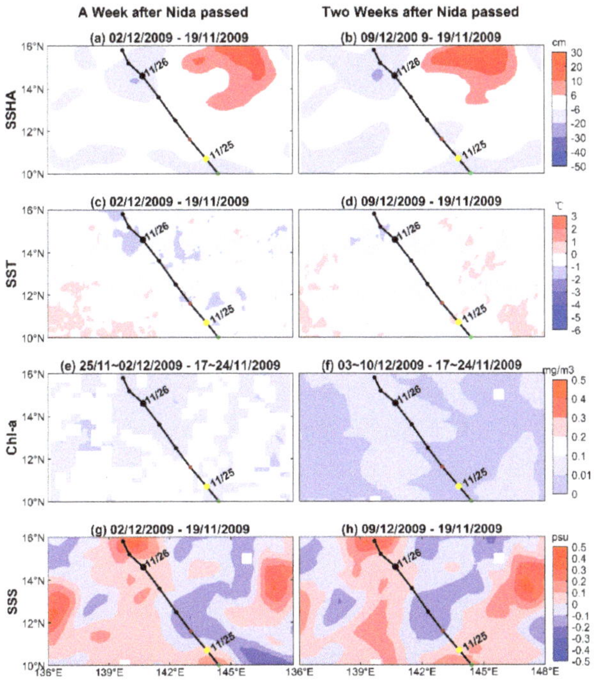

Figure 5. The straight-line stage. (**a**,**b**) The SSHA fields (shaded, cm) on 9 and 15 December 2009, indicating one week and two weeks after the passage of Nida, respectively. (**c**,**d**) The SST responses (shaded, cm) on 9 and 15 December 2009. (**e**,**f**) The chl-a responses (shaded, cm) several weeks after the passage of typhoon Nida. (**g**,**h**) The SSS responses (shaded, cm) on 9 and 15 December 2009.

In contrast, Figure 6 shows the long-memory ocean response one week and two weeks after Nida passed in the ST stage. The SSHA response (Figure 6a,b) indicates that the strong cold eddy with an SSHA decrease of 60 cm generated by Nida in the ST stage existed and strengthened for several weeks. Such an SSHA response delay after the passage of typhoon has also been noted in previous studies [43,44], and confirmed with Argo and SST data due to sustained upwelling [21,40]. However, the exact delay date might be uncertain [44], as the temporal resolution of the grid can be higher than two weeks [45,46]. Figure 6c,d show that Nida induced a maximum SST cooling exceeding 6 °C, which lasted more than two weeks (Figure 6a,b). Meanwhile, for several weeks after the passage of Nida, the cold eddy caused a long-lasting chl-a bloom exceeding 0.6 mg m^{-3} and an SSS increase of 0.6 psu, as shown in Figure 6e,f and Figure 6g,h, respectively. This played an important role in the upper marine ecological environment and marine transport of salinity, nutrients, and chlorophyll. As is well known, in the tropical oceans, limited nutrient concentrations in the upper ocean are not conducive to enhancing primary productivity [47–49]. Typhoon Nida induced the transport of richer nutrients from the subsurface to the surface layer, and together with the "presumed" sufficient sunlight and photosynthesis after the passing of the typhoon, there was a marked increase in the surface chl-a concentration, especially on the left-hand side of the typhoon track (five times that before the typhoon's passage). Lin and Oey [50] also showed that the typhoon's left-side rainfall can induce left-side blooming, but in the case of Nida the left-side blooming was dominated by the strong upwelling within the strong cold eddy. Therefore, typhoon Nida not only directly affected the upper-ocean environment during its passage, but also induced long-term non-negligible ocean responses by forcing the formation of the cold eddy.

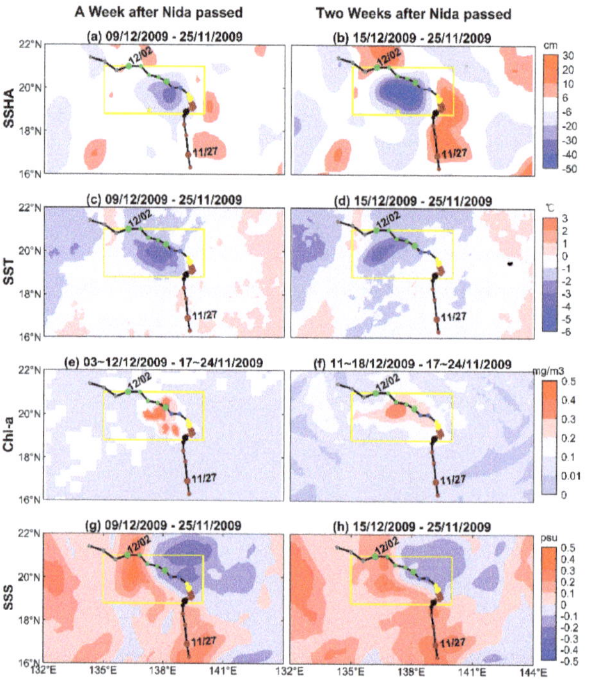

Figure 6. The sudden-turning stage. (**a**,**b**) The SSHA fields (shaded, cm) on 9 and 15 December 2009, indicating one week and two weeks after the passage of Nida, respectively. (**c**,**d**) The SST responses (shaded, cm) on 9 and 15 December 2009. (**e**,**f**) The chl-*a* and zooplankton responses (shaded, cm) several weeks after the passage of typhoon Nida. (**g**,**h**) The SSS responses (shaded, cm) on 9 and 15 December 2009. The yellow rectangle and yellow star indicate the cold eddy area and the location of Argo float 2900401, respectively.

Furthermore, we quantified the average ocean responses along the typhoon track in the two Category 5 stages (Table 1). The high-latitude MLD in the ST stage was slightly deeper than the low-latitude MLD in the SL stage. The average intensity of Nida for both stages was about 120 kts. However, the turning angle and speed of Nida were quite different in the two stages. In the ST stage, the turning angle and speed were 62.48° and 0.94 m s^{-1}, respectively, which were four times and one-quarter of the turning angle of 15.68° and speed of 4.39 m s^{-1} in the SL stage. Under the conditions of a similar ocean environment and typhoon intensity, the SST response, SSHA response and chl-*a* response in the ST stage were −4.15 °C, −10.29 cm, and 0.12 mg m^{-3}, which were three times, two times, and four times those in the SL stage, respectively. In summary, owing to the difference in turning angle and speed, the upper ocean responses during the ST stage were about three times more than those in the SL stage.

Table 1. Quantitative impact factors and biophysical responses in the two Category 5 stages.

	SL Stage	ST Stage
MLD (m)	33.41	62.08 (↑)
MSW speed (kts)	120.77	120.00 (=)
SSW speed (m s^{-1})	10.21	13.94 (↑)
Turning angle (°)	15.68	62.48 (↑)
Translation speed (m s^{-1})	4.39	0.94 (↓)
Rainfall (mm d^{-1})	264.58	601.02 (↑)
SST response (°C)	−1.44	−4.15 (↑)
SSHA response (cm)	−5.00	−10.29 (↑)
SSS response (psu)	0.06	0.00 (↓)
Chl-a response (mg m^{-3})	0.03	0.12 (↑)

Note. The MLD, SSW, rainfall, SST, SSHA, SSS and chl-*a* were averaged along the typhoon in the two stages: SL stage was from 24 November 06:00 to 27 November 12:00; ST stage was from 27 November 18:00 to 30 November 12:00. The signals up, down or equal mean the variables in the SL stage higher than, lower than or equal to the ST stage.

3.2.3. Subsurface Response

To investigate the subsurface response, we analyzed vertical profiles of temperature and salinity measured by Argo floats before and after typhoon Nida (Figure 7). Typhoon Nida affected floats 5901213 and 5902121 on 24 November. At this time, Nida moved fast and its intensity was weak, with an MSW speed of 65 kts. As shown in Figure 7a, the thickness of the mixed layer showed no obvious change after Nida passed. Nida affected float 5900976 from 26 to 27 November after it reached the maximum intensity, increasing the MLD by 35 m (Figure 7c). Float 2900401 was located at the sudden-turning track and was affected by Nida from 28 November to 1 December. During this stage, the MLD increased by 50 m (Figure 7e). At all three locations, the subsurface temperature responses showed a three-layer (decreasing–increasing–decreasing) vertical structure (Figure 7b,d,f). In addition, Figure 7g,h show the temporal evolution of temperature and salinity profiles measured by Argo float 2900401. First, Nida induced strong vertical mixing that not only cooled the surface but also heated the subsurface, which is the so-called 'heat pump' effect of the typhoon [10,51]. In the following, the cyclonic stress of the typhoon caused a strong uplift of the ocean thermocline, which acted as a 'cold suction' to the upper ocean [52,53]. Therefore, the upper-ocean response to typhoon Nida consisted of two consequential processes: strong vertical mixing and long-lasting upwelling. Consistent with the research on 14 typhoons presented by Hsu and Ho [54] through Glider data with higher temporal resolution, the two main effects occurring in the temperature profile of the typhoon response are cold suction due to upwelling and heat pumping owing to vertical mixing. The subsurface warming induced by Nida were showed not only at Category 5 (Figure 7d,f: floats 5900976 and 2900401), but also Category 1 (Figure 7b: floats 5901213 and 5902121), which is consistent with the observation in Hsu and Ho (2019): "significant subsurface warming for typhoons belonging to all categories". In addition, according the types of heat pump in Hsu and Ho [54], the subsurface temperature responses induced by Nida belonged to the second type, which was characterized by surface cooling and subsurface warming with additional layers of cooling and warming in the thermocline. The shift of the T, S profiles measured by the Argo floats may also have played a minor role, since only a minor part shifted within the eddy and the major part propagated with the eddy.

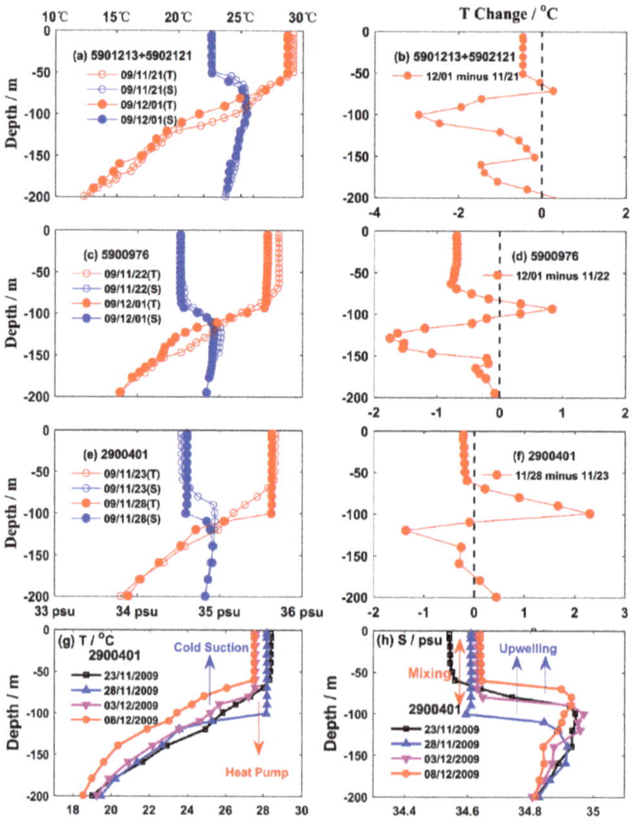

Figure 7. Temperature and salinity profiles, and temperature differences between the post-storm and pre-storm, measured by Argo floats 5901213 and 5902121 (**a**,**b**), 5900976 (**c**,**d**), and 2900401 (**e**,**f**). (**g**,**h**) Temporal evolution of temperature and salinity profiles measured by Argo float 2900401 before and after Nida.

3.3. Upper-Ocean Feedback to Typhoon Nida

To investigate the negative feedback of the cold wake on the typhoon intensity, Figure 8 shows the evolution of Nida's inner-core cold wake and inner-core enthalpy flux (sensible heat flux and latent heat flux) in the two Category 5 stages. In accordance with previous studies [14,55], we defined the typhoon inner core as the distance within a radius of 100 km from the typhoon center, indicated by the black solid circle in Figure 8. When Nida moved in a straight line in the SL stage, the cold wake induced by Nida lagged far behind the inner-core area because of Nida's fast speed (Figure 8a,b). Figure 8e,f indicate that the air–sea enthalpy flux supplied from the ocean to Nida was considerable, especially on 26 November, with the maximum enthalpy flux exceeding 1000 W m^{-2}. Therefore, Nida intensified rapidly in the SL stage. However, in the ST stage, the inner-core SST cooling was notable because the local cold wake no longer lagged behind the typhoon center, but completely overlapped the typhoon's inner core due to the sudden-turning track of the typhoon (Figure 8c,d). The strong inner-core SST cooling resulted in a sharp decrease of the enthalpy fluxes at the sea–air interface and the inner-core enthalpy flux decreased sharply from 681.9 W m^{-2} on 26 November to −101.2 W m^{-2} on 30 November, especially on the left-hand side of the track (Figure 8g,h). Furthermore, the sharp decrease of the inner-core enthalpy fluxes caused the energy supply from the ocean to the typhoon to cease, rapidly weakening the intensity of typhoon Nida from Category 5 to Category 2.

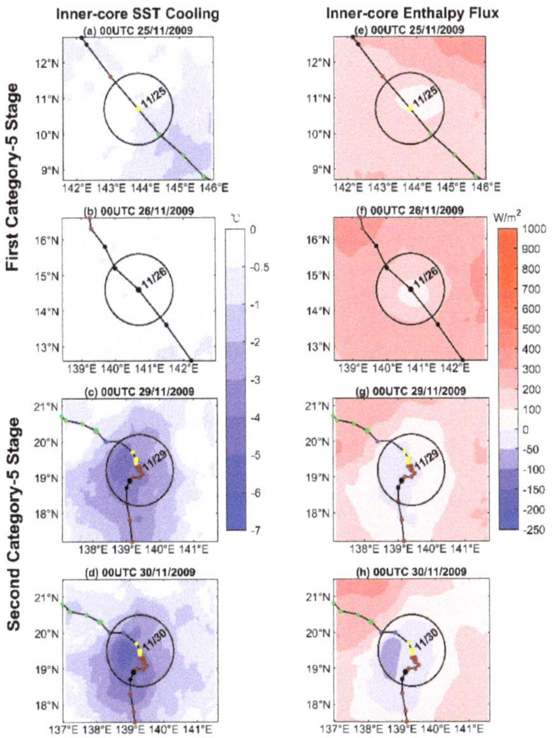

Figure 8. Evolution of the inner-core SST cooling (shaded, °C) and inner-core enthalpy flux (shaded, W m^{-2}) in the SL stage and ST stage. The black solid circle represents the radius of 100 km from the typhoon, indicating its inner core.

Figure 9 shows the change in the typhoon turning angle (a), translation speed (b), inner-core SST cooling and local ΔSST wake (c), enthalpy flux and frictional flux (d), input power and net power (e) during the lifecycle of typhoon Nida. Meanwhile, the feedback factor F_{sst} and the center pressure depression (ΔPc) of Nida [19] were calculated and the results are presented in Figure 9f,g. F_{sst} means the negative feedback level of cold wake to typhoon, the closer F_{sst} is to -1, the greater the negative feedback of cold wake is. During 22–24 November, Nida turned left (Figure 9a) and moved at a speed of about 2 m s^{-1} (Figure 9b). The local cold wake, inner-core SST cooling caused by Nida (Figure 9c), and the energy input from the ocean into the typhoon (Figure 9d,e) were very weak due to the weak SSW. The feedback factor F_{sst} was relatively strong because Nida moved slowly along the turning track (Figure 9f). Therefore, due to weak energy from the ocean and strong F_{sst}, Nida's intensity strengthened slowly (Figure 9g) during this time.

On 24 November, Nida strengthened to typhoon intensity and evolved into the SL stage. In the SL stage, the typhoon track of Nida was close to a straight line (Figure 9a) and the maximum translation speed reached 6.4 m s^{-1} (Figure 9b), as shown by the blue line segment in Figure 9. Because Nida moved along a straight-line track and the speed was fast, the local cold wake caused by the typhoon was very weak (Figure 9c). As the speed gradually slowed to 4 m s^{-1}, the typhoon-induced local cold wake gradually strengthened to 3 °C. Cione and Uhlhorn [14] showed that it took 23 h for the background water temperature to change into a cold wake. Assuming that the radius of the cold wake was 100 km, if the inner-core area was to feel the cold wake, the speed of the typhoon should be slower than 2 m s^{-1}. Therefore, when the speed of the typhoon slowed to 4 m s^{-1}, although the typhoon-induced local cold wake gradually strengthened, the inner-core SST cooling was still weak. The environment in the inner-core area still provided favorable conditions for the development of the typhoon. In this stage, the enthalpy flux supplied by the ocean to the typhoon increased to 1000 W m^{-2} (Figure 9d), and the corresponding input power (and net power) from the tropical ocean to Nida increased to 500 (200) W m^{-2} (Figure 9e). The feedback factor Fsst of Nida in this stage was very weak, with a value above −0.4 (Figure 9f). Therefore, because of the strong energy from the ocean and weak Fsst, Nida intensified rapidly to a Category 5 typhoon with a ΔPc of 99 hPa and MSW speed of 155 kts, and the intensification rate was about three times the definition of rapid intensification in previous studies (Figure 9g).

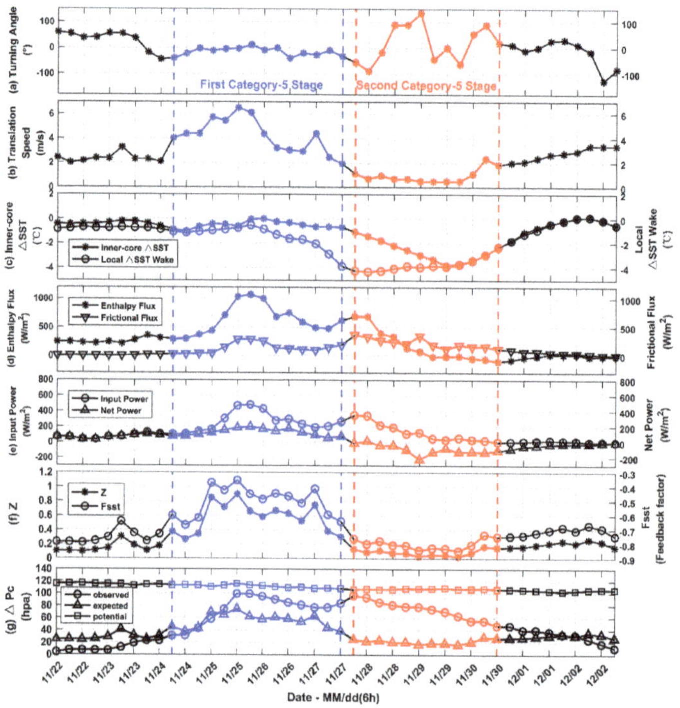

Figure 9. The (**a**) turning angle (°), (**b**) translation speed (m s^{-1}), (**c**) inner-core SST cooling and local ΔSST wake (°C), (**d**) enthalpy flux and frictional flux (W m^{-2}), (**e**) input power and net power (W m^{-2}), (**f**) feedback factor Fsst and reference value Z in Schade and Emanuel (1999), and (**g**) observed, expected and potential pressure depression in the eye of typhoon Nida during its passage. The blue and red segments indicate the straight-line stage and the sudden-turning stage of typhoon Nida, respectively.

On 27 November, Nida suddenly turned left and evolved into the ST stage. In the ST stage, the typhoon moved very slowly (about 1 m s^{-1}), especially during sudden-turning (Figure 9a,b), as shown by the red line segment in Figure 9. Owing to the cyclonic sharp turning and extremely slow speed, the typhoon caused a strong cold wake of 4 °C (Figure 9c). At the same time, the sharp turning and slow speed resulted in the typhoon center revisiting and staying over the cold wake, and the inner-core SST cooling began to drop rapidly to 4 °C. The decrease in the inner-core SST resulted in a rapid decrease in the enthalpy flux supplied by the ocean to Nida (Figure 9d), and the corresponding input power (and net power) decreased to 50 (−200) W m^{-2} (Figure 9e). The feedback factor Fsst strengthened rapidly to −0.85 because of Nida's extremely slow translation speed along the sudden-turning track (Figure 9f). As shown in Figure 9g, Nida weakened from

Category 5 with a ΔPc of 96 hPa, to Category 2 with a ΔPc of 55 hPa before and after the ST stage. In summary, owing to the different motions of typhoon Nida in the different stages, the ocean's responses and feedbacks to the typhoon were quite different in the different stages.

4. Discussion

This study used multiple satellite observations to study typhoon–ocean interactions: left bias ocean responses purely due to sharp-left sudden-turning of typhoon; resulting in more in-tense marine biological and physical responses and negative feedback; such negative feedback could only work under very restricted conditions with typical translation speed < 2 m s^{-1}. However, the coarseness of the spatial and temporal resolutions of these data might have limited our findings. Given the nature of the strong atmospheric forcing associated with the sharp gradient from a super typhoon, high temporal and spatial resolution analyses and comparisons are certainly preferable in order to investigate the oceanic responses and feedbacks in sufficient detail. This requires the development of new techniques to obtain higher spatial and temporal resolution data. NOAA already has a successful program of hurricane observation and in situ data collection from airplanes inside hurricanes that provide high-resolution data, so the new techniques have already been developed. The problem is the cost of it.

Although there were negative feedbacks to typhoon Nida's intensity, the statistical regression of feedback F_{sst} [19] was several times larger than the measured intensity change, as shown in Figure 9. The overestimation of the feedback F_{sst} might have resulted in overestimation of the negative feedback time. The feedback F_{sst} was obtained from statistical regression of the steady-state intensity of the hurricane in model simulations. According to numerical simulations [16,19], it takes a storm more than two to three days to settle into a statistically steady state in coupled models. This time in general is much longer than the interaction time between a typhoon and ocean cold wake. Thus, the negative feedbacks were too weak to be noted in observations.

Although the negative feedbacks were generally too weak to kill the typhoon, they may have suppressed the typhoon's intensity, as in Figure 9g. Typhoon Nida is not an isolated case. For typhoon Hagibis in 2007 [18], and typhoon Prapiroon in 2012 [56], their curved typhoon tracks caused them to undergo a similar process: generation and revisiting of a cold eddy, and the negative feedback process suppressed their intensity to below Category 1. The differences of the three typhoons are: 1. Hagibis and Prapiroon only reached the Category 1 typhoon intensity, while Nida reached the Category 5 typhoon intensity, which is more destructive; 2. The negative feedback effect of cold wake on Hagibis and Prapiroon is to suppress the development of typhoon, while the cold wake resulted in the rapid weakening of typhoon Nida, from Category 5 typhoon to Category 2 typhoon. The research of super typhoon Nida is more important for predicting the rapid weakening of super typhoons.

Besides, we considered which parameter might be dominant, although the negative feedback factor z includes many parameters, such as SST, ocean MLD, ocean stratification, typhoon translation speed, typhoon size, latitude, and so on [19]. This can be seen from Figure 9b,f, in which both curves have similar variation and their correlation is 0.85. This implies that the dominant parameter of negative feedback was translation speed in the case of typhoon Nida. This is consistent with the results of past studies [17], which indicate that mostly slow-moving TCs experience negative feedback that can be more or less pronounced depending on the ocean stratification (e.g., the depth of the 26 °C isotherm).

The sharp sudden-turning track is an effective way of slowing the translation speed. When the typhoon moves along a straight-line track, it needs to move at a very slow speed (<2 m s^{-1}) for a significant negative feedback, which is restricted by the radius of the typhoon core and cold wake. According to statistical analysis, in the North Atlantic, for about 15% of the duration of a hurricane, the speed is below 2 m s^{-1}, and only 3%

of the time is below 1 m s^{-1} [57]; in the Pacific Ocean, the proportion of time below 2 m s^{-1} may be as low as 10% [27]. According to our statistical analysis, only 6% of typhoons in the NWP moved slower than 2 m s^{-1} in the past 40 years from 1981 to 2018, while the proportion below 1 m s^{-1} was as low as 1% (Figure 10a). Therefore, for a typhoon that moves in a straight line at a typical speed (4–6 m s^{-1}), even if there is a decrease in SST, negative feedback barely exists [12]. However, when a typhoon moves along a sudden-turning track, the equivalent speed of the typhoon is greatly reduced, which is equivalent to the typhoon staying in a specific area for a longer time, increasing the probability of negative feedback. Therefore, sharp sudden-turning could effectively relax restrictions by making equivalent translation speed lower and air–sea interaction time longer. According to the angle statistics (Figure 10b), more than 60% of typhoons are close to straight (amplitude of turning angle <20°), and due to the influence of the NWP subtropical high pressure and large-scale circulation, more than 60% of typhoons turn right. Note in particular, the proportion of typhoons with a turning angle greater than 50° is about 10%, which can increase the probability that a typhoon will revisit the cold wake, thus enhancing the negative feedback effect that is difficult to trigger.

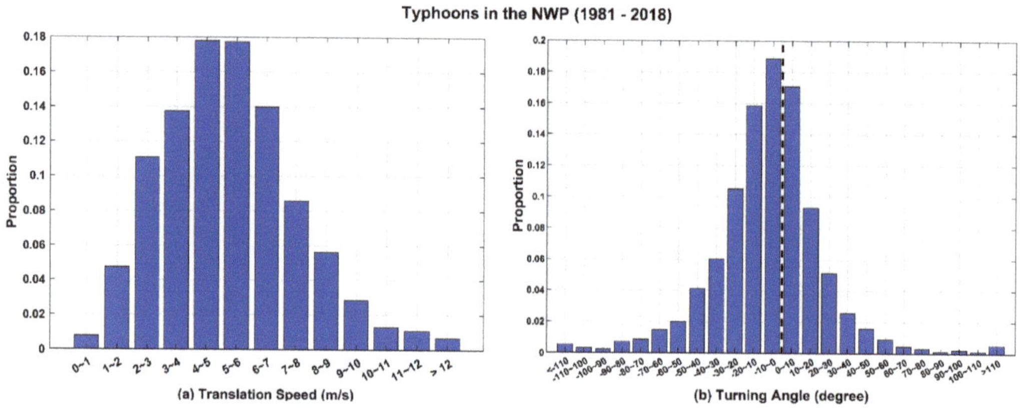

Figure 10. Distribution of typhoon (**a**) translation speed and (**b**) turning angle in the NWP during the 40 years from 1981 to 2018.

5. Conclusions

The ocean surface and subsurface biophysical responses and their feedbacks to super typhoon Nida were comprehensively investigated based on a substantial dataset of multiple-satellite observations, Argo profiles, and reanalysis data. Super typhoon Nida generated in the NWP in 2009 and consisted of two Category 5 stages. In the straight-line stage, Nida moved quickly along a straight-line track and rapidly intensified from Category 1 to Category 5. In the sudden-turning stage, Nida moved slowly along a sharp-left sudden-turning track and weakened from Category 5 to Category 2. Summarizing the regularity of typhoon–ocean interaction under different typhoon motions can provide a comprehensive and in-depth understanding of the typhoon–ocean interaction process. Based on this study, a schematic illustration of the interaction between the typhoon and the ocean, including the SL stage and the ST stage, is shown in Figure 11.

In the SL stage (24–27 November), Nida moved fast and along a straight northwest line, and the intensity strengthened rapidly from 65 to 155 kts. The strong winds induced strong vertical mixing and Ekman pumping. However, Nida caused only weak SST cooling because of the straight-line track and fast speed. As it slowed below 4 m s^{-1}, the ocean response gradually strengthened. During the whole stage, Nida induced an average SST cooling of 1.44 °C, an SSHA response of −5.00 cm, and a chl-a response of 0.03 mg m^{-3}. Meanwhile, Nida's inner-core SST cooling was weaker than 0.5 °C due to the fast speed along a straight-line track. Consequently, the enthalpy flux supplied from the ocean to the typhoon remained at a high level, with a maximum flux exceeding 1000 W m^{-2}, which intensified Nida from Category 1 to Category 5.

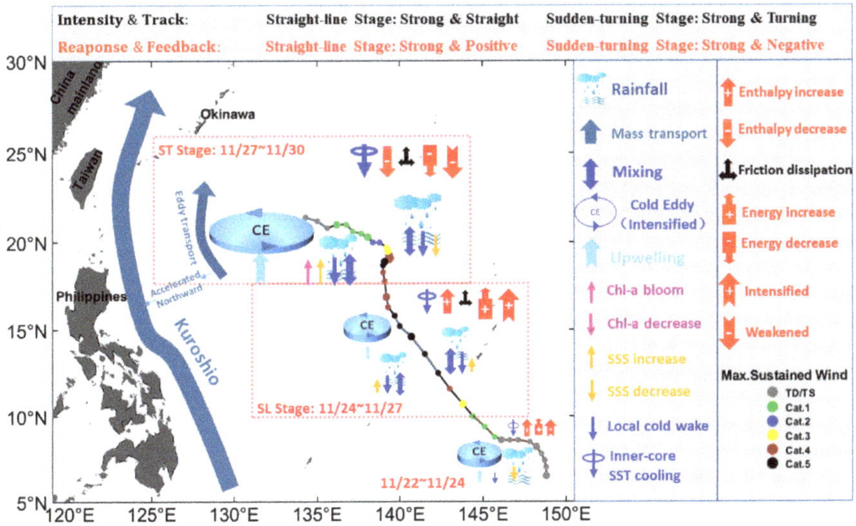

Figure 11. Schematic illustration of the interaction between super typhoon Nida and the ocean environment.

In the ST stage (27–30 November), Nida suddenly turned left, with a maximum left sudden-turning angle of 135°, and its translation speed was slower than 1 m s^{-1}. Owing to the *cyclonic* sudden-turning and longer air–sea interaction time on the left-hand side of the track, Nida induced strong left bias ocean responses, which was a new phenomenon not reported and explained before. The average SST response, SSHA response and chl-a response was −4.15 °C, −10.29 cm, and 0.12 mg m^{-3}, which was three times, two times, and four times more than those in the SL stage, respectively. Nida also induced a strong cold eddy (SSHA < −60 cm, maximum SST cooling > 6.68 °C), which caused a chl-a bloom exceeding 0.6 mg m^{-3} (more than five times) that lasted for weeks, which had significant impacts on the upper-ocean ecological environment. This strong cold eddy also potentially enhanced the Kuroshio Current as it approached the current, accelerating it by about 0.25 Sv. The strong oceanic responses in turn induced a negative feedback effect on Nida's intensity. The average inner-core SST cooling of Nida reached 4 °C, which resulted in a dramatic decrease in the sensible and latent heat fluxes at the sea–air interface from 681.9 to −101.2 W m^{-2}. Then, the sharp decrease in the enthalpy flux terminated the net energy supply from the ocean to the typhoon and further continually weakened the intensity of Nida from 150 (Category 5) to 90 kts (Category 2) within three days.

It is concluded that the ocean response is weak when the typhoon intensity is weak or the typhoon speed is fast (>6 m s^{-1}) along a straight-line track and the ocean response is much stronger when the typhoon intensity is stronger and the speed is slower (~4 m s^{-1}), which aligns with previous studies [2,17,58,59]. However, the ocean negative feedback to the typhoon intensity needs stricter conditions. According to the present case, the ocean negative feedback requires the typhoon to move at a slow speed (<2 m s^{-1}); otherwise, the inner core of the typhoon cannot feel the local SST cooling, and the negative feedback might be negligible. Comparing with the previously found slow translation speed (~4 m s^{-1}) of significant ocean response, the negative feedback requires even more restriction of translation speed (<2 m s^{-1}) and sharp sudden-turning could effectively relax restrictions by making equivalent translation speed lower and air-sea interaction time longer. In addition, the negative feedback of the ocean may weaken and/or suppress the typhoon intensity. However, such negative feedback may not be as strong as that suggested in previous studies. This might be due to the fact that SST is not the only factor affecting typhoon intensity. Other factors (including, for example, atmospheric conditions, the interaction between the typhoon and land) may also have roles [19].

Supplementary Materials: The following are available online at https://www.mdpi.com/2072-4292/13//2648/s1, Figure S1: The wind fields at 10 m (arrows) and the Ekman pumping velocity (shaded, positive value is upwelling) during the passage of Typhoon Nida (22 November–1 December 2009). The track line color indicated the max-sustained wind. Figure S2: Evolution of daily precipitation during the passage of typhoon Nida (22 November–1 December 2009). The track line color indicated the max-sustained wind. Figure S3: Evolution of daily SST (shaded, °C) during the passage of typhoon Nida (22 November–3 December 2009). The track line color indicated the max-sustained wind. Figure S4: Evolution of the altimeter-derived SSHA (shaded, cm) and surface geostrophic velocity (arrows) during the passage of typhoon Nida (22 November–3 December 2009). The track line color indicated the max-sustained wind of typhoon. Figure S5: Evolution of daily SSS (shaded, cm) during the passage of typhoon Nida. The track line color indicated the max-sustained wind of typhoon (22 November–3 December 2009). Figure S6: (a–d) Evolution of weekly Chlorophyll concentration (shaded, mg/m^3), (e–h) Evolution of weekly 2D fields of zooplankton (shaded, g/m^2) during the passage of typhoon Nida (17 November–18 December 2009). The track line color indicated the max-sustained wind of typhoon. Figure S7: Evolutions of mean SSHA within the cold eddy and the mass transport induced by the cold eddy on the left side of the track in the second Category 5 stage. Figure S8: Comparison of typhoon wind field and ocean condition on the left-hand and right-hand side if typhoon track in SL stage and ST stage respectively, extracted from Figures 3 and 4.

Author Contributions: Conceptualization, L.S.; data curation, J.L.; formal analysis, J.L. and L.S.; funding acquisition, L.S.; investigation, J.L.; methodology, J.L. and L.S.; writing—original draft, J.L.; writing—review and editing, J.L., Y.Y., G.W., H.C. and L.S. All authors have read and agreed to the published version of the manuscript.

Funding: This research was supported by the National Foundation of Natural Science of China (No. 41876013), the National Program on Global Change and Air-Sea Interaction (GASI-IPOVAI-04), the National Basic Research Program of China (2013CB430301 and 2013CB430303).

Institutional Review Board Statement: Not applicable.

Informed Consent Statement: Not applicable.

Data Availability Statement: Publicly available datasets were analyzed in this study. The daily SST data can be found here: [http://data.remss.com/] (accessed on 1 April 2021). The daily SSS data can be found here: [http://marine.copernicus.eu/services-portfolio/access-to-products/] (accessed on 1 April 2021). The daily SSHA and sea surface geostrophic velocity data can be found here: [https://www.aviso.altimetry.fr/en/data/products/] (accessed on 1 April 2021). The wind vector datasets can be found here: [http://data.remss.com/] (accessed on 1 April 2021). The daily precipitation product can be found here: [https://pmm.nasa.gov/data-access/downloads/gpm/] (accessed on 1 April 2021). Argo float profiles data can be found here: [http://www.argo.org.cn/] (accessed on 1 April 2021). Mixed-layer climatology and database can be found here: [http://mixedlayer.ucsd.edu/] (accessed on 1 April 2021). The daily and weekly ocean color datasets from GlobColour can be found here: [http://hermes.acri.fr/index.php?class=archive] (accessed on 1 April 2021).

Acknowledgments: We thank Remote Sensing Systems for the SST and sea surface wind data (http://www.remss.com/, (accessed on 1 April 2021)), GlobColour for chl-a product (http://hermes.acri.fr/index.php?class=archive, (accessed on 1 April 2021)), National Aeronautics and Space Administration's Precipitation Processing System for the Integrated Multi-satellite Retrievals for GPM final precipitation product (https://pmm.nasa.gov/data-access/downloads/gpm, (accessed on 1 April 2021)), Copernicus Marine and Environmental Monitoring Service Center for the SSS and zooplankton data (http://marine.copernicus.eu/services-portfolio/access-to-products/, (accessed on 1 April 2021)), Archiving, Validation and Interpretation of Satellite Oceanographic for the SSHA and Geostrophic Velocity data (https://www.aviso.altimetry.fr/en/data/products/, (accessed on 1 April 2021)), China Argo Real-time Data Center for Argo measurements (http://www.argo.org.cn, (accessed on 1 April 2021)), and Scripps Institution of Oceanography, UC San Diego for MLD data (http://mixedlayer.ucsd.edu/, (accessed on 1 April 2021)).

Conflicts of Interest: The authors declare that they have no known competing financial interest or personal relationships that could have appeared to influence the work reported in this paper.

References

1. Price, J.F. Upper ocean response to a hurricane. *J. Phys. Oceanogr.* **1981**, *11*, 153–175. [CrossRef]
2. Vincent, E.M.; Lengaigne, M.; Madec, G.; Vialard, J.; Samson, G.; Jourdain, N.; Menkes, C.; Jullien, S. Processes setting the characteristics of sea surface cooling induced by tropical cyclones. *J. Geophys. Res. Space Phys.* **2012**, *117*, 117. [CrossRef]
3. Walker, N.D.; Leben, R.; Balasubramanian, S. Hurricane-forced upwelling and chlorophyllaenhancement within cold-core cyclones in the Gulf of Mexico. *Geophys. Res. Lett.* **2005**, *32*. [CrossRef]
4. Wang, G.; Wu, L.; Johnson, N.C.; Ling, Z. Observed three-dimensional structure of ocean cooling induced by Pacific tropical cyclones. *Geophys. Res. Lett.* **2016**, *43*, 7632–7638. [CrossRef]
5. Lin, I.-I.; Liu, W.T.; Wu, C.-C.; Wong, G.T.F.; Hu, C.; Chen, Z.; Liang, W.-D.; Yang, Y.; Liu, K.-K. New evidence for enhanced ocean primary production triggered by tropical cyclone. *Geophys. Res. Lett.* **2003**, *30*. [CrossRef]
6. Liu, Y.; Tang, D.; Evgeny, M. Chlorophyll Concentration Response to the Typhoon Wind-Pump Induced Upper Ocean Processes Considering Air–Sea Heat Exchange. *Remote Sens.* **2019**, *11*, 1825. [CrossRef]
7. Fisher, E.L. Hurricanes and the Sea-Surface Temperature Field. *J. Meteorol.* **1958**, *15*, 328–333. [CrossRef]
8. Leipper, D.F. Observed Ocean Conditions and Hurricane Hilda, 1964. *J. Atmos. Sci.* **1967**, *24*, 182–186. [CrossRef]
9. Stramma, L.; Cornillon, P.; Price, J.F. Satellite observations of sea surface cooling by hurricanes. *J. Geophys. Res. Space Phys.* **1986**, *91*, 5031–5035. [CrossRef]
10. Emanuel, K.A. An Air-Sea Interaction Theory for Tropical Cyclones. Part I—Steady-State Maintenance. *J. Atmos. Sci.* **1986**, *43*, 585–604. [CrossRef]
11. Emanuel, K. 100 Years of Progress in Tropical Cyclone Research. *Meteorol. Monogr.* **2018**, *59*, 15.1–15.68. [CrossRef]
12. Change, S.W.; Anthes, R.A. The mutual response of the tropical cyclone and the ocean. *J. Phys. Oceanogr.* **1979**, *9*, 128–135. [CrossRef]
13. Sutyrin, G.; Khain, A. Effect of the Ocean-Atmosphere Interaction on the Intensity of a Moving Tropical Cyclone. *Atmos. Ocean. Sci. Lett.* **1984**, *20*, 697–702.
14. Cione, J.J.; Uhlhorn, E.W. Sea Surface Temperature Variability in Hurricanes: Implications with Respect to Intensity Change. *Mon. Weather. Rev.* **2003**, *131*, 1783–1796. [CrossRef]
15. Emanuel, K.A. Thermodynamic control of hurricane intensity. *Nat. Cell Biol.* **1999**, *401*, 665–669. [CrossRef]
16. Ma, Z. A Study of the Interaction between Typhoon Francisco (2013) and a Cold-Core Eddy. Part I: Rapid Weakening. *J. Atmos. Sci.* **2020**, *77*, 355–377. [CrossRef]
17. Lin, I.-I.; Pun, I.-F.; Wu, C.-C. Upper-Ocean Thermal Structure and the Western North Pacific Category 5 Typhoons. Part II: Dependence on Translation Speed. *Mon. Weather. Rev.* **2009**, *137*, 3744–3757. [CrossRef]
18. Sun, L.; Yang, Y.-J.; Xian, T.; Lu, Z.-M.; Fu, Y.-F. Strong enhancement of chlorophyll a concentration by a weak typhoon. *Mar. Ecol. Prog. Ser.* **2010**, *404*, 39–50. [CrossRef]
19. Schade, L.R.; Emanuel, K.A. The Ocean's Effect on the Intensity of Tropical Cyclones: Results from a Simple Coupled Atmosphere–Ocean Model. *J. Atmos. Sci.* **1999**, *56*, 642–651. [CrossRef]
20. Bruneau, N.; Wang, S.; Toumi, R. Long Memory Impact of Ocean Mesoscale Temperature Anomalies on Tropical Cyclone Size. *Geophys. Res. Lett.* **2020**, *47*, 47. [CrossRef]
21. Lu, Z.; Wang, G.; Shang, X. Strength and Spatial Structure of the Perturbation Induced by a Tropical Cyclone to the Underlying Eddies. *J. Geophys. Res. Oceans* **2020**, *125*, 125. [CrossRef]
22. Wang, X.; Wang, C.; Zhang, L.; Wang, X. Multidecadal Variability of Tropical Cyclone Rapid Intensification in the Western North Pacific. *J. Clim.* **2015**, *28*, 3806–3820. [CrossRef]
23. Zheng, Z.-W.; Ho, C.-R.; Kuo, N.-J. Importance of pre-existing oceanic conditions to upper ocean response induced by Super Typhoon Hai-Tang. *Geophys. Res. Lett.* **2008**, *35*. [CrossRef]
24. Wentz, F.J. Satellite Measurements of Sea Surface Temperature through Clouds. *Science* **2000**, *288*, 847–850. [CrossRef]
25. Jaimes, B.; Shay, L.K.; Uhlhorn, E.W. Enthalpy and Momentum Fluxes during Hurricane Earl Relative to Underlying Ocean Features. *Mon. Weather Rev.* **2015**, *143*, 111–131. [CrossRef]
26. Powell, M.D.; Vickery, P.J.; Reinhold, T.A. Reduced drag coefficient for high wind speeds in tropical cyclones. *Nat. Cell Biol.* **2003**, *422*, 279–283. [CrossRef]
27. Lin, I.; Pun, I.; Lien, C.-C. "Category-6" supertyphoon Haiyan in global warming hiatus: Contribution from subsurface ocean warming. *Geophys. Res. Lett.* **2014**, *41*, 8547–8553. [CrossRef]
28. Mei, W.; Lien, C.-C.; Lin, I.-I.; Xie, S.-P. Tropical Cyclone–Induced Ocean Response: A Comparative Study of the South China Sea and Tropical Northwest Pacific. *J. Clim.* **2015**, *28*, 5952–5968. [CrossRef]
29. Brennan, M.J.; Hennon, C.C.; Knabb, R.D. The Operational Use of QuikSCAT Ocean Surface Vector Winds at the National Hurricane Center. *Weather Forecast.* **2009**, *24*, 621–645. [CrossRef]

30. Sun, J.; Oey, L.-Y. The Influence of the Ocean on Typhoon Nuri (2008). *Mon. Weather. Rev.* **2015**, *143*, 4493–4513. [CrossRef]
31. Li, J.; Sun, L.; Yang, Y.; Cheng, H. Accurate Evaluation of Sea Surface Temperature Cooling Induced by Typhoons Based on Satellite Remote Sensing Observations. *Water* **2020**, *12*, 1413. [CrossRef]
32. Li, Q.-Y.; Sun, L.; Lin, S.-F. GEM: A dynamic tracking model for mesoscale eddies in the ocean. *Ocean Sci.* **2016**, *12*, 1249–1267. [CrossRef]
33. Liu, S.-S.; Sun, L.; Wu, Q.; Yang, Y.-J. The responses of cyclonic and anticyclonic eddies to typhoon forcing: The vertical temperature-salinity structure changes associated with the horizontal convergence/divergence. *J. Geophys. Res. Oceans* **2017**, *122*, 4974–4989. [CrossRef]
34. Sun, W.; Dong, C.; Tan, W.; He, Y. Statistical Characteristics of Cyclonic Warm-Core Eddies and Anticyclonic Cold-Core Eddies in the North Pacific Based on Remote Sensing Data. *Remote Sens.* **2019**, *11*, 208. [CrossRef]
35. Wang, G.; Ling, Z.; Wang, C. Influence of tropical cyclones on seasonal ocean circulation in the South China Sea. *J. Geophys. Res. Space Phys.* **2009**, *114*, 114. [CrossRef]
36. Zhang, Y.; Zhang, Z.; Chen, D.; Qiu, B.; Wang, W. Strengthening of the Kuroshio current by intensifying tropical cyclones. *Science* **2020**, *368*, 988–993. [CrossRef]
37. Wang, Z.-F.; Sun, L.; Li, Q.-Y.; Cheng, H. Two typical merging events of oceanic mesoscale anticyclonic eddies. *Ocean Sci.* **2019**, *15*, 1545–1559. [CrossRef]
38. Elsberry, S.A. Near-Inertial Ocean Current Response to Hurricane Frederic. *J. Phys. Oceanogr.* **1987**, *17*, 1249–1269.
39. Huang, S.-M.; Oey, L.-Y. Right-side cooling and phytoplankton bloom in the wake of a tropical cyclone. *J. Geophys. Res. Oceans* **2015**, *120*, 5735–5748. [CrossRef]
40. Sun, L.; Yang, Y.-J.; Xian, T.; Wang, Y.; Fu, Y.-F. Ocean Responses to Typhoon Namtheun Explored with Argo Floats and Multiplatform Satellites. *Atmos. Ocean* **2012**, *50*, 15–26. [CrossRef]
41. Li, C.; Zhang, Z.; Zhao, W.; Tian, J. A statistical study on the subthermocline submesoscale eddies in the northwestern Pacific Ocean based on Argo data. *J. Geophys. Res. Oceans* **2017**, *122*, 3586–3598. [CrossRef]
42. Sun, L.; Li, Y.-X.; Yang, Y.-J.; Wu, Q.; Chen, X.-T.; Li, Q.-Y.; Li, Y.-B.; Xian, T. Effects of super typhoons on cyclonic ocean eddies in the western North Pacific: A satellite data-based evaluation between 2000 and 2008. *J. Geophys. Res. Oceans* **2014**, *119*, 5585–5598. [CrossRef]
43. Mei, W.; Primeau, F.; McWilliams, J.C.; Pasquero, C. Sea surface height evidence for long-term warming effects of tropical cyclones on the ocean. *Proc. Natl. Acad. Sci. USA* **2013**, *110*, 15207–15210. [CrossRef] [PubMed]
44. Jansen, M.F.; Ferrari, R.; Mooring, T. Seasonal versus permanent thermocline warming by tropical cyclones. *Geophys. Res. Lett.* **2010**, *37*, 15207–15210. [CrossRef]
45. Pujol, M.-I.; Faugère, Y.; Taburet, G.; Dupuy, S.; Pelloquin, C.; Ablain, M.; Picot, N. DUACS DT2014: The new multi-mission altimeter data set reprocessed over 20 years. *Ocean Sci.* **2016**, *12*, 1067–1090. [CrossRef]
46. Ciani, D.; Rio, M.-H.; Nardelli, B.B.; Etienne, H.; Santoleri, R. Improving the Altimeter-Derived Surface Currents Using Sea Surface Temperature (SST) Data: A Sensitivity Study to SST Products. *Remote Sens.* **2020**, *12*, 1601. [CrossRef]
47. Lomas, M.; Moran, S.; Casey, J.; Bell, D.; Tiahlo, M.; Whitefield, J.; Kelly, R.; Mathis, J.; Cokelet, E. Spatial and seasonal variability of primary production on the Eastern Bering Sea shelf. *Deep. Sea Res. Part II Top. Stud. Oceanogr.* **2012**, *65–70*, 126–140. [CrossRef]
48. Liu, S.; Li, J.; Sun, L.; Wang, G.; Tang, D.; Huang, P.; Yan, H.; Gao, S.; Liu, C.; Gao, Z.; et al. Basin-wide responses of the South China Sea environment to Super Typhoon Mangkhut (2018). *Sci. Total Environ.* **2020**, *731*, 139093. [CrossRef] [PubMed]
49. Huang, J.; Xu, F. Observational Evidence of Subsurface Chlorophyll Response to Mesoscale Eddies in the North Pacific. *Geophys. Res. Lett.* **2018**, *45*, 8462–8470. [CrossRef]
50. Lin, Y.-C.; Oey, L.-Y. Rainfall-enhanced blooming in typhoon wakes. *Sci. Rep.* **2016**, *6*, 31310. [CrossRef]
51. Sriver, R.L.; Huber, M. Observational evidence for an ocean heat pump induced by tropical cyclones. *Nat. Cell Biol.* **2007**, *447*, 577–580. [CrossRef]
52. Chen, D.; Lei, X.; Han, G.; Wang, W.; Zhou, L.; Wang, G. Upper Ocean Response and Feedback Mechanisms to Typhoon. *Adv. Earth Sci.* **2013**, *28*, 1077–1086. [CrossRef]
53. Jullien, S.; Menkes, C.; Marchesiello, P.; Jourdain, N.; Lengaigne, M.; Koch-Larrouy, A.; Lefèvre, J.; Vincent, E.; Faure, V. Impact of Tropical Cyclones on the Heat Budget of the South Pacific Ocean. *J. Phys. Oceanogr.* **2012**, *42*, 1882–1906. [CrossRef]
54. Hsu, P.-C.; Ho, C.-R. Typhoon-induced ocean subsurface variations from glider data in the Kuroshio region adjacent to Taiwan. *J. Oceanogr.* **2018**, *75*, 1–21. [CrossRef]
55. Cione, J.J. The Relative Roles of the Ocean and Atmosphere as Revealed by Buoy Air–Sea Observations in Hurricanes. *Mon. Weather. Rev.* **2015**, *143*, 904–913. [CrossRef]
56. Li, Y.X.; Yang, Y.J.; Sun, L.; Fu, Y.F. The upper ocean environment responses to typhoon Prapiroon (2012). *SPIE Asia-Pac. Remote Sens.* **2014**, *9261*, 92610U. [CrossRef]
57. Yablonsky, R.M.; Ginis, I. Limitation of One-Dimensional Ocean Models for Coupled Hurricane–Ocean Model Forecasts. *Mon. Weather Rev.* **2009**, *137*, 4410–4419. [CrossRef]
58. D'Asaro, E.; Black, P.G.; Centurioni, L.R.; Chang, Y.-T.; Chen, S.S.; Foster, R.C.; Graber, H.C.; Harr, P.; Hormann, V.; Lien, R.-C.; et al. Impact of Typhoons on the Ocean in the Pacific. *Bull. Am. Meteorol. Soc.* **2014**, *95*, 1405–1418. [CrossRef]
59. Mei, W.; Pasquero, C.; Primeau, F. The effect of translation speed upon the intensity of tropical cyclones over the tropical ocean. *Geophys. Res. Lett.* **2012**, *39*, L07801. [CrossRef]

remote sensing MDPI

Article

Modulation Effect of Mesoscale Eddies on Sequential Typhoon-Induced Oceanic Responses in the South China Sea

Weifang Jin [1], Chujin Liang [1], Junyang Hu [1], Qicheng Meng [1], Haibin Lü [2], Yuntao Wang [1,*], Feilong Lin [1], Xiaoyan Chen [1] and Xiaohui Liu [1]

1 State Key Laboratory of Satellite Ocean Environment Dynamics, Second Institute of Oceanography, Ministry of Natural Resources, Hangzhou 310012, China; jinweif@sio.org.cn (W.J.); cjliang@sio.org.cn (C.L.); junyanghu@sio.org.cn (J.H.); q.meng@sio.org.cn (Q.M.); lfl@sio.org.cn (F.L.); chenxy@sio.org.cn (X.C.); xh_liu@sio.org.cn (X.L.)

2 Jiangsu Key Laboratory of Marine Bioresources and Environment/Jiangsu Key Laboratory of Marine Biotechnology, Jiangsu Ocean University, Lianyungang 222005, China; haibin_lv@jou.edu.cn

* Correspondence: yuntao.wang@sio.org.cn; Tel.: +86-571-8196-3629

Received: 3 July 2020; Accepted: 14 September 2020; Published: 18 September 2020

Abstract: The impacts of mesoscale eddies on the modulation of typhoon-induced oceanic responses are important for understanding ocean dynamics. Satellite observations identified prominent ocean surface temperature and chlorophyll changes over the regions with mesoscale eddies after two sequential typhoons, e.g., Linfa and Nangka, in the South China Sea. The impacts of typhoons on the ocean surface were more prominent within cyclonic eddies than within anticyclonic eddies. The wind speed (translation speed) of Linfa was much larger (slower) than that of Nangka; thus, the changes induced by Linfa were stronger. However, the second typhoon easily generated mixing through the weak stratification induced by the first typhoon and impacted the upper ocean. The strong chlorophyll enhancement induced by Nangka was identified at a cyclonic eddy. Using a combination of reanalysis data, the depth of water origin (DWO) was applied to quantify the depth to which a typhoon's impact could be exerted. Prominent changes were identified when the DWO reached the depth at which the temperature and nutrients differed from those within the mixed layer. This method can overcome the impacts of cloud coverage when examining a typhoon's influence with remotely sensed data and offers a quantitative approach to determine the mechanisms responsible for typhoon-induced ocean surface changes.

Keywords: mesoscale eddies; sea surface chlorophyll; typhoon; South China Sea; sea surface temperature; depth of water origin

1. Introduction

Tropical cyclones, which present as intense and devastating atmospheric events, are generated mainly near the tropics and propagate to midlatitudes [1]. The tropical cyclones generated in the northwestern Pacific Ocean, known as typhoons, account for more than 1/3 of all tropical cyclones in the Northern Hemisphere annually [2]. During their movement, typhoons can induce significant impacts on the upper ocean by driving the motions of waters beneath [3], e.g., large typhoon winds generate strong vertical mixing and upwelling within the ocean, which brings subsurface cold, high-nutrient water to the surface [4]. A cold surface is formed in front or at the wake of the typhoon path [5]. Consequently, the cold surface temperature modifies the typhoon intensity by blocking the following typhoons from gaining energy from the warm ocean surface [6,7]. Thus, single typhoons usually trigger substantial sea surface cooling and sea surface chlorophyll (CHL) blooms [8,9], but the impacts of sequential typhoons are much less pronounced [10].

Typhoon-induced ocean responses vary greatly depending on the typhoon's features (e.g., wind speed, pressure, and translation speed) and the preconditions of the upper ocean (e.g., mesoscale eddies, thermocline, mixed layer depth and strength of stratification) [1,11]. Strong winds during typhoons induce intensive mixing, which largely influences the upper ocean [12]. The wind stress curl associated with atmospheric cyclonic motions can drive substantial Ekman pumping, leading to upwelling beneath the typhoon center [13]. Both mixing and upwelling can occur simultaneously, and their relative importance is different in each case [2]. In general, strong-intensity and slow-moving typhoons are expected to induce large responses [14]. Numerous studies have investigated the impacts of typhoons on the ocean, including the spatial extent and temporal dependence of the typhoon's influence. For instance, typhoons can stimulate strong winds and wind–current resonance on the right side of the typhoon's track, resulting in stronger mixing than on the left side [4,9]. Moreover, the temperature changes instantly as the typhoon approaches, whereas CHL blooms are usually delayed [15]. This difference is attributed primarily to wind-induced mixing, which decreases temperature and elevates nutrients within a few hours [13], while the growth of phytoplankton takes several days [3]. On the other hand, CHL enhancement can be associated with cooling temperatures when typhoons bring subsurface CHL maxima directly to the surface [16].

The status of the ocean can also influence the oceanic responses to typhoons; previous studies generally described the changes near the surface, although the vertical structure of the water column and the depth of typhoon-induced mixing are particularly important [17]. Large ocean responses are found in regions with strong vertical gradients, and stratification is greatly reduced during the passage of a typhoon [18]. Because the temperature usually decreases monotonically with depth, stronger responses occur when mixing reaches greater depths [2]. The variability in CHL is more complex than that in temperature. In general, the nutrient content is low within the mixed layer depth (MLD) that the growth of phytoplankton is limited by the availability of nutrients, leading to low CHL levels [19]. As the nutrient content gradually increases with depth, CHL also slightly increases [20]. The nutrient level is much larger beneath the nutricline; however, the growth of phytoplankton is limited by the availability of light [21]. A subsurface CHL maximum (SCM) is usually identified within the euphotic zone around the nutricline, where both the nutrient level and the availability of light can barely support phytoplankton growth [8]. Thus, CHL is simultaneously modulated by nutrients and light, and the nutrients within the euphotic zone can be greatly elevated when typhoon-induced vertical mixing deepens the MLD [22]. Although the typhoon-induced mixing depth and upwelling velocity can be estimated [23], it is fundamentally important to gauge whether subsurface cold and high-nutrient water can be lifted to explicitly assess the responses of upper ocean.

As a ubiquitous dynamic feature, mesoscale eddies are characterized by intensive vertical dynamics and can modulate ocean responses to typhoons. Cyclonic (anticyclonic) eddies are characterized by upwelling (downwelling) within their interior [24]. A previous study revealed that the responses of the ocean to typhoons can be intensified (suppressed) by cyclonic (anticyclonic) eddies because the elevated (depressed) MLD via central upwelling (downwelling) can be enhanced (reduced) by typhoons [13]. Thus, pre-existing eddies can alter the intensity of temperature cooling and enhancement in CHL [25–27]. A previous study identified that a weak typhoon can still induce large and long-lasting oceanic responses in a pre-existing cyclonic eddy [13]. Other studies suggested that only 18% of typhoons can induce CHL enhancement, even with ubiquitous eddies [28]. Therefore, in addition to the importance of typhoon features and oceanic status, mesoscale eddies contribute to the large discrepancies among typhoons in different oceans, which requires further explanation.

Every year, a substantial number of typhoons pass through the South China Sea (SCS), a semienclosed ocean basin in the northwestern Pacific Ocean [16]. Because the nutrient level is extremely low at the ocean surface, the SCS is classified as a typical oligotrophic zone, although solar radiation is strong throughout the year with a euphotic zone that can extend up to 100 m [29]. A prominent seasonal cycle of typhoon occurrence is identified that strong and frequent typhoon activities occur from spring to fall [30]. During this period, the MLD is very shallow, ranging between

20 m and 40 m [31]. Moreover, the nutricline and SCM are highly consistent with each other, located near a depth of 60 m [32]. Numerous studies have been conducted in the SCS to investigate the impacts of typhoons on the ocean surface, and these studies revealed that typhoons can significantly change the ocean status [3]. Predominant mesoscale eddies are observed within the SCS, occupying approximately 25% of the ocean surface on average [33]. Previous studies have revealed that mesoscale eddies in the SCS can greatly modulate typhoon-induced ocean surface responses [11,22,34,35]. The SCS is a good study site with abundant typhoon events and complex oceanic dynamics, e.g., ubiquitous eddies, that can help to quantitatively disentangle typhoon-induced oceanic responses. Specifically, Chen and Tang [36] found that cyclonic eddies enhanced ocean responses during the passage of typhoon Linfa in 2009. This phenomenon mainly occurred because eddy-induced upwelling is enhanced by strong winds [37]. As eddies rapidly advect upwelled high-nutrient water to the surroundings, the induced phytoplankton bloom at the surface can last for a long period. However, a typhoon occurred immediately after Linfa, which was not identified in the former study [36]. The dynamics of sequential typhoons can be quite different from those of a single typhoon, and thus, this scenario should be further investigated.

Multiple observational methods, e.g., Argo floats, satellites and moorings, have been applied to study the responses of typhoon-impacted oceans. Argo profiles describe the changes in the vertical structure of the water column during typhoons, and satellites capture the features of the ocean surface [2,36]. Numerical modeling and reanalysis datasets are highly valuable for further resolving the three-dimensional processes and offer quantitative descriptions of typhoons' impacts [38]. Our understanding of ocean dynamics under the impact of a typhoon has been greatly improved. However, the situation can become very complex when different conditions are encountered simultaneously. For example, sequential typhoons passing over the ocean with mesoscale eddies are a perfect example to study ocean dynamics, but such events have rarely been identified. Hence, the tandem effects of typhoons and eddies on the sea surface in the SCS are an interesting topic that needs further exploration. In this study, the impacts of typhoons on the ocean surface in the SCS modulated by mesoscale eddies with different polarities are investigated. A universally applicable method based on a three-dimensional reanalysis dataset is proposed to quantitatively gauge typhoon-induced upper ocean responses. The remainder of this paper includes five sections: Section 2 describes the information on the typhoons and the materials and methods used in this study; the typhoon's impacts on the ocean in the SCS and the underlying eddy-induced dynamics are shown in Section 3; a comprehensive analysis of the typhoon-induced dynamics is presented in Section 4, followed by a summary in Section 5.

2. Data and Methods

The typhoon information was obtained from the Japan Meteorological Agency (JMA, http://www.jma.go.jp/jma/indexe.html). These data include the maximum sustained wind (MSW), category and location (longitude and latitude) of the typhoon center every 6 h. The category of typhoons is defined following the rule of the JMA that the MSWs of a tropical cyclone at category-3 and category-4 are between 17 m/s and 24 m/s and between 25 m/s and 32 m/s, respectively. The applied category in current study is different from the international categorization, which corresponds to tropical storm and severe tropical storm, respectively. The translation speed (Uh, m/s) of the typhoon was calculated as the ratio between the spatial distance traveled between two successive typhoon centers and the corresponding time span. The daily wind fields were obtained as the average of two satellite observational datasets. The first dataset was measured by the SeaWinds microwave scatterometer onboard the Quick Scatterometer (QuikSCAT, https://winds.jpl.nasa.gov/) satellite with a spatial distribution of 0.25° × 0.25°. To compensate for some observational gaps in the wind field, the other dataset from the Advanced Scatterometer (ASCAT), which was developed by the European Organisation for the Exploitation of Meteorological Satellites (EUMETSAT, http://www.remss.com/missions/ascat/), was applied.

Daily sea surface temperature (SST) data with a resolution of 0.25° × 0.25° was measured by the Advanced Very High-Resolution Radiometer (AVHRR) onboard the Polar-Orbiting Environmental Satellites, which was obtained from the National Centers for Environmental Information (NCEI, https://www.ncei.noaa.gov/). The corresponding climatology was calculated as the 15-year average ranging between 2002 and 2016 for each date over the year. Daily remotely sensed CHL concentrations with a spatial resolution of 4 km developed by the European Space Agency (ESA) Ocean Color Climate Change Initiative (OC-CCI, http://www.oceancolor.org/) are used to investigate the spatial and temporal variations in the phytoplankton biomass at the sea surface. Correspondingly, the daily CHL climatology is calculated on each day of the year using the same 15-year dataset.

Archiving, Validation and Interpretation of Satellite Oceanographic (AVISO) sea level anomaly (SLA) data with a spatial resolution of 0.25° × 0.25° was downloaded from the Copernicus Marine Service (CMEMS, http://marine.copernicus.eu/). The SLA data is used to derive the geostrophic current and to detect mesoscale eddies following Chelton et al. [24]. Specifically, pre-existing oceanic eddies are identified based on the sea surface height anomaly (SSHA) as follows: a local peak (trough) of SLA is first found, and the largest closed contours of SLA with a single regional maximum (minimum) are defined as anticyclonic (cyclonic) eddies. The anticyclonic eddies are characterized by a positive SLA (>5 cm) relative to the surroundings, while cyclonic eddies have a negative relative SLA (<−5 cm) [39].

A reanalysis dataset of the three-dimensional daily ocean temperature, salinity and derived density was extracted from the Global Ocean Forecasting System (GOFS) 3.1 reanalysis product based on the Hybrid Coordinate Ocean Model (HYCOM, http://apdrc.soest.hawaii.edu/). This dataset has a spatial resolution of 1/12° × 1/12° and 30 layers in the vertical direction. A detailed description can be found in Vidya et al. [40]. The product was forced by the Navy Operational Global Atmospheric Prediction System [41] and adjusted by the data assimilation system to produce a 3D variational scheme [42]. The assimilative data sources include satellite altimeter observations, satellite SST, and in situ vertical temperature and salinity profiles from expendable bathythermographs (XBTs), Argo floats and moored buoys. The MLD is calculated as the depth where the water temperature is 0.55 °C cooler than the corresponding temperature at the surface [12].

3. Results

3.1. Typhoons' Impacts on the Ocean Surface

Two sequential but weak typhoons, Linfa and Nangka, passed over the SCS in June 2009 (Figure 1). Typhoon Linfa was generated in the middle of the SCS on 17 June 2009; it remained near its origin for another day before moving northward on 19 June. The translation speed was approximately 2.3 m/s during this period. The typhoon was first classified as category 2 (tropical depression, MSW < 17 m/s) and was gradually increased to category 3 (tropical storm, 17 m/s < MSW < 24 m/s) and category 4 (severe tropical storm, 24 m/s < MSW < 32 m/s) before it entered the East China Sea on 21 June [43]. Typhoon Nangka originated in the western Pacific (10.8°N, 129.0°E) on 22 June with a wind speed of 18 m/s, after which it moved northwestward across the Philippines and entered the SCS on 24 June as a category-3 typhoon. In the following days, the typhoon moved rapidly towards the northwest at a maximal wind speed of 20.6 m/s until making landfall in Guangdong Province on 26 June. The averaged translation speed of Nangka was 6.2 m/s, which was substantially larger than that of Linfa. The wind speed for Linfa was slightly larger than that for Nangka, although both were characterized by relatively small MSW (less than 32 m/s). Both events passed over the SCS, as denoted by the dashed box in Figure 1a, and the occurrences of these typhoons were temporally separated by approximately a week. The water depth is more than 1000 m in the central SCS and gets shallower towards the coast. The corresponding climatological SST and CHL in June decreases and increases, respectively, from south to north (Figure 1b), although the spatial variance is actually small. In particular, for the region impacted by both typhoons, the SST is approximately 29.4 °C, and the CHL is close to 0.07 mg/m^3.

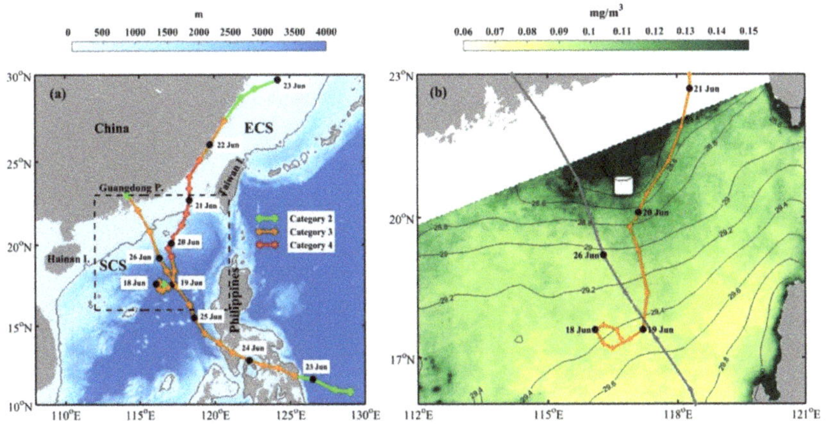

Figure 1. (**a**) Map of the South China Sea (SCS) and East China Sea (ECS) with 200-m isobath; and the trajectories of the two typhoons. The 00:00 time on each day is labeled at the locations of the typhoons. The typhoon category, which is based on JMA (Japan Meteorological Agency) categorization, is represented by the color. The study region is delineated by the dashed black square. (**b**) Climatological average of the sea surface chlorophyll (CHL, shading, mg/m^3) and sea surface temperature (SST, contours, °C) in June. The trajectories for typhoons Linfa and Nangka are shown in orange and gray, respectively. The values over the region close to the coast and island are masked. The SST and CHL data are respectively obtained from Ocean Color Climate Change Initiative (OC-CCI) and Advanced Very High-Resolution Radiometer (AVHRR).

The typhoons dominate the wind patterns throughout the study region during the period as they pass over the SCS (Figure 2). The wind speed intensity is roughly the same from 17 June to 20 June for Linfa; the wind fields on 19 and 20 June are shown to be associated with large Ekman pumping velocities (EPVs) within the typhoon center that are approximately 11.0×10^{-4} m/s and 15.7×10^{-4} m/s, respectively. For Nangka, the wind fields on 25 June and 26 June are shown since the typhoon rapidly traverse the study region, and the corresponding EPVs are 4.8×10^{-4} m/s and 8.1×10^{-4} m/s, respectively. Compared with those for Linfa, the wind intensity and EPV for Nangka are prominently smaller. The average typhoon radii of Linfa and Nangka with wind speeds of 15 m/s are 200 km and 250 km, respectively, when they are within the study region (Figure 3a,c).

Prominent SST cooling and CHL enhancement are identified along the typhoons' tracks within the typhoon-impacted area (Figure 3). The typhoon-induced ocean surface change is obtained as the average of several days' anomalies after the typhoons, where an anomaly is defined as the difference between the ocean surface parameter and the corresponding climatology. Because the times required for the SST and CHL to respond after a typhoon vary from each other [5,44], the selected time period should be different. In this study, the SST (CHL) is calculated over the period from 20 June to 22 June (20 June to 25 June) for Linfa and from 25 June to 27 June (26 June to 1 July) for Nangka. Some studies obtained the typhoon-induced changes as the difference between the time series after and before the typhoon [35]. Because the seasonal variability is prominent for entire period over two typhoons [22], the daily climatology is removed instead to obtain the anomalous field. The impacts of the typhoons on the SST are clearly captured, and prominent cooling can be found along their tracks, particularly on the right side. CHL enhancement can be found along the tracks as well, although the right-side bias is less prominent than with the SST. Because the CHL change is much larger in the coastal region than in the open ocean, the region within 200 km from the coast is masked to highlight the typhoon-induced CHL changes offshore. Linfa induces larger impacts on the SST and CHL than Nangka; e.g., the largest SST cooling is approximately −4 °C and −2 °C for Linfa and Nangka, respectively, and the corresponding maximum CHL enhancements are 0.5 mg/m^3 and 0.2 mg/m^3.

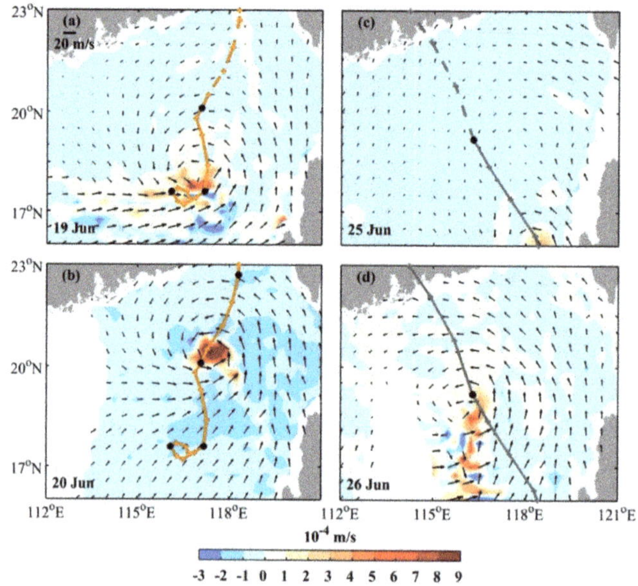

Figure 2. Wind field (vector, m/s) overlaid with the Ekman pumping velocity (shading, m/s) for typhoon Linfa on (**a**) 19 June and (**b**) 20 June and for typhoon Nangka on (**c**) 25 June and (**d**) 26 June. The typhoon trajectory after the depicted date is shown as a dashed line.

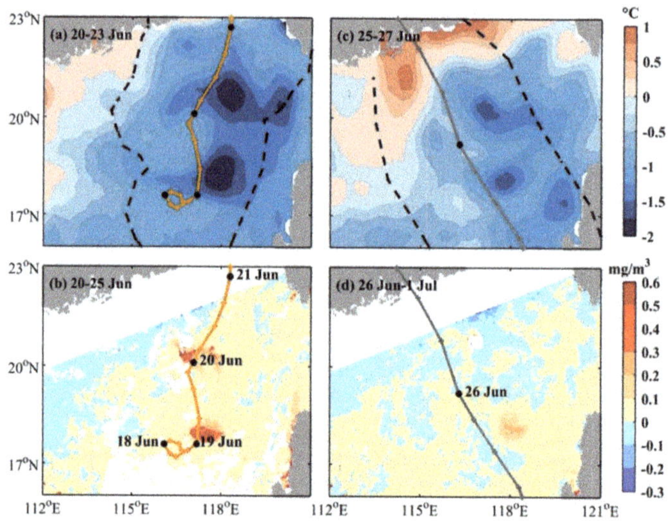

Figure 3. Changes induced by typhoons Linfa (**a**,**b**) and Nangka (**c**,**d**) in the (**a**,**c**) sea surface temperature and (**b**,**d**) sea surface chlorophyll. A change is defined as the average anomaly, which is the difference between the daily observations and corresponding climatology during the period denoted in each panel. The radius of the typhoon speed equal to 15 m/s is shown as dashed black lines.

3.2. Modulation Effect of Eddies on Typhoon-Induced Cooling

There are two zones featuring SST cooling greater than −2 °C centered at 17.7°N, 117.6°E and 20.1°N, 117.4°E, where strong responses are identified along the track of Linfa. The two areas are approximately circular with radii equal to 160 km and 110 km. As depicted in the mean SLA field between 18 June and 27 June, these areas correspond to the occurrence of two cyclonic eddies (Figure 4). Both eddies are similar in size, but the northern eddy (CE1) has a strong intensity and is less impacted by surrounding eddies; detailed information on both eddies during the passage of typhoons can be found in Table 1. Here, the relative SLA is defined as the difference between the maximum SLA within the eddy and the SLA averaged around the eddy perimeter, whose value is positive (negative) for anticyclonic (cyclonic) eddies. During the same period, two anticyclonic eddies are also located near the typhoons' tracks. The southern region (ACE2) is characterized by a strong intensity with a radius equal to 86.4 km and a relative SLA of approximately 10.7 cm. In contrast, the northern anticyclonic eddy (ACE1) is located between cyclonic eddies with a small radius and magnitude (Table 1). The longevity of ACE1 is much shorter than that of other eddies, as it fades on 27 June when immersed by CE2 (not shown). The eddies are generally located to the right of both typhoons' trajectories; the only exception is Nangka, which passes immediately over the southern anticyclonic eddy (ACE2). The four identified eddies pre-existed in the SCS before Linfa and remained after Nangka, and they have very important impacts on the vertical structure of the water column and typhoon-induced variation, which will be described later. Although some other eddies can be identified as well, such as the one centered at 18°N, 115.3°E, they are far from the typhoons and less prominently impact the ocean surface.

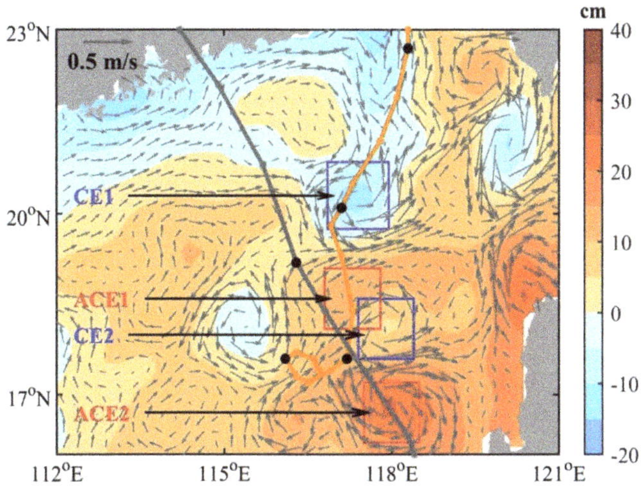

Figure 4. The averaged sea level anomaly (shading, cm) and geostrophic current (vector, m/s) between 18 June and 27 June. The selected locations for cyclonic eddies (blue) and anticyclonic eddies (red) are labeled along the typhoon's trajectories.

Table 1. Information on all eddies during the passage of typhoons (CE: cyclonic eddy; ACE: anticyclonic eddy; SLA: sea level anomaly; TY: typhoon).

		CE1	CE2	ACE1	ACE2
Period		10 June–31 July	19 June–16 July	10–27 June	18 May–28 July
Region of eddies		116.8–117.9°E 19.7–20.8°N	117.4–118.4°E 17.6–18.6°N	116.8–117.8°E 18.1–19.1°N	117.5–118.5°E 16.2–17.2°N
Area of eddies (km^2)		13,981	11,718	11,684	11,811
Linfa (17–23 June)	Averaged position	117.5°E 20.5°N	117.8°E 18.2°N	117.5°E 18.6°N	118.0°E 16.8°N
	Averaged radius (km)	63.1	54.8	57.2	86.4
	Maximum relative SLA (cm)	−7.8	−5.7	7.4	10.7
	Smallest distance to TY (km)	4.75	45.34	11.48	108.90
Nangka (24 June–2 July)	Averaged position	117.3°E 20.0°N	117.8°E 18.2°N	17.5°E 18.7°N	118.0°E 16.8°N
	Averaged radius (km)	74.9	78.2	40.0	68.8
	Maximum relative SLA (cm)	−7.4	−5.6	0.6	11.6
	Smallest distance to TY (km)	112.36	65.22	65.50	4.08

The temporal dependence of the typhoon-induced ocean surface responses is further investigated at the location of each eddy (Figure 5). The wind speed starts to increase before Linfa's arrival on 17 June and reaches a maximum, e.g., approximately 18 m/s at CE2, during its passage. The daily wind speed is obtained by averaging the wind within the area of the eddy; thus, the value is less than the MSW. The difference in time when the wind speed reaches its peak is clearly related to the locations of the eddies, which appear in the order of ACE2, CE2, ACE1 and CE1 from south to north (Table 2). While the peak value of the wind speed is related to the distance between the eddies and typhoon center, the wind in ACE2 is clearly weaker, e.g., approximately 15 m/s, because the eddy is farther from the typhoon track (Table 1). For Nangka, the wind speed starts to increase on 25 June upon entering the SCS and rapidly reaches a peak on 26 June before decreasing back to its normal speed. The wind speed during Nangka is prominently weaker than that during Linfa, with the former reaching a maximum of approximately 12 m/s. Comparing the wind speeds among the different eddy locations, their time series are highly similar because Nangka traverses the study region in a very short period. The winds during the time period between the two typhoons are characterized by slow speeds, and the corresponding values are comparable to the normal status.

The SST decreases prominently for all eddies simultaneously with the increases in the wind speed between 16 June and 21 June for all eddies. The cooling in the CE2 is more than −2 °C, which is more prominent than that in the anticyclonic eddies, with a decrease close to −1.5 °C. The SST cooling trend relaxes as the typhoon passes over the SCS on 21 June, but surface cooling can still be identified during the following three days. As Nangka arrives on 25 June, it instantly stimulates another surface cooling event. The SST depression is weak but recognizable compared with the cooling induced by Linfa, and the SST decreases are comparable among all the eddy locations. The SST increases back to its climatological value over the four days after Nangka passes through the study area. CHL enhancement is captured for all eddies after Linfa except ACE2; in contrast, after Nangka, a bloom is identified for only CE2. The CHL change is prominently delayed by a few days compared with the delay in the SST change.

From 23 to 24 June, 3 days after Linfa, a weak CHL bloom (e.g., 0.03 mg/m^3) can be identified for ACE1, while large CHL blooms (e.g., 0.2 mg/m^3) are found in the cyclonic eddies. Interestingly, after Nangka, another CHL peak is observed at CE2 with an enhancement of approximately 0.14 mg/m^3, but no CHL change is associated with the other eddies. This difference occurs because the three-dimensional processes cannot be captured from satellites. The CHL in the cyclonic eddies relaxes back to the climatological average on 2 July, approximately one week after Nangka. On the other hand, the CHL in the anticyclonic eddies is persistently higher than the climatological average during the study period, which may be due to the advection of surrounding water with a high CHL content from CE2. In summary, the modulation effect of eddies on typhoon-induced changes at the ocean surface is prominent, as will be discussed in the following section.

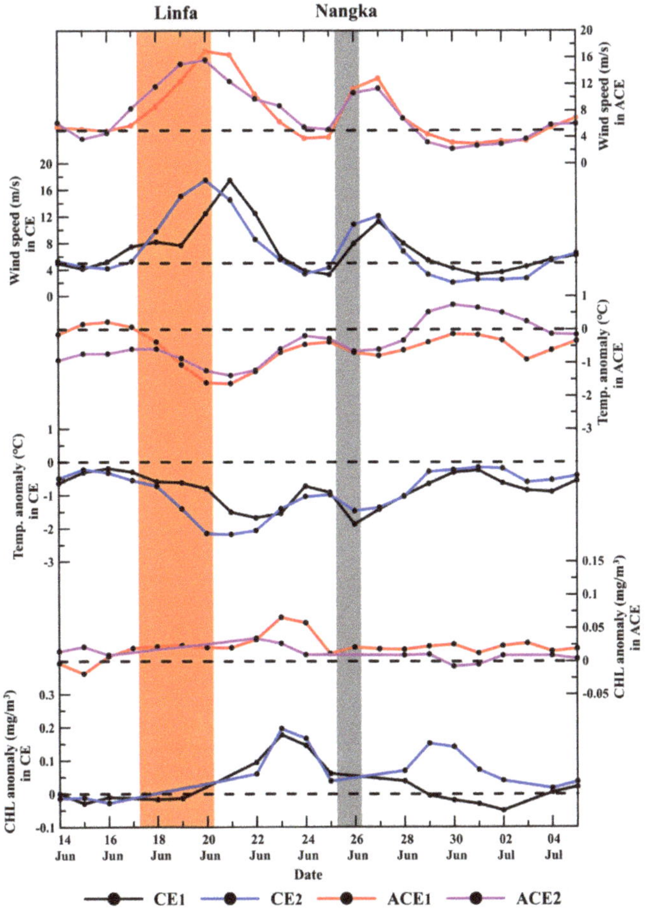

Figure 5. Time series of the wind speed (m/s), temperature anomaly (°C, AVHRR) and CHL anomaly (mg/m^3, OC-CCI) at CE1 (black), CE2 (blue), ACE1 (red) and ACE2 (magenta). The periods for Linfa and Nangka are indicated by the orange and gray bands, respectively. The averaged wind speed is shown as horizontal dashed lines. The anomaly field is obtained as the difference between the daily observations and the corresponding climatology between 8 June and 7 July.

Table 2. Typhoon-induced ocean changes at all eddy locations.

		CE1	CE2	ACE1	ACE2
Linfa	Period when typhoon passed over the eddy	18 on 19 June–00 on 20 June	00–06 on 19 June	06–12 on 19 June	06 on 17 June–00 on 19 June
	Maximum ΔT (°C)	−1.68	−2.17	−1.63	−1.38
	Time for maximum ΔT	22 June	21 June	21 June	21 June
	Maximum ΔCHL	0.17	0.20	0.07	0.03
	Time for maximum ΔCHL	23 June	23 June	23 June	22 June
Nangka	Period when typhoon passed over the eddy	00–06 on 26 June	18 on 25 June–00 on 26 June	12–18 on 25 June	06–12 on 25 June
	Maximum ΔT (°C)	−1.86	−1.46	−0.78	−0.65
	Time for maximum ΔT	26 June	26 June	27 June	26 June
	Maximum ΔCHL	0.04	0.15	0.03	0.01
	Time for maximum ΔCHL	28 June	29 June	30 June	29 June

The vertical sections of the water columns within the eddies are further investigated to understand the dynamics therein. Reanalyzed data from HYCOM are applied with the MLD depicted for each location in Figure 6. The initial MLDs are approximately 20 m at all locations, and the MLD starts to deepen when the typhoon arrives, associated with the occurrence of upwelling. The MLD and upwelling continue to increase until two days after the typhoon, characterized by prominent SST cooling. Subsequently, the temperature within the top 20 m increases, associated with shoaling of the MLD, which can rebound back to an even shallower depth than that under pre-typhoon conditions. The shallowest MLD is generally observed when Nangka arrives on 24 June, which induces further mixing. Although the wind speeds during Nangka are much weaker than those during Linfa, the maximum induced MLD, which also takes place two to three days after Nangka, is similar to or even larger than the maximum MLD after Linfa. Then, the MLD becomes very shallow and remains so for approximately four days, and after 2 July, the MLD deepens to a depth similar to the initial value.

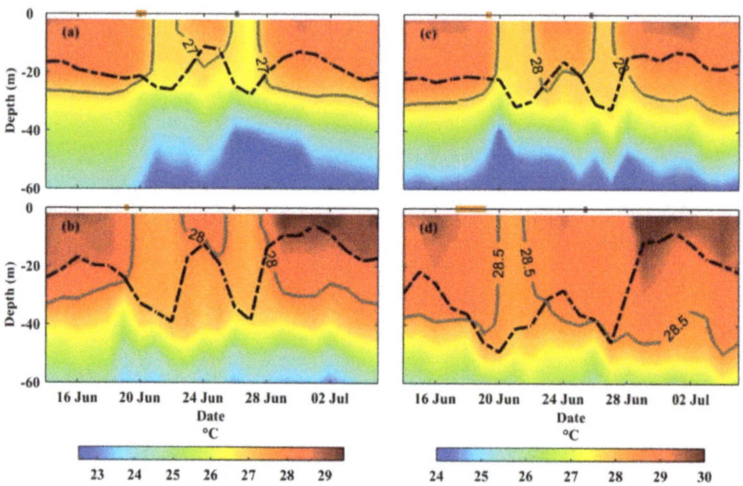

Figure 6. Time series of the temperature profile with isobaths of 27 °C, 28 °C, 28 °C and 28.5 °C for (**a**) CE1, (**b**) CE2, (**c**) ACE1 and (**d**) ACE2, respectively. The dashed black lines indicate the mixed layer depth. The horizontal bars represent the duration of each typhoon.

The climatological profiles of temperature and nutrients are averaged for June over the region covering all identified eddies (Figure 7b,c). The calculated MLD is 20 m, where the temperature is 0.55 °C less than the SST following Obata et al. [45]. The temperature is well mixed above the MLD with a value of approximately 29 °C and monotonically decreases beneath the MLD. On the other hand, the nutrient concentration is extremely low at the surface and persistently lower than 0.5 µM within 50 m of where the nutricline is identified [46], while a prominent nutrient increase is identified beneath the nutricline. Upon comparing the profiles at different eddy locations with each other, the patterns are highly similar within the top 50 m, which can be confirmed by the small standard deviations. The depth of water origin (DWO) is obtained by comparing the SST with the climatological temperature profile, where the upwelling origin is defined as the depth where the climatological water temperature is similar to the SST. If the SST is warmer than the climatological surface temperature, the DWO is defined as zero. Although the depth influenced by EPV and mixing can be estimated using wind stress, the upper ocean structure is largely different and can impact the influenced depth [40]. The advantage for the application of DWO is that the reanalyzed dataset offers a near real-time description of the ocean status [42] that considers the vertical variation. The DWO time series is shown for all eddy locations (Figure 7a), and the general patterns are similar to the variations in the MLD. For cyclonic eddies, the surface water always originates from the subsurface due to eddy-induced persistent upwelling. Before the arrival of Linfa, the DWO is approximately −12 m and −30 m for CE1 and CE2, respectively. As Linfa moves across the study region, the DWO gradually increases to greater depths, and the maximum depths are −30 m and −40 m for CE1 and CE2, respectively, on 22 June. The DWO subsequently shallows as the wind-induced mixing is reduced; however, the DWO deepens again when Nangka arrives on 24 June. The Nangka-induced DWO in CE1 is similar to the Linfa-induced DWO but is 5 m deeper in CE2. A very different feature is found for the anticyclonic eddies, in which the DWO is mostly near zero except during the period when a typhoon passes. The maximum DWO values for ACE1 and ACE2 are −17 m (−23 m) and −20 m (−8 m) during Linfa (Nangka), respectively.

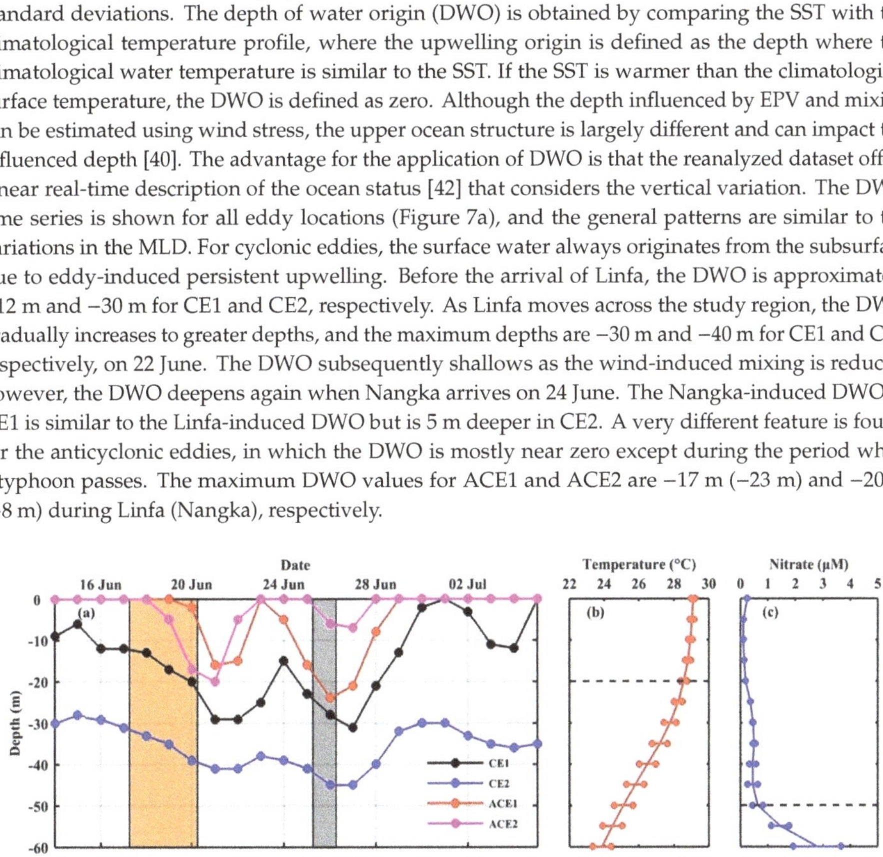

Figure 7. (**a**) Time series of the depth of water origin for CE1 (black), CE2 (blue), ACE1 (red) and ACE2 (magenta). The periods for Linfa and Nangka are indicated by the orange and gray bands, respectively. Climatological profiles for the average and standard deviation of the (**b**) temperature (°C) and (**c**) nitrate concentration (µM) for the regions including all eddies in June. The black dashed lines indicate the mixed layer depth and nutricline.

4. Discussion

Satellite data were used to capture prominent ocean surface responses induced by sequential typhoons in the SCS. Because both typhoons passed over the same region with pre-existing mesoscale eddies, this scenario offers a great opportunity to quantitatively analyze the mechanisms responsible for the typhoon-induced ocean changes and the modulation effect of eddies on those changes. The passage of typhoons drives strong mixing, which simultaneously transports heat downward and nutrients upward [47,48]. In the current study, the typhoons induced a maximum SST cooling equal to 4 °C, which is slightly weaker than many previous cooling intensities in the same region [3,10,20,34]. For example, Sriver and Huber [49] suggested that the largest typhoon-induced SST depression can reach 8 °C within cyclonic eddies. This disparity is due at least to the weak intensity of both typhoons (neither of them sustained wind speeds larger than 32 m/s), as large changes generally occur in response to typhoons with strong intensities [14]. In addition, both of the cyclonic eddies are characterized by small amplitude and short longevity (Table 1), making them less favorable for driving strong responses when compared with many former studies [13,24]. On the other hand, the impacts of the typhoons on CHL, with a maximum increase equal to 0.2 mg/m^3, are distinct, as depicted by the time series of surface CHL (Figure 5). This is particularly true for the SCS in summer when MLD is shallow [32] and surface water is oligotrophic [33], leading to low chlorophyll [22]. Some studies discovered that typhoon-induced CHL may have increased by a factor of 30 in response to a single typhoon [48], while others revealed that over 50% and 80% of all typhoons in the SCS [16] and Northwest Pacific [28], respectively, cannot even trigger increases in CHL. The selected typhoons in the current study can be considered to be highly representative for conducting investigations on typhoon-induced oceanic responses in the SCS.

Consistent with former studies [17,50,51], large SST and CHL responses are detected to the right of the typhoon tracks (Figure 2). Especially for the SST, typhoon-induced cooling predominantly affects the ocean surface to the right of Linfa and Nangka (Figure 3). Besides the influence of the occurrence of cyclonic eddies on the right of both typhoons (Figure 4), the winds and wind–current resonance on the right side are stronger than those on the left [4]. Huang and Oey [38] used a high-resolution biophysical coupled model and found that submesoscale recirculation cells induce restratification, which decreases turbulence to the right of typhoon tracks. Our results also reveal stronger wind speeds and EPV to the right side of the typhoon tracks to a large extent (Figure 2).

The comparison of the oceanic responses induced by Linfa and Nangka indicates that stronger changes, e.g., SST cooling, are found after the former. This difference is largely attributed to the typhoon features, such as the influence of the wind speed discussed above. The translation speed of Nangka is much higher than that of the majority of typhoons in the SCS [17], resulting in a short forcing time. Thus, the slow wind speed and fast translation speed of Nangka are not favorable for generating large responses [2]. Sun et al. [13] found that a weak typhoon classified as category 1 can still induce evident CHL increases, mainly because a slow translation speed results in a longer forcing time and increased MLD, which consequently contributes to a strong phytoplankton bloom. It is interesting to notice that although these features of Nangka are not favorable for driving oceanic response, prominent changes are also observed for SST (Figure 5), which is highly related with the unique sequential typhoons [35]. Linfa thoroughly mixed the surface ocean structure, and the short period between the two typhoons (Figure 6) does not allow the water to relax back to the initial condition [35]. Furthermore, typhoons can induce prominent surface ocean responses when the ocean is characterized by weak stratification [17]. Thus, another oceanic response would be expected if a sequential typhoon takes place, even if the second event has a weak intensity and fast translation speed (Figures 1 and 5).

The SST cooling and CHL enhancement distributions are not spatially uniform; instead, some locations are associated with large changes (Figure 2). The spatial locations of these large changes are highly coincident with those of cyclonic eddies (Figure 4), indicating that mesoscale eddies can modulate the responses of the ocean to typhoons [52]. Former studies revealed eddy-entrained suspended sediments, phytoplankton and colored dissolved organic matter in the surface ocean, implying that CHL can be largely elevated by typhoons [53]. Phytoplankton blooms then take place 3–4 days after the passage of typhoons associated with the rapid intensification of cyclonic eddies [25]. Zheng et al. [26] found that pre-existing cyclonic eddies in the western North Pacific significantly enhance the SST cooling trend in response to strong typhoons. The presence of warm ocean eddies can effectively block cold, nutrient-rich water from being entrained to the surface [28,54]. In our study, the prominent increases in CHL are highly consistent with pre-existing cyclonic eddies (Figure 2). Additionally, the occurrence of both cyclonic and anticyclonic eddies offers a great opportunity to evaluate their unique impacts, and clear distinctions are observed among them (Figure 4). Our results indeed indicate that typhoon-induced ocean changes are more prominent in cyclonic eddies than in anticyclonic eddies (Figure 5).

Because mesoscale eddies are ubiquitous in the global ocean [55,56], typhoons can encounter many eddies during their propagations [57]. It is crucially important to understand the ability of mesoscale eddies to regulate ocean responses to typhoons [58]. Pre-existing eddies greatly impact the vertical structure of the water column, and persistent upwelling can be found in cyclonic eddies where the surface water upwells from the subsurface (Figure 7a) [58]. On the other hand, convergence inhibits upwelling within anticyclonic eddies such that subsurface water outcrops to the surface only when a typhoon initiates intensive mixing (Figure 6), while the water column is restratified rapidly after a typhoon (Figure 6b,d). Liu and Tang [59] consistently found that the typhoon-induced CHL increase differs greatly among pre-existing oceanic conditions, with cyclonic and anticyclonic eddies increasing the CHL by approximately 35% and 6%, respectively. Furthermore, the magnitude of SST cooling varies only weakly between the two types of eddies (Figure 5) because typhoon-induced mixing is dominant, while upwelling within cyclonic eddies adds to the typhoon's impact and results in greater cooling at the surface than that within anticyclonic eddies [37].

The intensity of the eddy and the vertical structure of the water column can be influenced by typhoons [60], which in turn modify the subsequent typhoon and its associated ocean responses (Figure 1). Wu and Li [35] investigated sequential typhoons and found that the second typhoon can be weakened by the SST cooling induced by the first typhoon. Indeed, Linfa is intensified to category 4 during its propagation (Figure 1), whereas Nangka is persistently weak throughout its lifespan (Figures 1 and 3c,d). This phenomenon at least occurs because the cooling induced by typhoon Linfa inhibits typhoon Nangka from extracting energy from the ocean surface during its propagation [5]. It is particularly true for regions with pre-existing cyclonic eddies that typhoons intensify the strength of cyclonic eddies after their passage [59,61], leading to stronger upwelling and cooler SST (Figure 5). Besides the impact on the following typhoon, the substantial cooling can even reduce the intensity of the first typhoon as well [5]. In particular, the lingering of Linfa during its early stage can be highly impacted by the cooling induced by itself (Figure 3), which inhibits it from growing into a stronger typhoon (Figure 1).

The delayed CHL responses to the typhoons (Figure 5) indicate that the typhoon-induced nutrient supply is more important than the direct upwelling of subsurface CHL for driving CHL variability [62]. In the current study, the occurrence of a second peak induced by sequential typhoons is determined by the DWO and the nutrient availability at the corresponding depth. However, the second peak presents only if the induced DWO is deeper than that induced by the first typhoon (Figure 7) and there are extra nutrients remaining in the subsurface that are not mixed upward by the first typhoon (Figure 7a,c). Typhoon induced nutrient supply is particularly important to determine the chlorophyll bloom in the SCS as the nutricline is shallow and can be easily mixed up by typhoons [62]. It is important to emphasize that the calculation of DWO based on reanalyzed data is applicable for investigating a typhoon's impact because of the intensive dynamics that take place in a relatively short period [63]. The warming of upwelled water and usage of nutrients for phytoplankton growth can hardly occur during rapid mixing [10]. The proposed method is a valuable approach for quantitatively analyzing the impacts of typhoons on the upper ocean. Although DWO is acquired at the surface (Figure 7), the water does not necessarily need to outcrop at the surface to stimulate phytoplankton responses. For example, some extreme typhoons can induce subsurface CHL blooms even at depths near 100 m, which is greater than the depth of a surface bloom, by transporting nutrients to the euphotic layer [51].

A previous study revealed that oceanic responses, e.g., phytoplankton blooms, induced by Linfa can last for substantially long periods [33]. However, they ignored the occurrence of typhoon Nangka and the impacts of sequential typhoons (Figure 3). In the current study, we evaluate the responses of the ocean to both typhoons and improve our understanding of the corresponding dynamics. The CHL actually returns to its initial condition very rapidly (Figure 5), e.g., 5 days for each bloom, which can be attributed to regional dynamics and biological processes [59]. In particular, intense radiation warms the ocean within the following three days, and the MLD can be even shallower than that under pre-typhoon conditions (Figure 6). Additionally, restratification at the surface (Figure 5) reduces the nutrient supply and limits phytoplankton growth [16,64], and simultaneous microzooplankton grazing may consume phytoplankton [65], resulting in a rapid decrease in CHL (Figure 4). A comprehensive air–sea interaction model [57] equipped with a biogeochemical cycle can be used in the future to improve our understanding of the regional dynamics and ecological processes [66].

5. Conclusions

Satellite observations clearly reveal the role of eddies in modulating typhoon-induced SST cooling and CHL enhancement in the SCS, while the underlying dynamics are largely dependent of the three-dimensional ocean status. Two sequential typhoons passed over the SCS where four pre-existing eddies occurred; accordingly, this scenario offers an important and unique case study to improve our understanding of typhoon dynamics. Cyclonic eddies can enhance cooling and phytoplankton blooms, while anticyclonic eddies can depress the ocean response. The mechanism of typhoon-induced changes is determined by the relative depths of the MLD, thermocline and nutricline. The DWO is subsequently designed to quantitatively estimate typhoon-induced oceanic responses. The first typhoon can override the impact induced by the second typhoon unless the latter can induce vertical mixing to even greater DWO. The tandem occurrence of typhoons and eddies is greatly important for changing the nutrient supply to the ocean surface and modifying primary production. The proposed method combines satellite observations and reanalysis datasets to offer a comprehensive and quantitative description of typhoon-induced upper ocean responses.

Author Contributions: Visualization, W.J.; conceptualization, C.L., H.L.; investigation, J.H., Q.M., F.L.; data curation, X.C., X.L.; writing—original draft preparation, W.J., Y.W.; writing—review and editing, C.L., Y.W. All authors have read and agreed to the published version of the manuscript.

Funding: The study was funded by the National Natural Science Foundation of China [nos. 41806026], the Strategic Priority Research Program, Chinese Academy of Sciences [no. XDB42040202], Scientific Research Fund of the Second Institute of Oceanography, Ministry of Natural Resources [no. HYGG2002] and the Project of State Key Laboratory of Satellite Ocean Environment Dynamics, Second Institute of Oceanography, MNR [nos. SOEDZZ1902 and SOEDZZ1904].

Acknowledgments: We are very thankful to the Japan Meteorological Agency (JMA) for providing typhoon information, the National Aeronautics and Space Administration (NASA), the National Oceanic and Atmospheric Administration (NOAA), the European Space Agency (ESA) and the European Organisation for the Exploitation of Meteorological Satellites (EUMETSAT) for producing the satellite dataset, the Copernicus Marine Service (CMEMS) for sharing sea level anomaly (SLA) data and the Navy Coupled Ocean Data Assimilation (NCODA) system for providing the HYCOM dataset.

Conflicts of Interest: The authors declare no conflict of interest.

References

1. Price, J.F. Upper ocean response to a hurricane. *J. Phys. Oceanogr.* **1981**, *11*, 153–175. [CrossRef]
2. Lin, S.; Zhang, W.Z.; Shang, S.P.; Hong, H.S. Ocean response to typhoons in the western North Pacific: Composite results from Argo data. *Deep-Sea Res. Part I* **2017**, *123*, 62–74. [CrossRef]
3. Zhao, H.; Tang, D.; Wang, Y. Comparison of phytoplankton blooms triggered by two typhoons with different intensities and translation speeds in the South China Sea. *Mar. Ecol. Prog. Ser.* **2008**, *365*, 57–65. [CrossRef]
4. Babin, S.M.; Carton, J.A.; Dickey, T.D.; Wiggert, J.D. Satellite evidence of hurricane induced phytoplankton blooms in an oceanic desert. *J. Geophys. Res.* **2004**, *109*, C03043. [CrossRef]
5. Glenn, S.; Miles, T.; Seroka, G.; Xu, Y.; Forney, R.; Yu, F.; Roarty, H.; Schofield, O.; Kohut, J. Stratified coastal ocean interactions with tropical cyclones. *Nat. Commun.* **2016**, *7*, 10887. [CrossRef]
6. Cheung, H.F.; Pan, J.; Gu, Y.; Wang, Z. Remote-sensing observation of ocean responses to Typhoon Lupit in the northwest Pacific. *Int. J. Remote Sens.* **2013**, *34*, 1478–1491. [CrossRef]
7. Emanuel, K.A. Thermodynamic control of hurricane intensity. *Nature* **1999**, *401*, 665–669. [CrossRef]

8. Wang, S.; Li, S.; Hu, J.; Geng, B. Experiments in optimizing simulations of the subsurface chlorophyll maximum in the South China Sea. *J. Mar. Syst.* **2016**, *156*, 1–15. [CrossRef]
9. Wu, R.; Zhang, H.; Chen, D.; Li, C.; Lin, J. Impact of Typhoon Kalmaegi (2014) on the South China Sea: Simulations using a fully coupled atmosphere-ocean-wave model. *Ocean Model* **2018**, *131*, 132–151. [CrossRef]
10. Zhang, H.; Liu, X.; Wu, R.; Liu, F.; Yu, L.; Shang, X.; Qi, Y.; Wang, Y.; Song, X.; Xie, X. Ocean Response to Successive Typhoons Sarika and Haima (2016) Based on Data Acquired via Multiple Satellites and Moored Array. *Remote Sens.* **2019**, *11*, 2360. [CrossRef]
11. Gierach, M.M.; Subrahmanyam, B. Biophysical responses of the upper ocean to major Gulf of Mexico hurricanes in 2005. *J. Geophys. Res. Oceans* **2008**, *113*, C04029. [CrossRef]
12. Price, J.F.; Weller, R.A.; Pinkel, R. Diurnal cycling: Observations and models of the upper ocean response to diurnal heating, cooling, and wind mixing. *J. Geophys. Res.* **1986**, *91*, 8411–8427. [CrossRef]
13. Sun, L.; Yang, Y.; Xian, T.; Lu, Z.; Fu, Y. Strong enhancement of chlorophyll a concentration by a weak typhoon. *Mar. Ecol. Prog. Ser.* **2010**, *404*, 39–50. [CrossRef]
14. Zhao, H.; Pan, J.; Han, G.; Devlin, A.T.; Zhang, S.; Hou, Y. Effect of a fast-moving tropical storm Washi on phytoplankton in the northwestern South China Sea. *J. Geophys. Res. Oceans* **2017**, *122*, 3404–3416. [CrossRef]
15. Mei, W.; Lien, C.C.; Lin, I.I.; Xie, S.P. Tropical cyclone-induced ocean response: A comparative study of the South China Sea and Tropical Northwest Pacific. *J. Clim.* **2015**, *28*, 5952–5968. [CrossRef]
16. Pan, G.; Chai, F.; Tang, D.; Wang, D. Marine phytoplankton biomass responses to typhoon events in the South China Sea based on physical-biogeochemical model. *Ecol. Modell.* **2017**, *356*, 38–47. [CrossRef]
17. Zhang, H.; Wu, R.; Chen, D.; Liu, X.; He, H.; Tang, Y.; Ke, D.; Shen, Z.; Li, J.; Xie, J.; et al. Net modulation of upper ocean thermal structure by Typhoon Kalmaegi (2014). *J. Geophys. Res. Oceans* **2018**, *123*, 7154–7171. [CrossRef]
18. Han, G.; Ma, Z.; Chen, N. Hurricane Igor impacts on the stratification and phytoplankton bloom over the Grand Banks. *J. Mar. Syst.* **2012**, *100*, 19–25. [CrossRef]
19. Tang, S.; Dong, Q.; Liu, F. Climate-driven chlorophyll-a concentration interannual variability in the South China Sea. *Theor. Appl. Climatol.* **2011**, *103*, 229–237. [CrossRef]
20. Zhang, W.Z.; Wang, H.; Chai, F.; Qiu, G. Physical drivers of chlorophyll variability in the open South China Sea. *J. Geophys. Res. Oceans* **2016**, *121*, 7123–7140. [CrossRef]
21. Geider, R.J.; MacIntyre, H.L.; Kana, T. A dynamic regulatory model of phytoplanktonic acclimation to light, nutrients, and temperature. *Limnol. Oceanogr.* **1998**, *43*, 679–694. [CrossRef]
22. Yu, Y.; Xing, X.; Liu, H.; Yuan, Y.; Wang, Y.; Chai, F. The variability of chlorophyll-a and its relationship with dynamic factors in the basin of the South China Sea. *J. Mar. Syst.* **2019**, *200*, 103230. [CrossRef]
23. Pan, J.; Sun, Y. Estimate of Ocean Mixed Layer Deepening after a Typhoon Passage over the South China Sea by Using Satellite Data. *J. Phys. Oceanogr.* **2013**, *43*, 498–506. [CrossRef]
24. Chelton, D.B.; Schlax, M.G.; Samelson, R.M. Global observations of nonlinear mesoscale eddies. *Prog. Oceanogr.* **2011**, *91*, 167–216. [CrossRef]
25. Walker, N.D.; Leben, R.R.; Balasubramanian, S. Hurricane-forced upwelling and chlorophyll a enhancement within cold-core cyclones in the Gulf of Mexico. *Geophys. Res. Lett.* **2005**, *32*, 109–127. [CrossRef]
26. Zheng, Z.-W.; Ho, C.-R.; Zheng, Q.; Lo, Y.-T.; Kuo, N.-J.; Gopalakrishnan, G. Effects of preexisting cyclonic eddies on upper ocean responses to Category 5 typhoons in the western North Pacific. *J. Geophys. Res.* **2010**, *115*, C09013. [CrossRef]
27. Shi, W.; Wang, M. Satellite observations of asymmetrical physical and biological responses to Hurricane Earl. *Geophys. Res. Lett.* **2011**, *38*, 2–6. [CrossRef]
28. Lin, I.I. Typhoon-induced phytoplankton blooms and primary productivity increase in the western North Pacific subtropical ocean. *J. Geophys. Res.* **2012**, *117*, C03039. [CrossRef]
29. Lu, Z.; Gan, J.; Dai, M.; Zhao, X.; Hui, C.R. Nutrient transport and dynamics in the South China Sea: A modeling study. *Prog. Oceanogr.* **2020**, *183*, 102308. [CrossRef]
30. Wang, G.; Su, J.; Ding, Y.; Chen, D. Tropical cyclone genesis over the south China sea. *J. Mar. Syst.* **2007**, *68*, 318–326. [CrossRef]
31. Wang, Y.; Yu, Y.; Zhang, Y.; Chai, F. Distribution and variability of sea surface temperature fronts in the South China Sea. *Estuar. Coast. Shelf Sci.* **2020**, *240*, 106793. [CrossRef]
32. Hu, J.; Kawamura, H.; Hong, H.; Qi, Y. A Review on the Currents in the South China Sea: Seasonal Circulation, South China Sea Warm Current and Kuroshio Intrusion. *J. Oceanogr.* **2000**, *56*, 607–624. [CrossRef]

33. Xiu, P.; Chai, F.; Shi, L.; Xue, H.; Chao, Y. A census of eddy activities in the South China Sea during 1993–2007. *J. Geophys. Res.* **2010**, *115*, C03012. [CrossRef]
34. Sun, L.; Li, Y.X.; Yang, Y.J.; Wu, Q.; Chen, X.T.; Li, Q.Y.; Li, Y.-B.; Xian, T. Effects of super typhoons on cyclonic ocean eddies in the western North Pacific: A satellite data-based evaluation between 2000 and 2008. *J. Geophys. Res. Oceans* **2014**, *119*, 5585–5598. [CrossRef]
35. Wu, R.; Li, C. Upper ocean response to the passage of two sequential typhoons. *Deep Sea Res. Part I* **2018**, *132*, 68–79. [CrossRef]
36. Chen, Y.Q.; Tang, D.L. Eddy-feature phytoplankton bloom induced by a tropical cyclone in the South China Sea. *Int. J. Remote Sens.* **2012**, *33*, 7444–7457. [CrossRef]
37. Liu, F.; Tang, S. Influence of the interaction between typhoons and oceanic mesoscale eddies on phytoplankton blooms. *J. Geophys. Res. Oceans* **2018**, *123*, 2785–2794. [CrossRef]
38. Huang, S.M.; Oey, L.Y. Right-side cooling and phytoplankton bloom in the wake of a tropical cyclone. *J. Geophys. Res. Oceans* **2015**, *120*, 5735–5748. [CrossRef]
39. Jyothi, L.; Joseph, S. Surface and sub-surface ocean response to Tropical Cyclone Phailin: Role of pre-existing oceanic features. *J. Geophys. Res. Oceans* **2019**, *124*, 6515–6530. [CrossRef]
40. Vidya, P.J.; Santosh, D.; Mani Murali, R. Contrasting Chl-a responses to the tropical cyclones Thane and Phailin in the Bay of Bengal. *J. Mar. Syst.* **2017**, 103–114. [CrossRef]
41. Chassignet, E.P.; Hurlburt, H.E.; Smedstad, O.M.; Halliwell, G.R.; Hogan, P.J.; Wallcraft, A.J.; Baraille, R.; Bleck, R. The HYCOM (Hybrid Coordinate Ocean Model) data assimilative system. *J. Mar. Syst.* **2007**, *65*, 60–83. [CrossRef]
42. Cummings, J.A.; Smedstad, O.M. *Variational Data Assimilation for the Global Ocean. Data Assimilation for Atmospheric, Oceanic and Hydrologic Applications*; Springer: Berlin/Heidelberg, Germany, 2013; Volume II, pp. 303–343. [CrossRef]
43. Zhang, W.Z.; Hong, H.S.; Yan, X.H. Typhoons enhancing northward transport through the Taiwan Strait. *Cont. Shelf Res.* **2013**, *56*, 13–25. [CrossRef]
44. Sanford, T.B.; Price, J.F.; Girton, J.B. Upper ocean response to Hurricane Frances (2004) observed by profiling EM-APEX floats. *J. Phys. Oceanogr.* **2011**, *41*, 1041–1056. [CrossRef]
45. Obata, A.; Ishizaka, J.; Endoh, M. Global verification of critical depth theory for phytoplankton bloom with climatological in situ temperature and satellite ocean color data. *J. Geophys. Res.* **1996**, *101*, 20657–20667. [CrossRef]
46. Van Wambeke, F.; Ghiglione, J.; Nedoma, J.; Mevel, G.; Raimbault, P. Short scale variations in nutrients, ectoenzymatic activities and bottom-up effects on bacterial production and community structure during late summer-autumn transition in the open NW Mediterranean Sea. *Biogeosci. Discuss.* **2009**, *6*, 687–727. [CrossRef]
47. Emanuel, K.A. Contribution of tropical cyclones to meridional heat transport by the oceans. *J. Geophys. Res.* **2001**, *106*, 14771–14782. [CrossRef]
48. Lin, I.; Liu, W.T.; Wu, C.C.; Wong, G.T.F.; Hu, C.; Chen, Z.; Liang, W.D.; Yang, Y.; Liu, K.K. New evidence for enhanced ocean primary production triggered by tropical cyclone. *Geophys. Res. Lett.* **2003**, *30*, 1718. [CrossRef]
49. Sriver, R.L.; Huber, M. Observational evidence for an ocean heat pump induced by tropical cyclones. *Nature* **2007**, *447*, 577–580. [CrossRef]
50. Shibano, R.S.; Yamanaka, Y.; Okada, N.; Chuda, T.; Suzuki, S.; Niino, H. Responses of marine ecosystem to typhoon passages in the western subtropical North Pacific. *Geophys. Res. Lett.* **2011**, *38*, L18608. [CrossRef]
51. Ye, H.J.; Sui, Y.; Tang, D.L.; Afanasyev, Y.D. A Subsurface Chlorophyll a Bloom Induced by Typhoon in the South China Sea. *J. Mar. Syst.* **2013**, *128*, 138–145. [CrossRef]
52. Xu, F.; Yao, Y.; Oey, L.; Lin, Y. Impacts of pre-existing ocean cyclonic circulation on sea surface chlorophyll-a concentrations off northeastern Taiwan following episodic typhoon passages. *J. Geophys. Res. Oceans* **2017**, *122*, 6482–6497. [CrossRef]
53. Zheng, G.M.; Tang, D.L. Offshore and nearshore chlorophyll increases induced by typhoon winds and subsequent terrestrial rainwater runoff. *Mar. Ecol. Prog. Ser.* **2007**, *333*, 61–74. [CrossRef]
54. Lin, I.I.; Chun-Chieh, W.U.; Emanuel, K.A.; Lee, I.H.; Wu, C.R.; Pun, I.F. The interaction of super typhoon Maemi (2003) with a warm ocean eddy. *Mon. Weather Rev.* **2005**, *133*, 2635–2649. [CrossRef]

55. Gaube, P.; Chelton, D.B.; Samelson, R.M.; Schlax, M.G.; Oneill, L.W. Satellite Observations of Mesoscale Eddy-Induced Ekman Pumping. *J. Phys. Oceanogr.* **2015**, *45*, 104–132. [CrossRef]
56. Sun, W.; Dong, C.; Tan, W.; Liu, Y.; He, Y.; Wang, J. Vertical Structure Anomalies of Oceanic Eddies and Eddy-Induced Transports in the South China Sea. *Remote Sens.* **2018**, *10*, 795. [CrossRef]
57. Yue, X.; Zhang, B.; Liu, G.; Li, X.; Zhang, H.; He, Y. Upper Ocean Response to Typhoon Kalmaegi and Sarika in the South China Sea from Multiple-Satellite Observations and Numerical Simulations. *Remote Sens.* **2018**, *10*, 348. [CrossRef]
58. Ning, J.; Xu, Q.; Zhang, H.; Wang, T.; Fan, K. Impact of Cyclonic Ocean Eddies on Upper Ocean Thermodynamic Response to Typhoon Soudelor. *Remote Sens.* **2019**, *11*, 938. [CrossRef]
59. Liu, Y.; Tang, D.; Evgeny, M. Chlorophyll Concentration Response to the Typhoon Wind-Pump Induced Upper Ocean Processes Considering Air-Sea Heat Exchange. *Remote Sens.* **2019**, *11*, 1825. [CrossRef]
60. Lu, Z.; Wang, G.; Shang, X. Response of a preexisting cyclonic ocean eddy to a typhoon. *J. Phys. Oceanogr.* **2016**, *46*, 2403–2410. [CrossRef]
61. Yang, Y.J.; Sun, L.; Liu, Q.; Xian, T.; Fu, Y. The biophysical responses of the upper ocean to the typhoons Namtheun and Malou in 2004. *Int. J. Remote Sens.* **2010**, *31*, 4559–4568. [CrossRef]
62. Pan, J.Y.; Huang, L.; Devlin, A.T.; Lin, H. Quantification of Typhoon-Induced Phytoplankton Blooms Using Satellite Multi-Sensor Data. *Remote Sens.* **2018**, *10*, 318. [CrossRef]
63. Zhang, S.; Xie, L.; Hou, Y.; Zhao, H.; Qi, Y.; Yi, X. Tropical storm-induced turbulent mixing and chlorophyll-a enhancement in the continental shelf southeast of Hainan Island. *J. Mar. Syst.* **2014**, *129*, 405–414. [CrossRef]
64. Baranowski, D.B.; Flatau, P.J.; Chen, S.; Black, P.G. Upper ocean response to the passage of two sequential typhoons. *Ocean Sci.* **2014**, *10*, 559–570. [CrossRef]
65. Zhou, L.; Tan, Y.; Huang, L.; Huang, J.; Liu, H.; Lian, X. Phytoplankton growth and microzooplankton grazing in the continental shelf area of northeastern South China Sea after Typhoon Fengshen. *Cont. Shelf Res.* **2011**, *31*, 1663–1671. [CrossRef]
66. Chai, F.; Johnson, K.S.; Claustre, H.; Xing, X.; Wang, Y.; Boss, E.; Riser, S.; Fennel, K.; Schofield, O.; Sutton, A. Monitoring ocean biogeochemistry with autonomous platforms. *Nat. Rev. Earth Environ.* **2020**, 315–326. [CrossRef]

remote sensing MDPI

Article

Nature versus Humans in Coastal Environmental Change: Assessing the Impacts of Hurricanes Zeta and Ida in the Context of Beach Nourishment Projects in the Mississippi River Delta

Qiang Yao [1], Marcelo Cancela Lisboa Cohen [1,2,*], Kam-biu Liu [1], Adriana Vivan de Souza [2] and Erika Rodrigues [1,2]

1 Department of Oceanography and Coastal Sciences, College of the Coast and Environment, Louisiana State University, Baton Rouge, LA 70803, USA; qyao4@lsu.edu (Q.Y.); kliu1@lsu.edu (K.-b.L.); erikarodrigues@ufpa.br (E.R.)

2 Graduate Program of Geology and Geochemistry, Federal University of Pará, Av. Perimentral 2651, Terra Firme, Belém 66077-530, PA, Brazil; adriana.vivan@terra.com.br

* Correspondence: mcohen@ufpa.br

Abstract: Hurricanes are one of the most devastating earth surface processes. In 2020 and 2021, Hurricanes Zeta and Ida pounded the Mississippi River Delta in two consecutive years, devastated South Louisiana, and raised tremendous concerns for scientists and stakeholders around the world. This study presents a high-resolution spatial-temporal analysis incorporating planialtimetric data acquired via LIDAR, drone, and satellite to investigate the shoreline dynamics near Port Fourchon, Louisiana, the eye of Ida at landfall, before and after the beach nourishment project and recent hurricane landfalls. The remote sensing analysis shows that the volume of the ~2 km studied beachfront was reduced by 240,858 m^3 after consecutive landfalls of Hurricanes Zeta and Ida in 2020 and 2021, while 82,915 m^3 of overwash fans were transported to the backbarrier areas. Overall, the studied beach front lost almost 40% of its volume in 2019, while the average dune crest height was reduced by over 1 m and the shoreline retreated ~60 m after the two hurricane strikes. Our spatial-temporal dataset suggests that the Louisiana Coastal Protection and Restoration Authority's (CPRA's) beach nourishment effort successfully stabilized the beach barrier at Port Fourchon during the hurricane-quiescent years but was not adequate to protect the shoreline at the Mississippi River Delta from intense hurricane landfalls. Our study supports the conclusion that, in the absence of further human intervention, Bay Champagne will likely disappear completely into the Gulf of Mexico within the next 40 years.

Keywords: Hurricane Ida; Hurricane Zeta; LIDAR; Mississippi River Delta; satellite images; drone

1. Introduction

Hurricanes are one of the most devastating earth surface processes, causing over 10,000 deaths around the world each year [1]. For example, Hurricane Katrina (a Category 3 on the Saffir–Simpson Scale) made landfall near the Mississippi River Delta (MRD) on 29 August 2005 and killed over 1800 people in the USA [2]. In October 2020, another major hurricane, Zeta (Category 3), made landfall near the MRD. Ten months after Zeta, Hurricane Ida (Category 4) devastated the MRD again on the 16th anniversary of Hurricane Katrina. Along with Hurricane Laura (August 2020) and the Last Island Hurricane (1856), Ida was among the strongest hurricanes that struck the Louisiana coast in the last two centuries [3]. It also marks two major hurricane landfalls in Louisiana in two consecutive years (2020 and 2021) for the first time since ~1850s [4]. Moreover, instrumental records show that 90 hurricanes have made landfall near the Louisiana coast since the 1850s. Among them, 32 were major hurricanes (Categories 3–5) [4]. Thus, there is a vital and urgent need to document the geomorphological impacts of such extreme catastrophic events, particularly in populated coastal areas such as the MRD [5–7].

Furthermore, the combined effects of rising sea levels (~9.16 mm/year at the present) [8,9], subsidence [10,11], reduced sediment supply [12], and hurricanes are jeopardizing the stability of the MRD [13,14], whereas the shorelines are retreating at an alarming rate of up to 14 m/year [5,6,15,16]. The prevalent longshore currents carry much of these eroded sediments west from the MRD and deposit it along the Gulf coastlines in west Louisiana and east Texas [17]. To combat the coastal erosion and strengthen the Mississippi Delta shoreline, a series of restoration projects was implemented [18–20]. More recently, a beach nourishment project was undertaken by the Louisiana Coastal Protection and Restoration Authority (CPRA) in 2013 to replenish ~2.8×10^6 m^3 of sediments along ~10 km of shoreline near Port Fourchon (on the west side of the Caminada–Moreau headland) [21]. In addition, sand fences and dune vegetation (*Panicum amarum* and *Uniola paniculata*) were installed and planted along the supratidal zone to control the aeolian erosion and increase sand deposition [6,22,23]. Have these coastal restoration efforts been able to withstand the natural and anthropogenic disturbances and subsequentially protect the coastal wetlands (e.g., saltmarsh and mangrove) over the years, especially against major hurricanes such as Ida and Zeta? This question has not been adequately addressed, yet it has significant scientific implications, since, globally, sandy beach coastlines such as the one examined in this study have been under increasing threats of shoreline retreat and erosion [24,25]. Thus, large data gaps exist in the literature regarding the coastal stability at the MRD.

To fill these data gaps and provide the first look at Zeta and Ida's damage on the MRD, we investigated the coastal zones at Port Fourchon, the ground zero of Hurricane Ida at landfall (Figure 1). This study presents a high-resolution spatial-temporal analysis incorporating planialtimetric data acquired via LIDAR, drone, and satellite to investigate the shoreline changes and sand dune dynamics at Port Fourchon from 2002 to the present. The overarching goal of this study is to provide a robust baseline dataset for future coastal restoration efforts in hurricane-prone subtropical coastal regions. The two specific objectives are to (1) reveal the shoreline dynamics near Port Fourchon before and after the beach nourishment project; (2) investigate the impacts of Hurricane Ida and Zeta on the MRD shoreline.

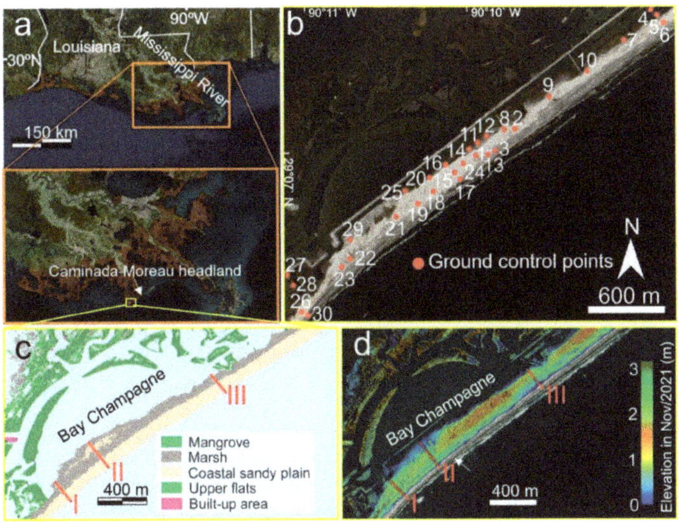

Figure 1. (**a**) Satellite image showing the study region; (**b**) Orthoimage marking the locations of the 30 ground control points; (**c**) Vegetation units and (**d**) Digital Elevation Model of the study area (November 2021). The red lines mark the cross-shore profile I–III. The GPS of each ground control point is listed in Table 1.

Table 1. Ground Control Points of the study area with longitude, latitude, elevation, and differences (m) between latitudes, longitudes, and elevation obtained by photogrammetry and those obtained in the field by a topographic survey.

GCP	X/Longitude	Y/Latitude	Z/Elevation (m)	X_{DIF}	Y_{DIF}	Z_{DIF}
1	−90.169318	29.117012	0.78	−0.014713	0.000967	0.04
2	−90.166061	29.119106	0.72	0.004428	0.004298	0.01
3	−90.167760	29.117485	0.45	−0.003994	0.000603	0.04
4	−90.154767	29.127904	0	0.066736	0.03	0.13
5	−90.154317	29.127531	0.45	0.088765	0.02	0.10
6	−90.153269	29.126278	0	0.088842	0.003677	0.038
7	−90.157144	29.124695	1.13	0.065826	0.009959	0.03
8	−90.166814	29.119035	0.6	0.034	0.303264	0.09
9	−90.164600	29.120375	1.18	0.018295	0.006667	0.03
10	−90.161136	29.122167	1.03	0.039914	0.006835	0.03
11	−90.168978	29.118352	0.27	−0.008730	0.008301	0.03
12	−90.168500	29.118745	0.21	−0.005413	0.009216	0.03
13	−90.168125	29.117093	0.20	−0.008472	−0.000703	0.02
14	−90.170295	29.117766	0.3	0.014279	−0.005867	−0.01
15	−90.171358	29.115810	1.0	0.002286	0.003633	−0.02
16	−90.172034	29.116723	0.61	0.003032	0.004055	0.003
17	−90.170875	29.115429	0.26	−0.000194	0.005523	−0.001
18	−90.173615	29.114195	0.75	0.002720	0.011527	−0.04
19	−90.176311	29.112635	0.78	0.006339	0.016190	0.003
20	−90.173905	29.115531	0.87	−0.022920	−0.040214	−0.002
21	−90.178845	29.110735	1.15	0.042	0.046	−0.02
22	−90.181842	29.108411	0.78	−0.636392	0.054	−0.04
23	−90.181485	29.108824	0.54	0.19	0.053	0.02
24	−90.169279	29.117025	0.78	0.008734	−0.000714	0.03
25	−90.176196	29.114711	0	−0.000529	−0.003827	0.05
26	−90.181666	29.109986	0	0.001820	−0.002098	0.08
27	−90.182765	29.111802	0	−0.006467	−0.008261	0.07
28	−90.182708	29.111594	0	0.005128	0.006270	0.03
29	−90.180144	29.111834	0.78	−0.130523	−0.783544	0.05
30	−90.181499	29.109870	0.58	0.769515	−0.681291	0.08

2. Geographic Background

2.1. Study Area

Our study area is located in Bay Champagne (between 29°09′–29°06′N and 90°11′–90°08′W) and the adjacent beach barrier (Figure 1). Bay Champagne is a shallow backbarrier lagoon situated to the southeast of Port Fourchon, ~80 km to the west of the Mississippi bird-foot delta. It is ~2 km^2 in size with a maximum depth of 2.5 m and an average salinity of ~30‰ [11]. The shoreline near Bay Champagne has retreated by ~2200 m since AD 1887 and has been retreating at ~10 m/year during the recent decades [16,26–28]. Currently, Bay Champagne is separated from the Gulf of Mexico (GOM) by a sand barrier of ~1–2 m above sea level (Figure 1d). The bay is surrounded by saltmarshes (*Spartina alterniflora*) and stunted (mostly < 1 m tall) black mangroves (*Avicennia germinans*), with a saltmarsh-mangrove ecotone on the north side of the lagoon [6] (Figure 1c). The beach barrier at Bay Champagne was regularly breached by storm surges caused by major hurricanes, including Lili (2002), Katrina (2005), Gustav (2008), Zeta (2020), and Ida (2021) [29,30].

2.2. Meteorological Data of Hurricane Ida and Zeta

Since the beginning of the 21st century, four major hurricanes have made landfall in close proximity to our study area [4]. The most recent one, Hurricane Ida, was formed in the southwest Caribbean Sea on 23 August 2021, and it rapidly intensified into a Category 1 hurricane on 27 August while making its way into the GOM across Cuba [3]. On 29 August 2021, the 16th anniversary of Hurricane Katrina, Ida made landfall at Port Fourchon, Louisiana as a Category 4 hurricane (Figure 2). At landfall, Ida had maximum sustained winds of ~240 km/h and caused up to 5 m of storm surges at Port Fourchon [3]. It was the 2nd and 5th strongest landfalling hurricane in Louisiana and the continental U.S. on the record. The severe winds, high storm surges, and extensive floods caused catastrophic damage to the coastal town of Port Fourchon and Grand Isle near the Mississippi Delta (Figure 2b–e). Ida also broke several rainfall records while moving across the USA Northeast. Overall, Ida caused over $65 billion in damages and a total of 115 deaths, including 33 total deaths in Louisiana [3].

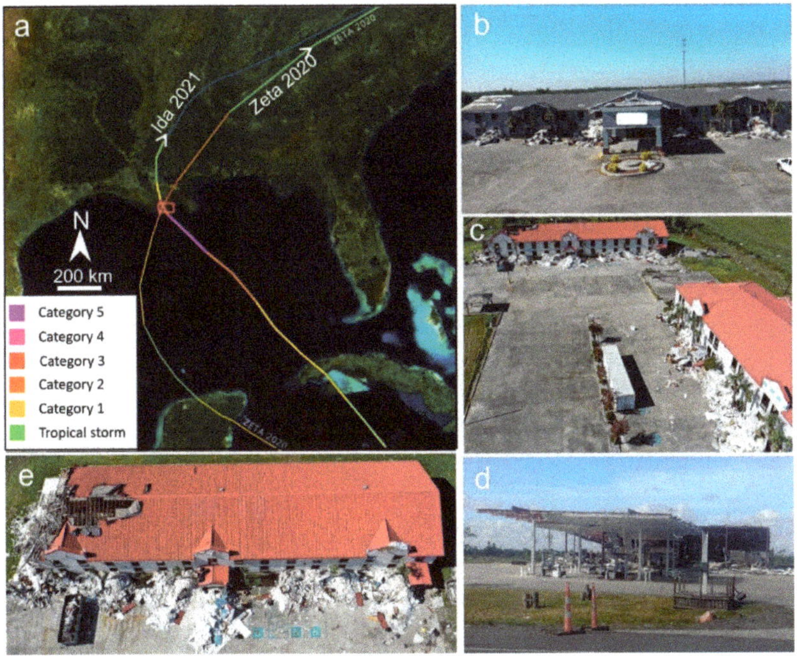

Figure 2. (**a**) Storm tracks (https://coast.noaa.gov/hurricanes, accessed on 10 May 2022) of Hurricane Ida and Zeta. The red box marks our study area, the coastal town of Port Fourchon. (**b**–**e**) Drone images showing Ida's impact on Port Fourchon.

Approximately ten months prior to Ida, another major hurricane, Zeta, made landfall at Cocodrie, Louisiana, ~50 km to the west of our study area, on 28 October 2020 (Figure 2). Zeta was a Category 3 hurricane at landfall, with maximum sustained winds of ~185 km/h. It caused ~3 m of storm surge inundation to the east of the MRD, especially in Mississippi and Alabama, and 1 to 2 m of storm surges at Port Fourchon [31].

3. Methods and Materials

3.1. Remote Sensing

The remote sensing analysis of the sand barrier dynamics at Bay Champagne utilized planialtimetric data and followed a three-phase methodology flow chart (Figure 3): (1) a spatial-temporal analysis based on satellite images and LIDAR data; (2) the production of a digital elevation model (DEM) based on ground-validated drone photogrammetry; (3) data integration to evaluate the impacts of Hurricane Ida on the beach barrier. All of the drone missions were conducted during low tide. More details regarding the three-phase methodology can be seen in [7].

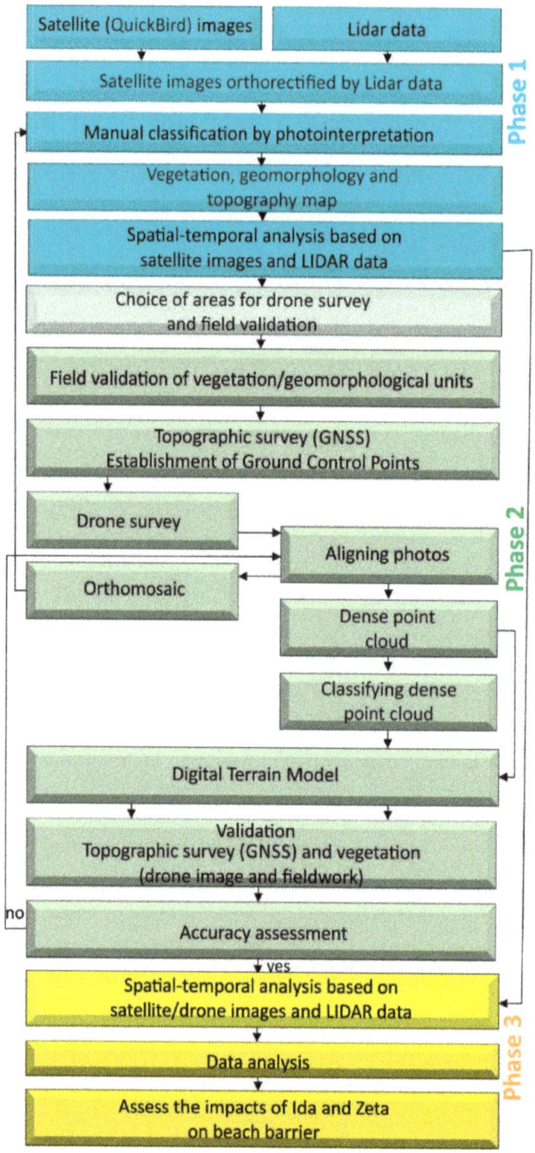

Figure 3. Methodological flow-chart of the spatial-temporal analysis.

Every remote sensing image was precisely orthorectified by the LIDAR model. QuickBird images (2.44 m multispectral ground pixel resolution with blue, green, and red band) of October/2002, April/2010, March/2013, October/2020, and September/2021 were processed with Agisoft Metashape Professional v.1.6.2. The LIDAR data for years 2002 (NOAA/NASA/USGS), 2010 (JALBTCX), and 2013 (USGS/NASA) were retrieved from the Atlas website from Louisiana State University (https://atlas.ga.lsu.edu/) (accessed on 21 November 2021) and National Oceanic and Atmospheric Administration (NOAA) website (https://coast.noaa.gov/dataviewer/#/) (accessed on 21 November 2021), and they had a horizontal and vertical accuracy of 73–100 cm and 10–15 cm, respectively. LIDAR data provided planialtimetric data to categorize the study area into ground, water, and vegetation.

3.2. Spatial-Temporal Analysis

The spatial-temporal analysis of years 2010 (before beach nourishment), 2013 (after beach nourishment), 2019 (before Zeta), 2020 (after Zeta), and 2021 (after Ida) was processed in the Global Mapper v.22.1 (Blue Marble Geographics, Hallowell, ME, USA). The planialtimetric analysis was completed with high-resolution images (2.6 cm/pixel) of November/2019 and November/2021 using Phantom 4 Advanced drone. Ground surveys were conducted during the drone surveys to validate the topography, vegetation heights/types, and shoreline dynamics between November/2019 and November/2021. Topographic surveys on the ground allowed the determination of the supratidal and intertidal zone limits during that time interval. The planialtimetric cross-shore profiles I–III (Figure 1c,d) were retrieved by using an electronic Self-Leveling Horizontal Rotary Laser, a Trimble Catalyst receiver, and a differential GPS system with decimeter correction (horizontal and vertical precision $\pm$ 10 cm). These absolute planialtimetric data, based on GEOID18, were used as Ground Control Points (GCPs, a total of 30 points) to improve and evaluate the image positions and to calibrate the DEM (Table 1). Sediment volumes were measured within a designated zone according to the elevation grid generated for each drone and Lidar survey, relative to a baseline, defined as mean sea-level (0 m) using Global Mapper software version 22.1. Volumetric calculations were executed by dividing the area of interest into small rectangular pieces following a uniform grid and then calculating the sum volume of the small 3D rectangles (Volume = Height $*$ Pixel Size) between terrain models and the cut surface [32].

The vertical differences (Z_{dif}) between low and high tide were less than 13 cm, revealing a vertical margin of error of $\pm$13 cm for the 3D models. The planimetric differences (latitude and longitude) were <0.78 m (Table 1). Considering the X_{dif}, Y_{dif}, and Z_{dif} values, margins of error were estimated at $\pm$0.079 m^3 and $\pm$0.15 m^3 for the volume calculations based on drone and Lidar data (vertical and horizontal accuracy of 15 and 100 cm), respectively.

GPSs of landmarks (e.g., bridges and houses) were used as stable reference lines for spatial-temporal analysis. The contour line of 0 m above the mean sea level was used as the lower limit of the intertidal zone to calculate the shoreline dynamics. Distance measurements were obtained by Global Mapper v.22.1 on georeferenced satellite and drone images. Considering the ground pixel resolution of 2.44 m of the QuickBird images, a margin of error of 2.5 m was estimated for the planimetric data obtained from these images. The vegetation and geomorphological features were manually classified by photointerpretation using the Global Mapper software. The dataset of sites with a previously identified land cover was applied to define the image features (multispectral digital numbers) related to the texture of drone orthoimages (spectral information between 380 and 710 nm) of each land cover type. Drone images (resolution of 2.6 cm) permitted the identification of mangroves, saltmarshes, and the sandy coastal barrier. Drone panoramic aerial photos were also used to identify the vegetation and geomorphological units. More details regarding drone image processing and 3-D model construction are described in the Supplementary Content Text S1.

4. Results

4.1. Beach Barrier Dynamics between 2019 and 2021

Three cross-shore profiles were established to compare the beach barrier dynamics during the study period (Figure 4a,b). Each profile was marked between two fixed reference points to calculate the vertical and horizontal shoreline change. Substantial changes were observed along all three cross-shore profiles, especially after the landfalls of Hurricanes Zeta and Ida (Figure 4c,d).

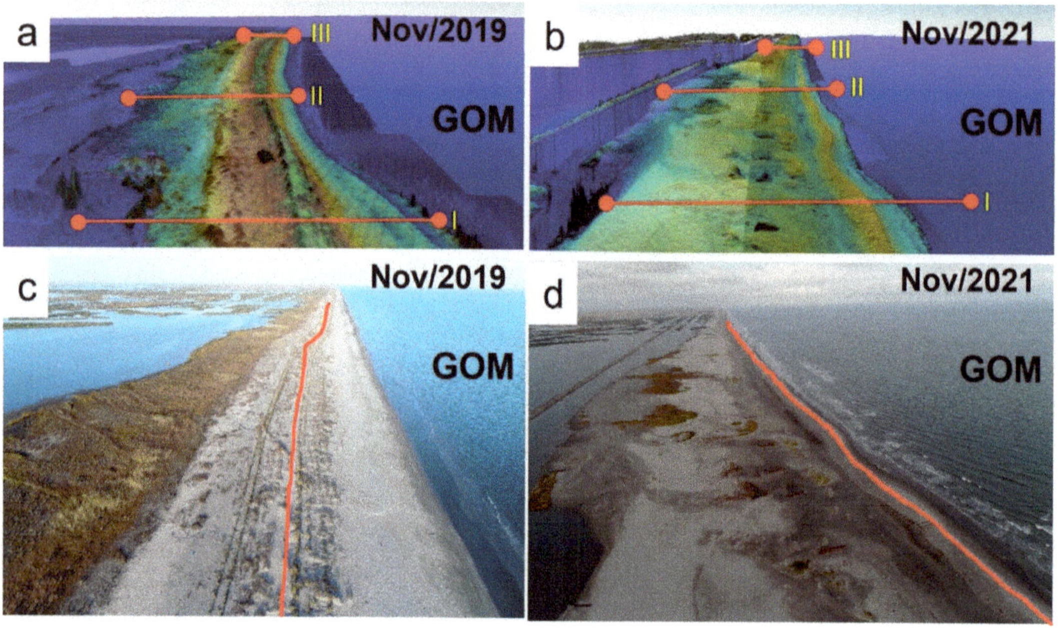

Figure 4. DEMs of the beach barrier in (**a**) 2019 and (**b**) 2021 (see location in Figure 1d). Drone images (taken during low-tide) of the studied beach barrier in (**c**) 2019 and (**d**) 2021. The red dots in (**a**,**b**) mark the fixed reference points that establish the three cross-shore profiles. The red lines in (**c**,**d**) mark the dune crest position in 2019.

Planialtimetric data (LIDAR and drone) show significant vertical and horizontal displacements of the dune crests between 2019 and 2021 along all three cross-shore profiles. The maximum dune heights in 2019 along the profiles I, II, and III were ~2.3, 2.2, and 2.1 m, but these numbers were reduced to ~1, 0.9, and 1 m in 2021, respectively (Figure 5). The average dune height was reduced by over 1 m after the two consecutive hurricane landfalls (Figure 5). Meanwhile, substantial erosion occurred along the sandy intertidal flats at the seaward edge of the beach barrier. For instance, the shoreline retreated for 62, 25, and 64 m along the profiles I, II, and III between 2019 and 2021 (Figure 5). Subsequently, the intertidal and supratidal portions of the cross-shore profiles (sum of the red areas in Figure 5) were reduced by 108,434 m^2 after Hurricane Zeta (between November/2019 and October/2020) and by 151,030 m^2 after Hurricane Ida (between October/2020 and September/2021). At the other end of the beach, the sandy backbarrier flats gained ~34,000 m^2 between November/2019 and October/2020 and ~100,872 m^2 between October/2020 and September/2021 (sum of the yellow areas in Figure 5).

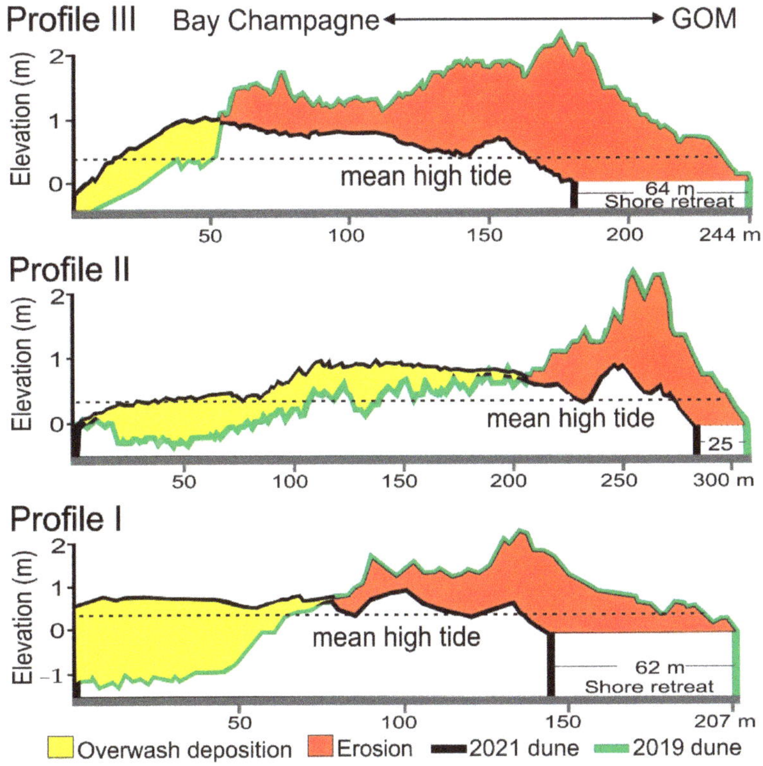

Figure 5. Planialtimetric cross-shore profiles I, II, and III (see location in Figure 1d) based on drone data. The dune crest positions along the three profiles were marked by green line in November/2019 and by black line in November/2021. The red and yellow areas mark the erosion and redeposition of the beach barrier sediments.

4.2. Spatial-Temporal Analysis of the Beach Barrier between 2002 and 2021

The total sediment volume of the beach barrier in our study area fluctuated between 2002 and 2021 (purple boxes in Figure 6). Before the CPRA beach nourishment project, the total floor area (the surface area of the studied beach barrier in the purple box) and the volume of the studied barrier were 32 ha and 329,628 m^3, respectively, in 2002 (Figure 6a). These numbers changed to 50.1 ha and 275,166 m^3 in 2010 (Figure 6b). Overall, the total floor area of the studied barrier gained 18.1 ha (+56.6%), but its volume was reduced by 54,462 m^3 (−16.5%) between 2002 and 2010. In particular, the average dune crest height was reduced by over 1 m (Figure 6a,b). The three red lines in Figure 6 mark the distance between three fixed reference points and the seaward edge of the beach barrier at the low tide. These lines were used to measure and compare the shoreline dynamics along the three cross-shore profiles during the study period. Between 2002 and 2010, the shoreline at cross-shore profiles I, II, and III retreated 142 m (−17.2%), 157 (−13.2%), and 210 m (−23.5%), respectively (Figure 6a,b). On average, the shoreline in the study area was retreating at a rate of 21.2 m/year between 2002 and 2010, over 50% higher than the previous estimation of 12−14 m/year between 1983 and 2014 (Yao et al., 2018; Cohen et al., 2021a; Dietz et al., 2018). Thus, the shoreline retreat was rapidly accelerating during the past decade.

At the start of the beach nourishment effort, the overall floor area and volume of the studied barrier were 41 ha and 304,583 m^3 in 2013 but then increased to 62.3 ha (+52%) and 646,533 m^3 (+112.3%) in 2019, respectively (Figure 6c,d). Meanwhile, the shoreline at profiles I, II, and III advanced 42 m (+6.2%), 56 m (+5.6%), and 40 m (+6.1%) between 2013 and

2019, respectively (Figure 6d). After being directly hit by two major hurricanes (Zeta and Ida) in consecutive years, these numbers decreased to 50.6 ha (−18.8%) and 405,675 m^3 (−37.3%) between 2019 and 2021, while the shoreline retreated 62 m (−9.1%), 24 m (−2.4%), and 64 m (−10.9%) at profiles I, II, and III, respectively (Figure 6e). On average, the shoreline was retreating at a rate of 25 m/year between 2019 and 2021, 18.4% higher than that prior to the beach nourishment effort.

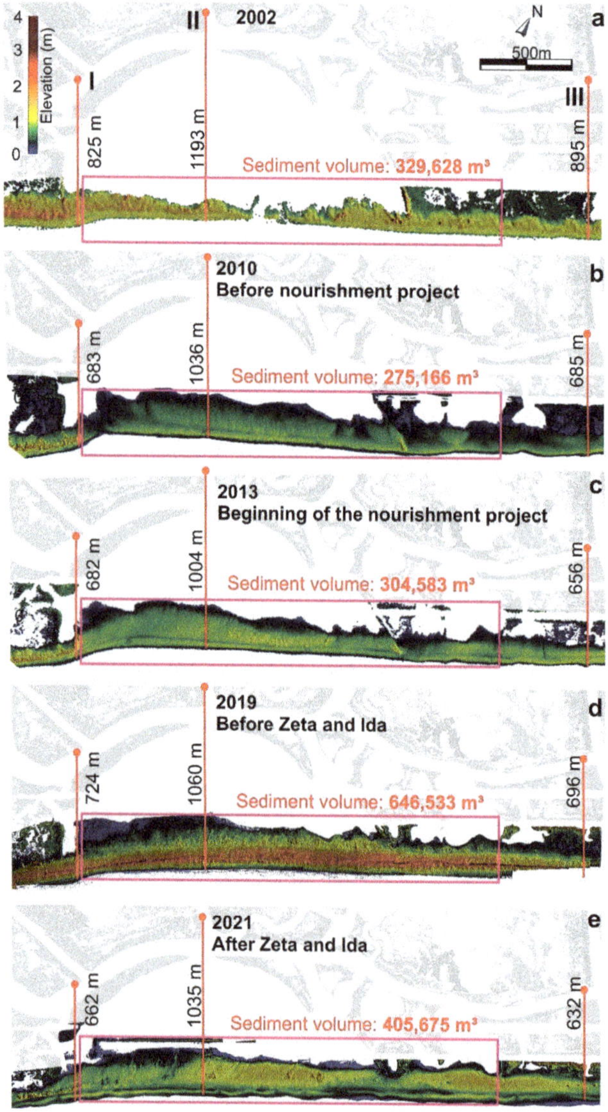

Figure 6. Spatial-temporal analysis of the beach barrier dynamics based on satellite and drone images between 2002 and 2021 (**a**–**e**). The purple boxes mark the areas used as references for sediment volume calculation. The red dots mark the three fixed reference points that were used to calculate the shoreline dynamics at planialtimetric cross-shore profiles I, II, and III. The three red lines in each figure mark the distance between the fixed reference points and the seaward edge of the beach barrier at the low tide.

5. Discussion

5.1. Shoreline Dynamics at Bay CHAMPAGNE Prior to Hurricanes Ida and Zeta

An analysis of the remote sensing images revealed dramatic shoreline fluctuations near Bay Champagne (Figure 6). Prior to the beach nourishment project, the shoreline in the study area retreated ~170 m (−18%) between 2002 and 2010 (average values of profiles I, II and III). However, after the beach nourishment project in 2013, the shoreline advanced (on average) ~46 m (6%) between 2013 and 2019. Thus, the remote sensing analysis indicates that the restoration effort was effective during the study period.

From a long-term perspective, historic maps from the Louisiana State University Cartographic Information Center show that Bay Champagne was a circular, inland lake with a direct tidal inlet connecting to the Gulf of Mexico in the late 19th century (Figure 7a). At that time, the lake was located approximately one kilometer inland from the GOM. By 1953, the shoreline had retreated to the edge of the lake, while the tidal inlet disappeared (Figure 7b). By 1978, nearly half of the original lake area had disappeared, and the shoreline moved landward for 500 m (Figure 7c). Over the next 32 years, two-thirds of the original lake area had disappeared, and the shoreline retreated another 240 m (Figure 7d). Thus, although the above shoreline movements were estimated based on a single reference point, it is clear that shoreline erosion was a continuous phenomenon along the studied beach barrier since the late 19th century, in line with previous studies of the Caminada–Moreau headland [15,16,28].

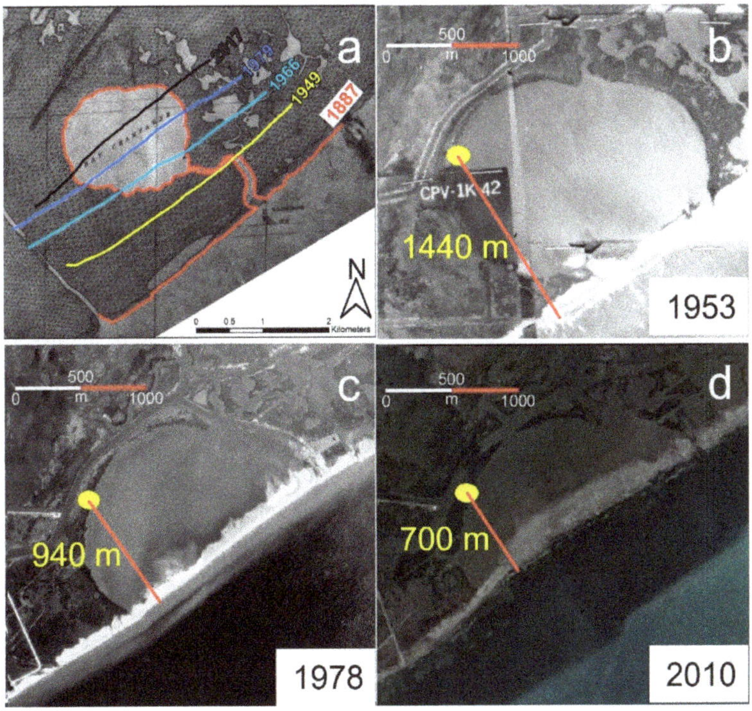

Figure 7. (**a**) Historic map (Dietz et al., 2022), (**b**,**c**) aerial photographs, and (**d**) satellite image showing rapid shoreline retreat at Bay Champagne since the late 19th century. The red lines perpendicular to the Shoreline show the distance between a fixed reference point (yellow circle) and the seaward edge of the beach on every satellite image. These lines were used as references to measure the shoreline retreat. Historic map and aerial photographs were retrieved from the Louisiana State University Cartographic Information Center.

Located to the west of the current Mississippi River Delta, the Caminada–Moreau headland was once part of the active Mississippi River Delta that was abandoned ~500 years ago when the delta lobe switched to the east [33–35]. Since then, erosional processes have been persistent and have caused significant land loss at the Caminada–Moreau headland [15,36], making it the region with the highest rate of shoreline retreat around the globe [36–38]. However, as our spatial-temporal analysis indicates (Figure 6c,d), CPRA's beach nourishment project (including sand replenishment, sand fence installation, and dune vegetation planting) had successfully stopped coastal erosion and stabilized the shoreline near Bay Champagne between 2013 and 2019, perhaps for the first time since the delta lobe was abandoned.

5.2. Hurricane Effects on Beach Barrier Dynamics

Before CPRA's beach nourishment project, the volume of this ~2 km long beachfront was disappearing at a rate of ~6808 m^3/year as a result of human and natural disturbances, and the shoreline was retreating at a rate of ~21 m/year (the average change of the three cross-shore profiles) between 2002 and 2010 (purple box in Figure 6a,b), when two intense hurricanes, Katrina (Category 4, 2005) and Gustav (Category 3, 2008), made landfalls nearby [29,39,40]. After the beach nourishment effort in 2013, the beachfront volume and shoreline width were increasing at a rate of ~56,992 m^3/year and ~15 m/year between 2013 and 2019 (Figure 6c,d), despite the frequent impacts of minor tropical storms but no major hurricane landfalls [4].

Over the past two decades, the major physical mechanisms driving coastal erosion at Bay Champagne, such as the relative sea-level rise and deltaic sediment compaction, have remained relatively constant [16,28]. Moreover, Atlantic hurricane activity has fluctuated during the 20th and early 21st centuries [41–44], and periods of frequent landfalling hurricanes are known to accelerate coastal erosion near the MRD [16]. Therefore, Hurricanes Katrina and Gustav have likely contributed to the rapid coastal erosion in our study area during the first decade of the 21st century (Figure 6a,b). However, the sand fences and planted dune vegetation have evidently reduced aeolian erosion and increased sand deposition after 2013. Thus, CPRA's restoration effort in our study area has effectively stabilized the coastal zone during the hurricane-quiescent years between 2013 and 2019.

Furthermore, after the two hurricane strikes by Hurricanes Zeta and Ida in consecutive years (2020 and 2021), the volume of the beachfront was reduced by 240,858 m^3, almost 40% of its volume in 2019, while the shoreline retreated ~60 m (Figure 8). Moreover, the average height of the dune crest across the studied beachfront was reduced by over 1 m (Figure 4c,d), while the transport of overwash sand (82,915 m^3) to backbarrier environments was recorded after both Zeta and Ida (Figure 8). All the evidence points to the removal and redistribution of beach sediments during these hurricane strikes [6,45]. Thus, it is clear that this beach nourishment effort was not able to withstand the direct landfall of intense hurricanes. If we consider the ~20-year return period for major (Categories 3−5) hurricane strikes in this area (NOAA, 2021b) and 40% loss of the beach volume after 2 major hurricane strikes, it is reasonable to assume that the entire beach barrier will be removed after several more major hurricane strikes. Moreover, Bay Champagne has lost over 45% of its surface since the 1970s as a result of long-term shoreline retreat [16]. Thus, our estimates support the conclusion that, at the current rate of shoreline recession averaged over the past four decades, Bay Champagne will disappear completely into the Gulf of Mexico within the next 40 years, if no further human intervention is undertaken [16,28]. Hence, it is possible to conclude that on a decadal timescale, the current restoration effort may not be adequate to compensate for the coastal erosion in this part of the MRD due to the likelihood of more frequent major hurricane strikes in the future [1].

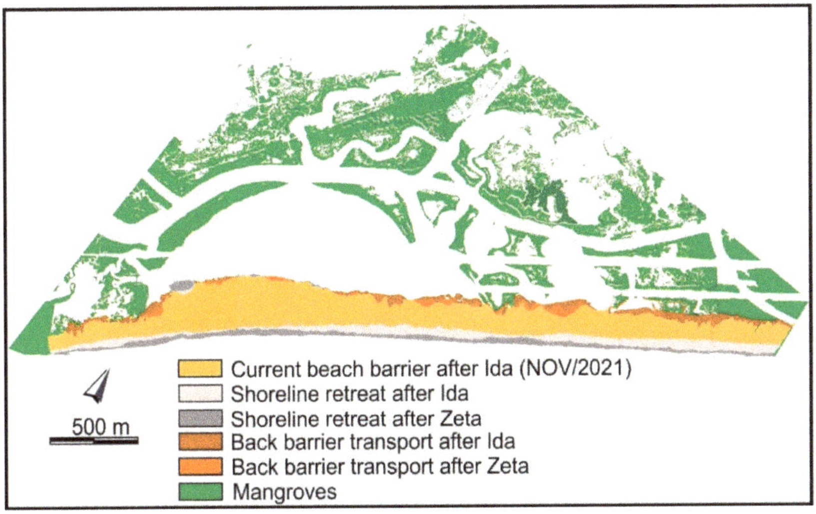

Figure 8. Composite diagram showing beach barrier dynamics associated with Hurricane Ida and Zeta.

5.3. Implications for Future Coastal Restoration Efforts

At the MRD, anthropogenic activities have become a primary factor of the land loss problem in the region. The construction of >50,000 dams and levees upstream of the Mississippi River not only has effectively stopped the natural overbank flooding, but also prevented river sediments from replenishing the wetlands [13,46]. Building of waterways and canals to support the petroleum industries has also led to increased saltwater intrusion and wetland fragmentation [39,47,48]. Moreover, the impoundments and other engineering activities such as soil erosion controls, bank revetments, and meander cutoffs have led to ~70% reduction of the fluvial sediment load since the 1850s [12,49–52] As a result of these human activities, the MRD has lost ~5000 km^2 of coastal wetlands and ~75% of the land area since the 1930s [20].

Figure 9 compares the beach barrier before and after Hurricanes Ida and Zeta. It is clear that the sand fences and dune vegetation were not able to withstand the two major hurricane landfalls (Figure 9). As the beach nourishment efforts were demolished by storms, Port Fourchon would once again become increasingly vulnerable to future hurricane impacts. In particular, when the beach nourishment efforts were intact, only direct landfalls of major hurricanes would have breached the beach barriers and transported marine water and sediment into Bay Champagne, as we discussed in the previous sections. After the average dune crest height was reduced by over 1 m (Figures 5 and 7), Bay Champagne would have become more susceptible to marine inundation through storm surge overwash processes [53]. As this cycle continues, our study area is expected to become more susceptible to hurricanes and human impacts, further accelerating the rate of shoreline retreat. What is the alternative solution for the rapidly retreating coast in the MRD? The short-term answer will be periodically repeating the beach nourishment efforts, especially after intense hurricane strikes. However, as the root of this coastal land loss issue is anthropogenic activities in the face of climate change and the rapid sea-level rise (~9.16 mm/year) in the area [9], repetitive beach nourishment efforts would become increasingly expensive and futile. Thus, Louisiana's coastal restoration effort is facing a dilemma that may ultimately lead to a tipping point.

Figure 9. Georeferenced drone images comparing the beach barriers near the three cross-shore profiles before and after Hurricane Ida and Zeta. The red circles mark the sand fences and planted dune vegetation prior to the two hurricanes.

An ongoing natural process that may help to reduce the rate of coastal erosion is the proliferation of mangroves in this part of the MRD. As demonstrated by previous studies, mangroves in the MRD are very resilient in withstanding intense hurricane impacts and in recovering from winter freeze damages [6,7,54,55]. More importantly, research from across the world has demonstrated that mangroves can stabilize the coastal zones by attenuating the wind and waves from storms and increasing vertical accretion in the coastal zone by trapping more sediments [56–58]. In fact, significant mangrove expansion has been reported near the MRD since the early 21st century due to warmer winters [7,59,60], and mangrove planting has already taken place near the MRD as a means of shoreline protection [61]. Thus, in the long term, such eco-engineering might be a viable solution to help save the Louisiana coast or other rapidly recessional shorelines across the tropics and subtropics.

6. Conclusions

This study indicates that the Louisiana CPRA's beach nourishment project (including sand replenishment, sand fence installation, and dune vegetation planting) has successfully stopped coastal erosion and stabilized the shoreline near Bay Champagne during the hurricane-quiescent years between 2013 and 2019. However, this beach nourishment effort was not able to withstand the landfalls of Hurricanes Zeta and Ida in two consecutive years. Thus, over the long term, the current restoration effort may not be adequate to compensate for the costal erosion in this part of the MRD due to the likelihood of more frequent major hurricane strikes in the future. However, we do acknowledge that this study was conducted immediately after the landfall of Hurricane Ida, whereas it is possible—albeit unlikely—that the studied beach barrier will be stabilized again with the recovery of the planted dune vegetation. Thus, future studies are needed to keep monitoring this dynamic shoreline near the Mississippi River Delta, particularly focusing on the post-hurricane recovery of the beach barrier. Our three-phase methodology (Figure 3) could provide a viable method for post-disaster assessment and an efficient approach to monitor the effectiveness of coastal restoration projects around the globe.

Supplementary Materials: The following supporting information can be downloaded at: https://www.mdpi.com/article/10.3390/rs14112598/s1. Text S1: 3-D Models.

Author Contributions: Q.Y. designed the study, led the sample collection, and wrote the paper. M.C.L.C. led the remote sensing analysis, contributed to writing and editing, and assisted in the fieldwork and vegetation survey. K.-b.L. directed the project and fieldwork and contributed to data interpretation, writing, and editing. A.V.d.S. assisted in the remote sensing analysis. E.R. assisted in field sampling and laboratory analysis. All authors have read and agreed to the published version of the manuscript.

Funding: This research was funded by the U.S. National Science Foundation (grant # 1759715), the Brazilian National Council for Technology and Science (CNPq 07497/2018−6, 403239/2021−4), Research Funding Agency of the State of São Paulo (FAPESP 2020/13715−1).

Data Availability Statement: Datasets produced in this article will be stored at the Neotoma Paleoecology Database (https://www.neotomadb.org (accessed on 26 May 2022)) and accessible to the public for free, upon the publication of this study.

Acknowledgments: We thank the Louisiana State University Cartographic Information Center for providing the historic map and aerial photographs of the study area. We also thank LSU Libraries for their Open Access Author fund.

Conflicts of Interest: The authors declare no conflict of interest.

References

1. Emanuel, K. Increasing destructiveness of tropical cyclones over the past 30 years. *Nature* **2005**, *436*, 686–688. [CrossRef] [PubMed]
2. Blake, E.S.; Landsea, C.; Gibney, E.J. The Deadliest, Costliest, and Most Intense United States Tropical Cyclones from 1851 to 2010 (and Other Frequently Requested Hurricane Facts). Available online: https://repository.library.noaa.gov/view/noaa/6929 (accessed on 27 April 2022).
3. National Oceanic and Atmospheric Administration (NOAA); National Hurricane Center. Hurricane IDA Advisory Archive. Available online: https://www.nhc.noaa.gov/archive/2021/IDA.shtml? (accessed on 20 October 2021).
4. National Oceanic and Atmospheric (NOAA). Historical Hurricane Tracks. Available online: https://coast.noaa.gov/hurricanes/#map=4/32/-80 (accessed on 28 October 2021).
5. Yao, Q.; Liu, K.-B.; Ryu, J. Multi-proxy Characterization of Hurricanes Rita and Ike Storm Deposits in the Rockefeller Wildlife Refuge, Southwestern Louisiana. *J. Coast. Res.* **2018**, *85*, 841–845. [CrossRef]
6. Cohen, M.; de Souza, A.; Liu, K.-B.; Rodrigues, E.; Yao, Q.; Pessenda, L.; Rossetti, D.; Ryu, J.; Dietz, M. Effects of Beach Nourishment Project on Coastal Geomorphology and Mangrove Dynamics in Southern Louisiana, USA. *Remote Sens.* **2021**, *13*, 2688. [CrossRef]
7. Cohen, M.C.; de Souza, A.V.; Liu, K.-B.; Rodrigues, E.; Yao, Q.; Ryu, J.; Dietz, M.; Pessenda, L.C.; Rossetti, D. Effects of the 2017–2018 winter freeze on the northern limit of the American mangroves, Mississippi River delta plain. *Geomorphology* **2021**, *394*, 107968. [CrossRef]
8. Jankowski, K.L.; Törnqvist, T.E.; Fernandes, A.M. Vulnerability of Louisiana's coastal wetlands to present-day rates of relative sea-level rise. *Nat. Commun.* **2017**, *8*, 14792. [CrossRef]
9. National Oceanic and Atmospheric (NOAA). Tides and Currents. Available online: https://tidesandcurrents.noaa.gov/sltrends/sltrends_station.shtml?id=8761724 (accessed on 18 October 2021).
10. Zou, L.; Kent, J.; Lam, N.S.-N.; Cai, H.; Qiang, Y.; Li, K. Evaluating Land Subsidence Rates and Their Implications for Land Loss in the Lower Mississippi River Basin. *Water* **2015**, *8*, 10. [CrossRef]
11. Coleman, J.M.; Roberts, H.H.; Stone, G.W. Mississippi river delta: An overview. *J. Coast. Res.* **1998**, *14*, 698–716.
12. Blum, M.D.; Roberts, H.H. Drowning of the Mississippi Delta due to insufficient sediment supply and global sea-level rise. *Nat. Geosci.* **2009**, *2*, 488–491. [CrossRef]

13. Lam, N.S.-N.; Cheng, W.; Zou, L.; Cai, H. Effects of landscape fragmentation on land loss. *Remote Sens. Environ.* **2018**, *209*, 253–262. [CrossRef]
14. Zhang, W.; Xu, Y.J.; Guo, L.; Lam, N.S.-N.; Xu, K.; Yang, S.; Yao, Q.; Liu, K.-B. Comparing the Yangtze and Mississippi River Deltas in the light of coupled natural-human dynamics: Lessons learned and implications for management. *Geomorphology* **2021**, *399*, 108075. [CrossRef]
15. Penland, S.; Zganjar, C.; Westphal, K.A.; Connor, P.; List, J.; Williams, S.J. *Shoreline Changes in the Caminada-Moreau Headland and Grand Isle—1887 to 1996 Lafourche and Jefferson Parishes, Louisiana*; United States Geological Survey: Lafayette, LA, USA, 2003.
16. Dietz, M.E.; Liu, K.-B.; Bianchette, T.A. Hurricanes as a Major Driver of Coastal Erosion in the Mississippi River Delta: A Multi-Decadal Analysis of Shoreline Retreat Rates at Bay Champagne, Louisiana (USA). *Water* **2018**, *10*, 1480. [CrossRef]
17. Roberts, H.H. Dynamic Changes of the Holocene Mississippi River Delta Plain: The Delta Cycle. *J. Coast. Res.* **1997**, *13*, 605–627.
18. Kolb, C.R.; van Lopik, J.R. *Geology of the Mississippi River Deltaic Plain, Southeastern Louisiana: US Army Corps of Engineers*; Technical Report 2; U.S. Army Engineer Waterways Experiment Station: Vicksburg, MI, USA, 1958.
19. Reed, D.J.; Wilson, L. Coast 2050: A New Approach to Restoration of Louisiana Coastal Wetlands. *Phys. Geogr.* **2004**, *25*, 4–21. [CrossRef]
20. Couvillion, B.R.; Beck, H.; Schoolmaster, D.; Fischer, M. *Land Area Change in Coastal Louisiana (1932 to 2016)*; Technical Report No. 3381; US Geological Survey: Lafayette, LA, USA, 2017. [CrossRef]
21. Coastal Protection and Restoration Authority. *Louisiana's Comprehensive Master Plan for a Sustainable Coast*; CPRA: Baton Rouge, LA, USA, 2017.
22. Li, B.; Sherman, D.J. Aerodynamics and morphodynamics of sand fences: A review. *Aeolian Res.* **2015**, *17*, 33–48. [CrossRef]
23. Bennett, R.; Chavarria, A. BA-171 Caminada Headland Back Barrier Marsh Creation Project. Available online: https://mississippiriverdelta.org//files/2017/03/CWPPRA-Caminada-headlands-back-barrier-marshrestoration.pdf (accessed on 19 April 2022).
24. Vousdoukas, M.I.; Ranasinghe, R.; Mentaschi, L.; Plomaritis, T.A.; Athanasiou, P.; Luijendijk, A.; Feyen, L. Sandy coastlines under threat of erosion. *Nat. Clim. Chang.* **2020**, *10*, 260–263. [CrossRef]
25. Luijendijk, A.; Hagenaars, G.; Ranasinghe, R.; Baart, F.; Donchyts, G.; Aarninkhof, S. The State of the World's Beaches. *Sci. Rep.* **2018**, *8*, 6641. [CrossRef]
26. Williams, S.J.; Penland, S.; Sallenger, J.A.H. *Atlas of Shoreline Changes from 1853 to 1989: Louisiana Barrier Island Erosion Study*; Technical Report No. 2150-A; United States Geological Survey: Baton Rouge, LA, USA, 1992. [CrossRef]
27. Byrnes, M.R.; Britsch, L.D.; Berlinghoff, J.L.; Johnson, R.; Khalil, S. Recent subsidence rates for Barataria Basin, Louisiana. *Geo-Mar. Lett.* **2019**, *39*, 265–278. [CrossRef]
28. Dietz, M.E.; Liu, K.-B.; Bianchette, T.A.; Ryu, J.; Yao, Q. Linking backbarrier lacustrine stratigraphy with spatial dynamics of shoreline retreat in a rapidly subsiding region of the Mississippi River Delta. *Geomorphology* **2021**, *397*, 108008. [CrossRef]
29. Liu, K.; Li, C.; Bianchette, T.; McCloskey, T.; Yao, Q.; Weeks, E.; Furmanczyk, K.; Giza, A.; Terefenko, P. Storm deposition in a coastal backbarrier lake in Louisiana caused by Hurricanes Gustav and Ike. *J. Coast. Res.* **2011**, *64*, 1866–1870. [CrossRef]
30. Dietz, M.E.; Liu, K.-B.; Bianchette, T.A.; Smith, D. Differentiating hurricane deposits in coastal sedimentary records: Two storms, one layer, but different processes. *Environ. Res. Commun.* **2021**, *3*, 101001. [CrossRef]
31. Blake, E.; Berg, R.; Hagen, A. *Hurricane Zeta*; Tropical Cyclone Report; National Hurricane Center: Miami, FL, USA, 2021.
32. Global Mapper User's Manual. Available online: https://www.globalmapper.it/helpv11/Help_Main.html (accessed on 9 April 2022).
33. Harper, J. Sediment Dispersal Trends of the Caminada-Moreau Beach-Ridge System. *Gulf Coast Assoc. Geol. Soc. Trans.* **1977**, *27*, 283–289.
34. Penland, S.; Ritchie, W. Short Term Morphological Changes along the Caminada-Moreau Coast, Louisiana. *Gulf Coast Assoc. Geol. Soc. Trans.* **1979**, *29*, 342–346.
35. DeWitt, N.T.; Flocks, J.G.; Hansen, M.; Kulp, M.; Reynolds, B. *Bathymetric Survey of the Nearshore from Belle Pass to Caminada Pass, Louisiana: Methods and Data Report*; Data Series 312; US Geological Survey: Lafayette, LA, USA, 2007. [CrossRef]
36. Penland, S.; Suter, J.R.; Moslow, T.F. *Inner-Shelf Shoal Sedimentary Facies and Sequences: Ship Shoal, Northern Gulf of Mexico*; SEPM Society for Sedimentary Geology: Tulsa, OK, USA, 1986.
37. McBride, R.A.; Byrnes, M.R. Regional Variations in Shore Response along Barrier Island Systems of the Mississippi River Delta Plain: Historical Change and Future Prediction. *J. Coast. Res.* **1997**, *13*, 628–655.
38. Henry, K.M.; Twilley, R.R. Soil Development in a Coastal Louisiana Wetland during a Climate-Induced Vegetation Shift from Salt Marsh to Mangrove. *J. Coast. Res.* **2013**, *29*, 1273. [CrossRef]
39. Day, J.W., Jr.; Boesch, D.F.; Clairain, E.J.; Kemp, G.P.; Laska, S.B.; Mitsch, W.J.; Orth, K.; Mashriqui, H.; Reed, D.J.; Shabman, L.; et al. Restoration of the Mississippi Delta: Lessons from Hurricanes Katrina and Rita. *Science (1979)* **2007**, *315*, 1679–1684. [CrossRef]
40. Naquin, J.D.; Liu, K.-B.; McCloskey, T.A.; Bianchette, T.A. Storm deposition induced by hurricanes in a rapidly subsiding coastal zone. *J. Coast. Res.* **2014**, *70*, 308–313. [CrossRef]
41. Elsner, J.B.; Kara, A.B.; Owens, M.A. Fluctuations in North Atlantic Hurricane Frequency. *J. Clim.* **1999**, *12*, 427–437. [CrossRef]
42. Webster, P.J.; Holland, G.J.; Curry, J.A.; Chang, H.-R. Changes in Tropical Cyclone Number, Duration, and Intensity in a Warming Environment. *Science (1979)* **2005**, *309*, 1844–1846. [CrossRef]
43. Mann, M.E.; Woodruff, J.D.; Donnelly, J.P.; Zhang, Z. Atlantic hurricanes and climate over the past 1500 years. *Nature* **2009**, *460*, 880–883. [CrossRef]

44. Yao, Q.; Liu, K.; Rodrigues, E.; Bianchette, T.; Aragón-Moreno, A.A.; Zhang, Z. A Geochemical Record of Late-Holocene Hurricane Events From the Florida Everglades. *Water Resour. Res.* **2020**, *56*. [CrossRef]
45. Johnson, C.L.; Chen, Q.; Ozdemir, C.E. Lidar time-series analysis of a rapidly transgressing low-lying mainland barrier (Caminada Headlands, LA, USA). *Geomorphology* **2019**, *352*, 106979. [CrossRef]
46. Xu, Y.J. Rethinking the Mississippi River diversion for effective capture of riverine sediments. *Proc. Int. Assoc. Hydrol. Sci.* **2015**, *367*, 463–470. [CrossRef]
47. Bass, A.S.; Turner, R.E. Relationships between Salt Marsh Loss and Dredged Canals in Three Louisiana Estuaries. *Coast. Res.* **1997**, *13*, 895–903.
48. Ko, J.-Y.; Day, J.W. A review of ecological impacts of oil and gas development on coastal ecosystems in the Mississippi Delta. *Ocean Coast. Manag.* **2004**, *47*, 597–623. [CrossRef]
49. Keown, M.P.; Dardeau, E.A.; Causey, E.M. Historic Trends in the Sediment Flow Regime of the Mississippi River. *Water Resour. Res.* **1986**, *22*, 1555–1564. [CrossRef]
50. Kesel, R.H. The decline in the suspended load of the Lower Mississippi River and its influence on adjacent wetlands. *Environ. Earth Sci.* **1988**, *11*, 271–281. [CrossRef]
51. Maloney, J.M.; Bentley, S.; Xu, K.; Obelcz, J.; Georgiou, I.Y.; Miner, M.D. Mississippi River subaqueous delta is entering a stage of retrogradation. *Mar. Geol.* **2018**, *400*, 12–23. [CrossRef]
52. Meade, R.H.; Moody, J.A. Causes for the decline of suspended-sediment discharge in the Mississippi River system, 1940-2007. *Hydrol. Process.* **2009**, *24*, 35–49. [CrossRef]
53. Liu, K.-B. Paleotempestology: Geographic Solutions to Hurricane Hazard Assessment and Risk Prediction. In *WorldMinds: Geographical Perspectives on 100 Problems*; Springer: Dordrecht, The Netherlands, 2004; pp. 443–448. [CrossRef]
54. Osland, M.J.; Day, R.H.; Hall, C.T.; Brumfield, M.D.; Dugas, J.L.; Jones, W.R. Mangrove expansion and contraction at a poleward range limit: Climate extremes and land-ocean temperature gradients. *Ecology* **2016**, *98*, 125–137. [CrossRef]
55. Osland, M.J.; Feher, L.C.; Anderson, G.H.; Vervaeke, W.C.; Krauss, K.W.; Whelan, K.R.; Balentine, K.M.; Tiling-Range, G.; Smith, T.J.; Cahoon, D.R. A Tropical Cyclone-Induced Ecological Regime Shift: Mangrove Forest Conversion to Mudflat in Everglades National Park (Florida, USA). *Wetlands* **2020**, *40*, 1445–1458. [CrossRef]
56. Barbier, E.B.; Koch, E.W.; Silliman, B.R.; Hacker, S.D.; Wolanski, E.; Primavera, J.; Granek, E.F.; Polasky, S.; Aswani, S.; Cramer, L.A.; et al. Coastal Ecosystem-Based Management with Nonlinear Ecological Functions and Values. *Science (1979)* **2008**, *319*, 321–323. [CrossRef] [PubMed]
57. Anthony, E.J.; Gratiot, N. Coastal engineering and large-scale mangrove destruction in Guyana, South America: Averting an environmental catastrophe in the making. *Ecol. Eng.* **2012**, *47*, 268–273. [CrossRef]
58. Phan, L.K.; van Thiel de Vries, J.S.M.; Stive, M.J.F. Coastal Mangrove Squeeze in the Mekong Delta. *J. Coast. Res.* **2015**, *300*, 233–243. [CrossRef]
59. Michot, T.C.; Day, R.H.; Wells, C.J. Increase in Black Mangrove Abundance in Coastal Louisiana. *Louisiana Natural Resources News*, 27 February 2010; pp. 4–5.
60. Yao, Q.; Cohen, M.C.L.; Liu, K.B.; Fan, D.D.; Rodrigues, E.; Maiti, K.; de Souza, A.V.; Aragón-Moreno, A.A.; Rohli, R.; Yin, D.X.; et al. Mangrove Expansion at Poleward Range Limits in North and South America: Late-Holocene Climate Variability or Anthropocene Global Warming? *Catena* 2022, *in press*.
61. Madison, M.J.; Mack, S.K.; Lane, R.R.; Day, J.W. Wetland Restoration Using Mangroves in Southern Louisiana. In Proceedings of the SPE Americas E&P Health, Safety, Security and Environmental Conference, Galveston, TX, USA, 18–20 March 2013. [CrossRef]

remote sensing

MDPI

Article

Automatic Creation of Storm Impact Database Based on Video Monitoring and Convolutional Neural Networks

Aurelien Callens [1,*], Denis Morichon [2], Pedro Liria [3], Irati Epelde [3] and Benoit Liquet [1,4]

1 LMAP, Université de Pau et des Pays de l'Adour, E2S UPPA, CNRS, 64000 Pau, France; benoit.liquet@univ-pau.fr
2 SIAME, Université de Pau et des Pays de l'Adour, E2S UPPA, 64600 Anglet, France; denis.morichon@univ-pau.fr
3 AZTI, Marine Research Division, KOSTARISK, 20110 Pasaia, Spain; pliria@azti.es (P.L.); iepelde@azti.es (I.E.)
4 Department of Mathematics and Statistics, Macquarie University, Sydney, NSW 2006, Australia
* Correspondence: aurelien.callens@univ-pau.fr

Abstract: Data about storm impacts are essential for the disaster risk reduction process, but unlike data about storm characteristics, they are not routinely collected. In this paper, we demonstrate the high potential of convolutional neural networks to automatically constitute storm impact database using timestacks images provided by coastal video monitoring stations. Several convolutional neural network architectures and methods to deal with class imbalance were tested on two sites (Biarritz and Zarautz) to find the best practices for this classification task. This study shows that convolutional neural networks are well adapted for the classification of timestacks images into storm impact regimes. Overall, the most complex and deepest architectures yield better results. Indeed, the best performances are obtained with the VGG16 architecture for both sites with F-scores of 0.866 for Biarritz and 0.858 for Zarautz. For the class imbalance problem, the method of oversampling shows best classification accuracy with F-scores on average 30% higher than the ones obtained with cost sensitive learning. The transferability of the learning method between sites is also investigated and shows conclusive results. This study highlights the high potential of convolutional neural networks to enhance the value of coastal video monitoring data that are routinely recorded on many coastal sites. Furthermore, it shows that this type of deep neural network can significantly contribute to the setting up of risk databases necessary for the determination of storm risk indicators and, more broadly, for the optimization of risk-mitigation measures.

Keywords: convolutional neural networks; storm impact database; transfer learning; video monitoring

1. Introduction

Databases containing information on past storm characteristics and their impacts on the coast are essential for the disaster-risk-reduction process. They enable scientists and coastal stakeholders to better understand the storm hazard in a specific area, to identify potential trends, and most importantly to assess coastal risks (present or future) through their use in the development and validation of early warning systems [1,2].

In these databases, storm impact is mostly represented as a qualitative variable with different categories. The different categories of storm impact are called "regimes" and are defined according to the Sallenger's scale [3]. This scale was originally derived to classify storm impact intensity based on the relation between wave-induced maximum water level and topographic elevations of the different sections of a natural beach. Recently, this approach has been extended to the estimation of storm impact intensity at an engineered beach backed by a seawall [4].

Due to the extreme and episodic nature of storms, databases covering a long period of time are necessary. Observed data about storm characteristics such as tide, wave, and wind are abundant and have been collected routinely for decades. In addition, numerous reanalyses and hindcasts are available for these variables. On the contrary, data on storm impacts are more sparse and mostly come from archives [5–7] or insurance data [8,9]. A few examples of storm impact databases are: the RISC-KIT database, which contains storm impact information for nine study sites in Europe [10]; the SurgeWatch database [5] for the UK; and a database for the Basque coast [7]. Even though archives and insurance provide information, there are some limitations including the heterogeneity, the incompleteness of the data sources, and the consequent amount of work needed [7]. A solution to routinely create a storm impact database could be to use images provided by coastal monitoring stations that are now widely used worldwide to survey and study coastal processes.

In recent decades, video monitoring systems have proven to be valuable assets in the study of the coastal zone due to their cost-efficiency and their ability to a provide continuous stream of data including intense storm conditions. Video monitoring systems are generally composed of one or several cameras operated by a monitoring software such as Argus [11], HORUS, Kosta (www.kostasystem.com, accessed on 1 May 2021) or Sirena [12]. The reader is referred to the work of Nieto et al. [12] for a comparison of the cited monitoring systems. These systems generate different types of images that can be applied to study coastal processes such as beach morphology changes, wave runup, and coastal currents [13,14]. Among the different types of imagse generated by the video monitoring system, timestacks images represent the time-varying pixel intensities along a particular cross-shore transect in the camera's field of view. They are used to perform wave runup parametrization [15–17], wave breaking detection [18], or intertidal topography [19] and also to estimate wave characteristics [20,21], sea level [22], and bathymetry [23,24]. Timestacks images have also been employed in the study of storm impact. In the work of Thuan et al. [25], they quantify the impact of two typhoons on the longshore-averaged shoreline changes based on the analysis of a series of timestacks images. To our knowledge, timestacks have not been used to directly measure storm impact regimes as defined previously. Image processing techniques are usually employed to transform the information contained in the images into quantitative measurements (runup elevation, wave height, shoreline). In this article, we propose extracting storm impact regimes (qualitative data) directly from the timestacks.

The storm impact regimes can be extracted from timestacks using two methodologies. The first methodology can be qualified as deterministic; it relies on image processing techniques and consists of two steps. First, the water line position is found by segmenting the image. The storm impact regime is then deduced by comparing the position of the waterline with the position of the defense infrastructure in the timestack. Different methods can be used to extract the waterline from timestacks images [26,27]. For example, Otsu's method [26] divides the pixels into two groups depending on their intensity values. It is not always robust and depends on the quality and lighting of the images. Most of the time, it requires rigorous and tedious human verification and correction [16]. The second methodology, presented in this article, relies on deep learning with convolutional neural networks (CNNs). CNNs are a class of deep neural networks, specializing in imagery analysis, that perform well on specific problems such as image classification and segmentation. First, timestack images are classified into storm impact regimes by human operators. Then, the CNN is trained to classify timestacks into storm impact regimes using the annotated dataset. During the training process, the CNN learns to simultaneously classify the images and which features to detect in order to achieve the best classification accuracy. Once the neural network has learned on the training dataset, it can be used to routinely analyze the timestacks produced by the video monitoring system and therefore create incrementally a qualitative storm impact database. This second methodology, based on a self-learning algorithm (CNN), allows for more automation compared to the first methodology because it does not require site-specific calibration [17].

Another advantage of CNNs is their transfer learning potential. Transfer learning consists in using knowledge gained on a first task, which usually involves a large number of data, and applies it to a different task [28] with smaller number of data. This technique has already been employed in coastal engineering domain [14,29] and usually results in faster training and better accuracy. In the case of storm impact recognition, where images of extreme storm impact regimes are rare by nature, this method can significantly improve the performances of CNNs. Moreover, it is reasonable to think that knowledge acquired at one site can be used to improve the performances on another site. This could be a non-negligible asset for the application of the method to a new site.

This paper aims to demonstrate the high potential of CNN methods to constitute a storm impact database using timestacks images provided by coastal video monitoring stations. Different methods are tested using images collected at two study sites. The best practice and the transferability of knowledge gained at one site to another are studied. In the following sections, the study sites and the video dataset are first described. The main features of the CNN implementation procedure are then shown in Section 3. Results and transferability of the CNN between the study sites are presented in Section 4 and discussed in Section 5. The main results are finally presented in the conclusion, Section 6.

2. Study Sites and Data

In this study, the storm impact intensity is classified into three storm impact regimes (Figure 1) derived from the Sallenger's scale [3]. The following three categories have been adapted for the timestack images:

- Swash regime: all the waves in the timestack are confined to the beach;
- Collision regime: at least one wave in the timestack collides with the bottom of the seawall ;
- Overwash regime: at least one wave in the timestack completely overtops the seawall.

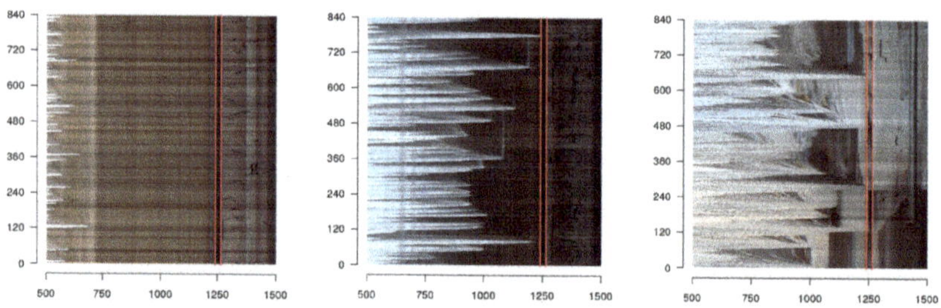

(a) (b) (c)

Figure 1. The three categories of storm impact regimes estimates from timestack images. y-axis represents the time in seconds, x-axis represent the pixel index (cropped images), and red lines represent the sea wall bottom and top positions. (**a**) Swash regime. (**b**) Collision regime. (**c**) Overwash regime.

The CNNs were trained on timestacks images collected by video monitoring stations operating on two sites along the basque coast (Figure 2), namely the Grande Plage of Biarritz (GPB), and the Zarautz beach (ZB). The use of two sets of data acquired from sites with different geological and morphological characteristics and distinct responses to oceanic forcing makes it possible to assess objectively the ability of CNN to detect storm impact regimes.

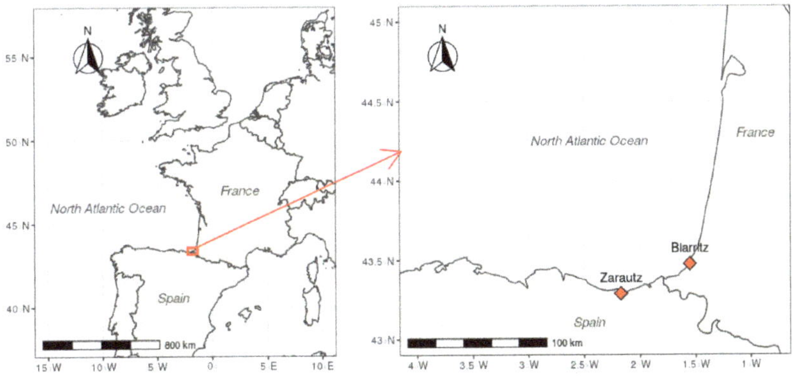

Figure 2. Map showing the locations of the two study sites.

2.1. Grande Plage de Biarritz

2.1.1. Site Characteristics

The Grande Plage of Biarritz (GPB) is an urban embayed beach that is 1.2 km long, located on the southern Aquitanian coast of France (Figure 2). It has a high socio-economic importance for the city of Biarritz due to its tourist appeal, its historical heritage, and its location near the city center. In terms of characteristics, the GPB is an intermediate-reflective beach with typically a steep foreshore slope of 8–9% and a gentle nearshore slope of 2–3% [30]. It is a mesotidal beach with 4.5 m spring tidal range around a mean water level of 2.64 m. This narrow beach is backed by a seawall with an alongshore elevation varying between 7 and 8 m. This seawall serves as defense infrastructure for back beach buildings.

The beach is predominantly exposed to waves coming from the WNW direction. The offshore wave climate is moderately to highly energetic. The annual average significant wave height and peak period are, respectively, H_s = 1.5 m and T_p = 10 s [30]. In this region, an event is qualified as a storm event when H_s and T_p are, respectively, greater than 3.5 m and 13.8 s. Such events correspond to 7.24% of the offshore wave climate [31] and are responsible for several overwash events each year.

This site has been equipped with a coastal video monitoring station since 2017. The station includes 4 cameras with different lenses to ensure the coverage of the entire beach with a sufficient spatial resolution. The cameras are operated by the open source software SIRENA [12]. For this site, one transect is monitored by the camera pointing to the beach and seawall location (transect Stack-Id01 in Figure 3). The timestack images correspond to pixel intensities recorded along this transect over 14 min with a sampling frequency of 1Hz. Among the 70,000 images of this database, only 8172 images were kept to be part of the ground truth dataset. Indeed, the timestacks generated in summer months were excluded as the human activities negatively affect the quality of the images. The images where the tide level was below 2.8 m were excluded as they corresponded to timestacks images without visible swash.

2.1.2. Timestack Image Preprocessing

The ground truth dataset was built by labeling the 8172 images. There are two methods to annotate the images: by hand or in a semi-automatic way. The annotation by hand is the most straightforward but also the most time-consuming method. The semi-automatic method consists of two steps. First, the position of the waterline is extracted automatically by segmenting the image using Otsu's thresholding method [16]. Then, the storm impact regime is identified by comparing the position of the waterline with the one of the defense infrastructure. This method is faster than the annotation by hand; however, it still requires an operator because it is not always robust and highly depends on the lightning conditions of the image. To employ this method, the position of the defense infrastructure in the image must be known. This is the case for the Grande Plage de Biarritz; therefore, semi-automatic annotation was performed.

After verification and correction by an operator, the result of the annotation was 7907/211/54 (Swash/Collision/Overwash). The classes are highly imbalanced, and this could have some effect on the classification accuracy of the CNN. Methods to deal with this problem are presented below. Before the training process, the images were resized to fit to the input dimensions of the CNNs tested in this study (224 × 224).

(a)

(b)

Figure 3. (**a**) Satellite view of the site of Biarritz from Google Earth with red lines representing the transects on the site. (**b**) Transect on the Grande Plage de Biarritz.

2.2. *Zarautz*

2.2.1. Site Characteristics

The beach of Zarautz is an embayed beach of 2.3 km long located on the Basque coast (northern Spain) in the SE Bay of Biscay, approximately 70 km southwest of GPB. The beach, facing north (345 degrees), can be divided into two parts (Figure 4): 30% of the beach in the eastern part presents a large and well-preserved dune system, with a maximum height of approximately 10 m above the minimum astronomic tide. The remaining 70% is an engineered urban beach, backed by a concrete seawall and the village of Zarautz.

In terms of characteristics, the beach of Zarautz is an intermediate-dissipative [32] and mesotidal beach with a 4 m spring tidal range. It is composed of fine–medium sand with a mean slope of around 2%. The annual average significant wave height and peak period are, respectively, 1 m and 9 s. Like the GPB, the beach of Zarautz is also exposed to highly energetic waves and storms coming from the WNW and NW directions. The seawall backing 70% of the beach has an along-shore elevation varying between 6.5 m in the western part and 8 m in the center of the beach. This seawall serves as a defense infrastructure for the buildings near the beach, and overtopping events are common at high tide during winter storms.

A video monitoring station, like the one used on the GPB site was installed in 2010. The station has 4 cameras of 1.4 Megapixels. Two of the cameras are equipped with 12 mm lens and have a panoramic view, and another 2 equipped with 25 mm lens cover with more resolution the mean high and low tide coastline positions. For the Zarautz dataset, 4 transects are monitored by the camera covering the supra-tidal beach with higher resolution (Figure 4). The transects are perpendicular to the seawall and are named corresponding to the elevation of the seawall in the point of intersection (i.e., transect 65 corresponds to the part of the seawall with 6.5 m elevation).

(a) (b)

Figure 4. (**a**) Satellite view of the site of Zarautz from Google Earth, with the red point representing the location of the video monitoring station. (**b**) Positions of the transects on the site of Zarautz.

2.2.2. Timestacks Images Preprocessing

Images from the site of Zarautz were annotated by hand. This method of annotation was preferred over semi-automatic annotation because: (i) the position of the seawall varied between timestack images, making the semi-automatic method more laborious, and (ii) the presence of strong winds and gust negatively impacted the quality of certain images, making the semi automatic method less robust. A simple web application was developed to facilitate the annotation for the operator and is accessible in a public GitHub repository (link in Data Availability Statement section). After classification by hand, the result of the annotation was 19,596/2776/162 (Swash/Collision/Overwash). Like the images of Biarritz, images of Zarautz present class imbalance, and they were resized to fit to the input dimensions of the CNNs before the training.

3. Convolutional Neural Networks

3.1. General Concept

CNNs are a type of neural networks widely used to perform tasks related to imagery analysis such as image segmentation, classification, or object detection. For classification problems, a CNN takes as input images with three channels (RGB), from which they output probabilities of belonging to specified categories, in our case storm impact regimes. Like a classical neural network, a CNN is a stacking of neurons that are organized in different layers. The structure of a CNN can be divided into two parts. The first part contains mostly convolutional and pooling layers and aims to learn specific features that help to classify the images correctly. The second part contains fully connected layers and the output layer. It uses the specific features extracted in the first part to output probabilities of belonging to specified categories.

In the feature extraction part, the convolutional layers detect features inside an image. They convolve their input with one or more filters, which results in one or more feature maps (one for each filter). The feature maps represent the activation of a specific filter at every spatial position of the input image. During the learning process, the network will learn filters that activate when they see specific visual features that help to correctly classify the training images. Usually, convolutional layers are stacked inside a CNN. The early layers detect simple features such as edges, whereas the deeper layers can detect more complex features.

Pooling layers are commonly found between convolutional layers. These layers also rely on convolutional operations and aim to reduce the dimensionality of the feature maps in order to increase the learning speed of the network and to control the overfitting of the CNN. If a CNN is overfitted, it would indicate that the network has learned exactly the characteristics of the training images and cannot generalize to new data. By stacking several convolutional and pooling layers inside a CNN, the complexity of the extracted information increases as we go deeper in the network with more feature maps with smaller dimensions.

The output of these specific layers serves as input to the second part of the network, which aims to classify the image into the correct category. In the classification part, neurons are organized in layers and are connected to the previous layers through weights (hence the name fully connected layers). To prevent overfitting, drop-out regularization can be applied on these layers. This method randomly ignores neurons during the training process, making the network learn more sparse and robust representation of the data. Finally, the output layer estimates the probabilities of belonging to the specified categories for the input image with a "softmax" activation function.

The CNNs are trained with backpropagation in the same manner as classical neural networks: the weights in the convolutional and fully connected layers are updated iteratively to minimize the errors between the prediction of the network and the ground truth. The ground truth dataset for such a network is made by annotating images. Details on the annotation of the timestacks can be found in the "Study Sites" section. Only the general ideas about CNN have been presented above; for a detailed description on CNN and their training, the reader is referred to the work of Bengio et al. [33].

There are many CNN architectures, each with different complexity and characteristics. In order to keep the computation time reasonable, it was decided to limit the comparison of performances between four architectures of increasing depth and complexity:

- A custom architecture inspired by the work of LeCun et al. [34] adapted for bigger images. The architecture is presented in the appendix (Table A1).
- AlexNet [35], which won the ImageNet challenge in 2012. Its architecture contains more convolutional layers and dense layers (Table A2). The number of filters is also larger than that of the custom architecture.
- VGG16 [36], which is a very deep CNN that uses 13 convolutional layers and three dense layers (Table A3).
- Inception v3, an improved version of the GoogleNet from Szegedy et al. [37] which won the ILSVRC in 2014. It relies on inception modules, which perform convolutions with filters of multiple size and concatenate their results (Table A4). In addition, the convolution operation with filters of large size inside an inception module are made by using $1 \times n$ filters to reduce computational cost. This results in deeper networks with significantly fewer parameters to learn.

3.2. *Training the CNN*

3.2.1. Data Processing

The datasets of both sites were divided into training, validation and testing sets containing, respectively, 65%, 15%, and 20% of the data (common proportions in the literature). Stratified random sampling was used to ensure that each part contains the same class proportions. The training set is used to fit the CNN. The validation set is used to stop the training for the CNN (early stopping). At last, the test part is used to evaluate the performance of the neural network on unseen data (not used in the training step).

During the training, each training image is seen multiple times by the CNN. This can be a problem as the network can learn exactly the characteristics of the training images and might not generalize to new data. To avoid this problem, called overfitting, data augmentation is employed during the training of the CNN. This method consists in making small changes to images in the training set before feeding them to the CNN. By generating modified images, this method artificially increases the number of images in the minority classes and makes the models more robust to overfitting.

The following changes were made:

- Random vertical flip: new timestack with inverted time;
- Random shift in the RGB image color to decrease the dependence on lighting conditions;
- Normalization of pixel values to 0–1 for faster training

3.2.2. Class Imbalance Problem

The datasets from both sites suffer from the class imbalance problem. Indeed, the distributions of storm impact regimes are highly imbalanced. For the Biarritz site, 96.8% of the images display swash regimes, 2.6% collision regimes, and 1% of overwash regimes. For Zarautz, 87% of the images are swash regimes, 12% collision regimes, and 1% overwash regimes. This class imbalance problem was expected as we are studying rare events.

It has been proven that class imbalance can negatively affect the performances of machine learning models in classification tasks [38]. Methods to deal with this problem are well known [38–40] and can be divided into two categories: data-level methods and classifier-level methods.

The data-level methods aim to modify the class distribution in order to reduce the imbalance ratio between classes. The most popular methods in this category are oversampling and undersampling. Oversampling consists in replicating random samples from minority classes until the imbalance problem is removed. In contrast, undersampling consists in removing random samples from the majority class until the balance between classes is reached.

The classifier-level methods aim to modify the training or the output of the machine learning algorithm. They include cost-sensitive learning, which is a method that gives more weights during learning to examples belonging to minority classes, and the thresholding method, which adjusts the output probabilities by taking into account the prior class probabilities [39].

3.2.3. Transfer Learning

For complex and deep CNN, it is common to use transfer learning to speed up the learning process and to improve performances. Transfer learning methods consist in using knowledge gained on a specific task to solve a different task. There are different methods of transfer learning for CNN; for an exhaustive listing, readers are referred to the work of Pan and Yang [28]. The method used in this article is "pre-training". It consists of using the weights of a CNN trained on a first task as initialization weights for a second CNN that will perform on a second task. The efficiency of pre-training was tested by using the pre-trained weights of VGG16 and Inceptionv3 on the ImageNet dataset, which is one of the largest labeled image dataset [41]. Then, transfer learning was performed between sites to see if the knowledge gained on one site is beneficial for the learning on the second site.

3.2.4. Application to the Datasets

The workflow for this study is presented in Figure 5. For each site, the four CNNs with different architectures were fitted without and with the two methods related to the class imbalance problem: oversampling and cost-sensitive learning (class weights). Transfer learning was used on the more complex architectures (VGG16, Inceptionv3) and only for the best performing method to cope with class imbalance. Data augmentation was used during the training of all the CNNs.

The networks were trained on a laptop equipped with a GPU (Quadro RTX 4000) using Keras (tensorflow GPU 1.12.0/Keras 2.3.1/Python 3.6.1), an open-source python library designed for building and training neural networks. The scripts used in this article are available on a public GitHub repository (link in Data Availability Statement section). The optimizer used is Mini-Batch gradient descent algorithm with batch size of 32 and a learning rate of 0.001 that decays by a factor of 2 every 10 epochs. The training is stopped at 100 epochs or earlier when the value of the validation loss does not decrease over 10 epochs (early stopping).

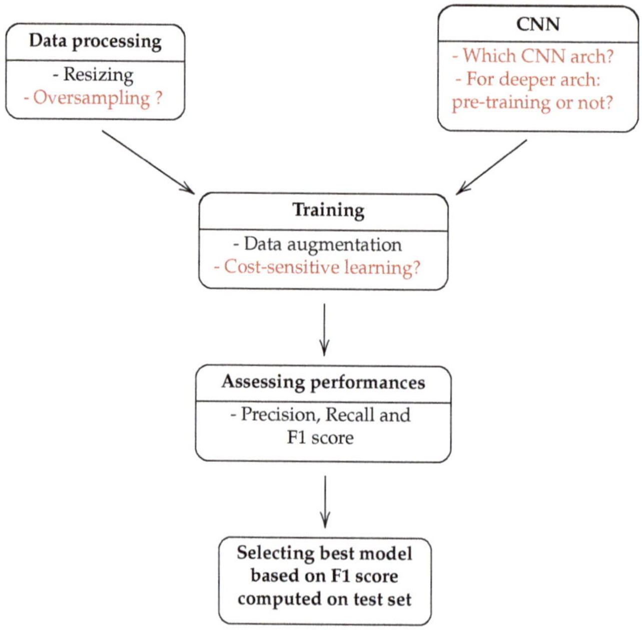

Figure 5. Workflow for this study. Items inside the boxes that are highlighted in red represent the choices tested in this study, whereas the items in black are the methods applied in every cases.

3.3. CNN Accuracy Assessment

To compare the performance of the different networks, the F_1-score is computed with the following formula:

$$F_1 = \frac{Precision \times Recall}{Precision + Recall}$$

where

$$Precision = \frac{True\ positives}{True\ positives + False\ positives}$$

and

$$Recall = \frac{True\ positives}{True\ positives + False\ negatives}.$$

The precision, recall, and F_1-score are computed for each storm impact regime and are averaged in order to have one global metric for each CNN. The F_1-score varies between 0 and 1, with 1 representing the best value. Unlike the global accuracy (number of correct predictions divided by the total number of predictions), the F_1 metric is not biased when data present a class imbalance.

4. Results

The results are organized into four subsections. Firstly, the performances of the different combinations of CNN architectures, methods to cope with class imbalance, and transfer learning are compared. Secondly, the prediction errors of the best CNN for each site are investigated. Thirdly, we present results related to transferability between sites. Finally, a sensitivity analysis is presented for the site of Zarautz.

4.1. CNN Performances

4.1.1. Architectures

Table 1 regroups the training time, number of epochs, and also the performance metrics (accuracy, recall, F_1-score) for different CNN architectures, methods to cope with class imbalance, and with or without pre-training with ImageNet dataset for both sites. For both sites and for every methods used to cope with class imbalance (or not), CNNs with deeper and more complex architectures yielded better results (higher values of precision, recall, and F_1-score). Indeed, these kind of architectures tend to learn more complex features that lead to better performance in harder tasks. The downside of these networks is the training time, which is significantly higher than simpler and shallower models.

Table 1. CNN performances for both sites. Best models are in bold font.

(a) Biarritz						
CNN	**Training Time (min)**	**Epochs**	**Time per Epoch (s)**	**Precision**	**Recall**	**F_1-Score**
Baseline						
Custom CNN	16.1	100	9.7	/	0.333	0.328
AlexNet	17.2	100	10.3	/	0.333	0.328
VGG16	81.4	89	54.9	/	0.481	0.476
Inception v3	40.0	69	34.8	0.721	0.714	0.713
Class weights						
Custom CNN	4.6	28	9.9	/	0.603	0.474
AlexNet	3.8	21	10.9	0.568	0.777	0.609
VGG16	43.4	46	56.6	0.574	0.832	0.645
Inception v3	23.2	39	35.8	0.563	0.798	0.631
Oversampling						
Custom CNN	10.6	26	24.6	0.642	0.880	0.718
AlexNet	11.8	27	26.1	0.716	0.885	0.777
VGG16	69.6	28	149.1	0.783	0.851	0.813
VGG16 Transfer	**49.9**	**20**	**149.6**	**0.869**	**0.865**	**0.866**
Inception v3	59.6	38	94.1	0.679	0.767	0.717
Inception v3 Transfer	34.5	21	98.6	0.777	0.786	0.780
(b) Zarautz						
CNN	**Training Time (min)**	**Epochs**	**Time per Epoch (s)**	**Precision**	**Recall**	**F_1-Score**
Baseline						
Custom CNN	22.5	49	27.5	/	0.637	0.616
AlexNet	24.0	48	30.0	/	0.628	0.616
VGG16	202.1	72	168.4	/	0.635	0.617
Inception v3	108.7	64	101.9	/	0.630	0.614
Class weights						
Custom CNN	11.6	26	26.7	0.666	0.846	0.720
AlexNet	22.7	45	30.3	0.671	0.817	0.716
VGG16	81.9	30	163.7	0.680	0.844	0.732
Inception v3	89.3	53	101.1	0.654	0.838	0.710
Oversampling						
Custom CNN	38.7	36	64.5	0.769	0.804	0.783
AlexNet	22.6	19	71.3	0.756	0.797	0.775
VGG16	146.6	22	399.8	0.775	0.812	0.792
VGG16 Transfer	**86.5**	**13**	**399.1**	**0.897**	**0.834**	**0.858**
Inception v3	97.7	24	244.2	0.777	0.801	0.784
Inception v3 Transfer	65.3	16	245.0	0.869	0.835	0.849

4.1.2. Class Imbalance

Without coping with class imbalance problem, CNNs tend to predict all the images as the majority class, resulting in poor classification results. Between the two methods tested, oversampling seems to perform better, with F_1-scores on average 30% better than the ones obtained with cost-sensitive learning (class weights). The superior performance of oversampling method on this dataset might be due to the fact that the CNNs see more images during training in the oversampling method than in the cost sensitive learning method, resulting in better classification accuracy.

4.1.3. Pre-Training

Finally, models using pre-trained weights (transfer learning) train faster (fewer epochs) and yield better classification results than models trained from scratch. Indeed, the F1-scores obtained with the pre-trained models are, respectively, 6 to 8% higher for the VGG16 and Inceptionv3 models. Even though the images from the ImageNet dataset have different characteristics than the timestack images that are being classified, the pre-trained weights might contain knowledge about general features that are helpful to better classify the timestacks.

4.1.4. Best Models

For GPB, the best model is the pre-trained VGG16 with an F1-score of 0.866. The pre-trained Inception model trains faster but shows a lower F1-score (0.780). For the Zarautz site, the best model is also the pre-trained VGG16 with an F1-score of 0.858, but this time the performance of the pre-trained Inception v3 model was very close with an F1-score of 0.849.

4.2. *Investigating the Errors*

The confusion matrices on the test sets are presented in Table 2. In general, the minority classes tended to have higher error rate. This is expected as minority classes contain fewer examples than majority classes.

Table 2. Confusion matrices obtained by the best models for both sites.

(a) Biarritz (best model: OV VGG16 Transfer)				
			Predicted	
		Swash	**Collision**	**Overwash**
Observed	Swash	1576	7	0
	Collision	4	34	2
	Overwash	1	2	9
(b) Zarautz (best model: OV VGG16 Transfer)				
			Predicted	
		Swash	**Collision**	**Overwash**
Observed	Swash	4265	40	0
	Collision	13	617	8
	Overwash	0	25	30

The prediction errors made by the CNN were manually inspected to gain an understanding of common error types. Prediction errors made on the GPB test set are presented in Figure 6 and in the appendix (Table A5). Among the 16 errors made on the test set, five errors came from human misclassification, five errors may have been caused by the presence of specific features such as vertical lines usually associated with collision regimes, two errors were made on images that were displayed in between category of storm impact regimes. The remaining errors correspond to images that were hard to classify, where lighting and meteorological conditions were poor. The misclassification errors were corrected for the test and validation sets, and the best network was trained once again, resulting in slightly lower results but this time without human misclassification error (Table 3).

Table 3. Classification results after correcting the misclassification in the test and validation sets.

CNN	Time (min)	Epochs	Time per Epoch (s)	Precision	Recall	F1 Score
Biarritz						
Best model before corr.	49.9	20	149.6	0.869	0.865	0.866
Best model after corr.	60	24	150	0.895	0.833	0.860
Zarautz						
Best model before corr.	86.5	13	399.1	0.897	0.834	0.858
Best model after corr.	88.5	13	408.5	0.917	0.859	0.883

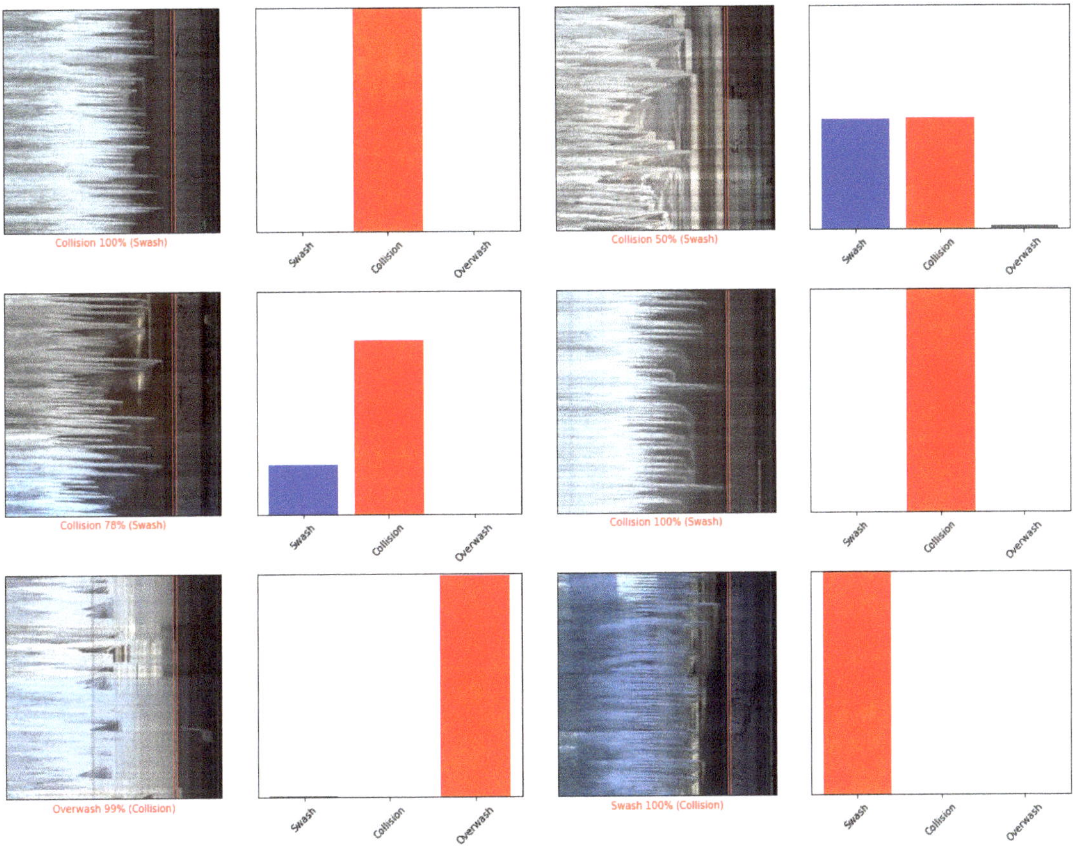

Figure 6. Prediction errors on the test set of Biarritz by the best CNN. The storm impact regime predicted is written under each timestack, and the ground truth is written in parentheses. The probabilities of belonging predicted by the CNN are represented on the right side of each timestack (red = prediction made by CNN, blue = ground truth). The red line in the timestack represents the position of the seawall.

The errors made on the Zarautz dataset were also analyzed (Table A6). A large number of errors were made on images that were in between the categories of storm impact regimes: either the images displayed a swash regime, which was very close to the collision regime, or the images displayed a regime impact, with one small overtopping of the wall. Some misclassification errors were made. The rest of the errors may have come from lighting conditions (large horizontal band, lighter in the images). The misclassification errors were corrected for the test and validation, and the best network was trained once again, resulting in a slightly better result for this site (Table 3).

4.3. Transferability between Sites

The interest in transfer learning for CNN training has been highlighted in Section 4.1.3. The models using pre-trained weights from ImageNet trained faster (fewer epochs) and yielded better classification results. In this section, we investigate if the knowledge acquired on one site (CNN weights) could be transferred to another site with pre-training. Pre-training is the most common way to transfer knowledge between tasks. It consists in using the weights of a neural network trained on one site as initialization weights for the training on the second site. The weights of the best CNN for each site are used for the pre-training of a CNN on the other site. The performances of these CNN are presented in Table 4.

Table 4. Performances of CNN learning from scratch, pre-trained with ImageNet, or pre-trained with the other site for Biarritz and Zarautz. "OV" stands for oversampling.

CNN	Time (min)	Epochs	Time per Epoch (s)	Precision	Recall	F_1-Score
Biarritz						
VGG16 (OV)	69.6	28	149.1	0.783	0.851	0.813
VGG16 (OV) Pretraining with ImageNet	49.9	20	149.6	0.869	0.865	0.866
VGG16 (OV) Pretraining with Zarautz data	47	19	148.4	0.826	0.832	0.823
Zarautz						
VGG16 (OV)	81.9	30	163.7	0.680	0.844	0.732
VGG16 (OV) Pretraining with ImageNet	86.5	13	399.1	0.897	0.834	0.858
VGG16 (OV) Pretraining with Biarritz data	92	14	394.2	0.909	0.867	0.885

The weights of the best model on Zarautz data (VGG16 transfer) were used as initialization weights for the training on Biarritz data. This resulted in classification results better than the learning from scratch with a higher precision and F_1-score (Table 4). However, the values of precision, recall, and F_1-score obtained with pre-training on Zarautz data remained slightly lower than the ones obtained with pre-training on ImageNet data.

Pre-training method was also applied on Zarautz data, where the weights of the best model on the Biarritz site were used as initialization weights. The classification results were better than learning from scratch and learning with pre-trained weights from ImageNet with higher F_1-score (Table 4).

4.4. Sensitivity Analysis

A sensitivity analysis was performed on the dataset of Zarautz to highlight the effect of the size of the training images dataset on the classification accuracy. The dataset of Zarautz was divided into three smaller datasets. Each of these datasets was divided into the training/validation/test sets with the proportions described in Section 3.2.1. Finally, a CNN model was trained on each smaller datasets (VGG16 with transfer learning ImageNet and oversampling).

The averaged F1-metric for these three models was 0.805. This value is slightly lower than the one obtained with the full dataset, which is 0.858 (Table 1). These results confirm what was already known in the literature: CNN performances tend to increase with the training set size [42].

5. Discussion

Even though we showed the strong potential of CNN to automatically generate storm impact regime database, the proposed methodology can be improved in several ways. More attention could be paid to the choice of CNN architectures and hyperparameters. Other CNN architectures need to be tested, especially recent architectures such as ResNet,

MobileNet, or Xception. They could perform better than the architectures presented in this work. For instance, the ResNet architecture contains skip connections between layers, which allows the training of much deeper and performant networks. Hyperparameters are parameters whose values are specified by the user before the training process begins; they affect the structure of a CNN and how well it trains. They have a non negligible impact on the final results. Several optimization algorithms such as Bayesian optimization could be employed to select the optimal hyperparameters [43], which has not been done in this study.

In addition to hyperparameter tuning, other methods of data augmentation could be used to improve the performances of the CNN. The analysis of prediction errors can help in the choice of other data augmentation methods. In our case, many errors were related to lighting conditions; it would be wise to test various data augmentation methods affecting the lighting or brightness of the images. This could make the CNN more robust to lighting conditions and therefore improve its performances.

It is worth noting that the performances of a CNN model implemented at a given site are expected to increase with time as more timestacks are collected by the video monitoring system. With more training images, the minority classes will contain more images, and this will lead to less classification errors for these classes. Moreover, if enough timestacks are collected, intermediate storm impact regime classes could be created. These classes could reduce the errors on the images displaying impact regimes not corresponding to the three regimes presented in this work.

One very interesting feature of CNNs models is their transferability. We showed that using the knowledge acquired from another site can lead to improved classification results when using pre-training (especially for Zarautz site). The weights of the best CNNs for both sites are available in a GitHub repository (link in Data Availability Statement section) and could be used as initialization weights for a CNN applied to a new site. The only requirement is to annotate timestacks from the new site, which will serve as training data.

Despite the promising performance, this methodology has some limitations, mainly related to the image annotation, an obligatory step for CNN training. The first limitation of this method is the lack of knowledge about its sensitivity. We showed for the site of Zarautz that CNNs yield lower performances when trained on a smaller training set. However, we do not know the minimum number of timestacks to annotate for a new site in order to have satisfactory accuracy. A sensitivity analysis should be performed to find this minimum threshold and to make some recommendations on the use of this method in the case of new sites with a small number of timestacks.

The second limitation of this method is the annotation process itself, which is tedious and time-consuming. An alternative solution could be to use the domain-adaptation approach presented in the work of Ganin and Lempitsky [44]. They propose a specific CNN architecture that can be trained simultaneously on a large number of labeled data from a source domain (one site) and unlabeled data from a target domain (new site). At the end of the training, the CNN is able to classify correctly images from both sites even though only images from one site have been labeled.

Finally, the performances of the proposed method must be compared objectively with human-level performance and other methodologies. Assessing the human-level performance on this task is essential and would give precious insights into how to improve further the CNN performances [45]. For example, a CNN performance lower than the human-level performance could indicate the presence of a bias, which can be avoided by using deeper models or by training more slowly and for longer. It would be of great interest to compare this methodology based on CNNs with methodologies based only on traditional imagery analysis. As stated in the introduction, a possible methodology could be to first extract the waterline position using Otsu's segmentation [16,26] or using the radon transform [27] and then compare its position with the position of defense infrastructure to define the storm impact. Another methodology could be based on the analysis of pixel intensity such as the works of Simarro et al. [46] and Andriolo et al.

[47]. The methodologies based on simple image processing algorithms could have some advantages over CNNs. Indeed, they would not require the building and training of a CNN structure, which is time-consuming, and the whole decisional process is known to be contrary to CNN, which can be considered as a "black box". However, these methodologies would need to be adapted for each site by indicating the position of the defense infrastructure, which is not needed with CNN. In addition, the simple image processing algorithms could be more affected by the erratic brightness of the timestacks than CNNs, which are trained with data augmentation.

This work is a first step in the analysis of storm impact with video monitoring. Numerous extensions can be envisaged, particularly on the type of information extracted and the type of image analyzed. Indeed, the CNN could be used to count the number of collision or overwash events in one timestack. This technique could be also extended to analyze other types of images produced by video monitoring systems such as oblique and/or rectified images. Finally, it can be employed to analyze images from already existing cameras such as surfcam [48,49]. This could constitute a low-cost monitoring method with a large spatial coverage for the qualitative study of storm impact. Many questions arose with this work, especially about the minimum number of images to annotate to have satisfactory accuracy or the lack of comparison with the current method or human level performance. More questions will arise during the operational implementation and use of the CNNs concerning the verification of predictions, the prediction error handling, or how often we need to re-train the neural networks with the newly classified images.

6. Conclusions

In this paper, we presented an innovative methodology based on convolutional neural networks and coastal imagery that could be used to collect storm impact data routinely. We described the methodologies associated with CNNs, including the annotation of the dataset, the training of the networks, or transfer learning. We also introduced the problem of class imbalance, which is due to the extreme nature of the storm impact regimes, and we proposed and compared different solutions such as oversampling or cost-sensitive learning.

The proposed methodology was tested on two sites: Biarritz and Zarautz. We showed that convolutional neural networks are well adapted for the classification of timestacks into storm impact regimes. Overall, we found that more complex and deeper architectures yielded better results. Best performances were achieved with the VGG16 architecture for both sites with F-scores of 0.866 for the site of Biarritz and 0.858 for the site Zarautz. For the class imbalance problem, the method of oversampling showed better classification accuracy than the cost-sensitive learning method, with F-scores on average 30% higher. Finally, we showed that the method can be easily applied to a new site with optimal efficiency using transfer learning. Indeed, training a CNN using pre-trained weights (ImageNet or weights of another site) resulted in better accuracy than training a CNN from scratch (F-scores on average 6 to 8% higher).

With convolutional neural networks, we can take full advantage of the large number of data produced by video monitoring systems. We showed that they are able to transform images into usable qualitative data about storm impact. Even if the data are not continuous (only day time and winter months), this method could be, without a doubt, a real asset in the future for coastal researchers and stakeholders by routinely collecting storm impact data, which are rare at present. These data are essential in the disaster risk reduction chain, and they have many uses. They can serve as validation data for impact models or early warning systems based on numerical modeling. They can also be used to train early warning system based on Bayesian networks [50,51]. Finally, statistical analysis can be performed to find relationships between observed storm impact regimes and local conditions such as wave characteristics, tide, or meteorological conditions.

Author Contributions: Conceptualization, A.C., D.M. and B.L.; methodology, A.C. and B.L.; software, A.C.; validation, A.C.; formal analysis, A.C.; investigation, A.C.; resources, D.M., P.L. and I.E.;

data curation, A.C.; writing—original draft preparation, A.C.; writing—review and editing, A.C., D.M., P.L. and B.L.; visualization, A.C.; supervision, D.M. and B.L.; project administration, D.M. and B.L.; funding acquisition, D.M. and B.L. All authors have read and agreed to the published version of the manuscript.

Funding: Funding was provided by the Energy Environment Solutions (E2S UPPA) consortium and the BIGCEES project from E2S-UPPA ("Big model and Big data in Computational Ecology and Environmental Sciences").

Data Availability Statement: The images used to train the CNN are not publicly available due to the large size of the files. SIAME laboratory owns the Biarritz images data; they can be provided upon request. Images of Zarautz's site used in this work are from Azti and the Zarautz Town Council (shared). They can also be provided upon request. The python scripts and weights of the best CNN are available here: https://github.com/AurelienCallens/CNN_Timestacks, accessed on 1 May 2021. The web application used to label the images of Zarautz site (R programming language) is available here: https://github.com/AurelienCallens/Shiny_Classifier, accessed on 1 May 2021.

Acknowledgments: The authors gratefully acknowledge European POCTEFA Program funding under the research project MARLIT EFA344/19 and the complementary program of OCA. The authors would like to thank Zarautz Town Council and the Directorate of Emergencies and Meteorology of the Basque Government for their continuous support of Zarautz's video station.

Conflicts of Interest: The authors declare no conflict of interest.

Abbreviations

Abbreviations

The following abbreviations are used in this manuscript:

CNN	Convolutional neural networks
GBP	Grande Plage of Biarritz

Appendix A. CNN Architectures

Appendix A.1. Custom CNN

Table A1. Architecture of the "custom" CNN.

Layer (Type)	Output Shape	Param
Block1 Conv (Conv2D)	(None, 111, 111, 32)	896
Block1 Pool (MaxPooling2D)	(None, 55, 55, 32)	0
Block2 Conv2d (Conv2D)	(None, 53, 53, 64)	18,496
Block2 Pool (MaxPooling2D)	(None, 26, 26, 64)	0
Block3 Conv2d (Conv2D)	(None, 24, 24, 128)	73,856
Block4 Pool (MaxPooling2D)	(None, 12, 12, 128)	0
Block5 Conv2d (Conv2D)	(None, 10, 10, 256)	295,168
Block5 Pool (MaxPooling2D)	(None, 5, 5, 256)	0
Flatten (Flatten)	(None, 6400)	0
Dense1 (Dense)	(None, 512)	3,277,312
Dropout1 (Dropout)	(None, 512)	0
Dense2 (Dense)	(None, 256)	131,328
Dropout2 (Dropout)	(None, 256)	0
Dense3 (Dense)	(None, 128)	32,896
Dropout3 (Dropout)	(None, 128)	0
Output (Dense)	(None, 3)	387

- Total params: 3,830,339
- Trainable params: 3,830,339
- Non-trainable params: 0

Appendix A.2. AlexNet

Table A2. Architecture of the AlexNet.

Layer (Type)	Output Shape	Param
Block1 Conv (Conv2D)	(None, 54, 54, 96)	34,944
Block1 Pool (MaxPooling2D)	(None, 27, 27, 96)	0
Block2 Conv (Conv2D)	(None, 17, 17, 256)	2,973,952
Block2 Pool (MaxPooling2D)	(None, 8, 8, 256)	0
Block3 Conv (Conv2D)	(None, 6, 6, 384)	885,120
Block3 Conv (Conv2D)	(None, 4, 4, 384)	1,327,488
Block4 Conv (Conv2D)	(None, 2, 2, 256)	884,992
Block4 Pool (MaxPooling2D)	(None, 1, 1, 256)	0
Flatten (Flatten)	(None, 256)	0
Dense1 (Dense)	(None, 4096)	1,052,672
Dropout1 (Dropout)	(None, 4096)	0
Dense2 (Dense)	(None, 4096)	16,781,312
Dropout2 (Dropout)	(None, 4096)	0
Output (Dense)	(None, 3)	12,291

- Total params: 23,952,771
- Trainable params: 23,952,771
- Non-trainable params: 0

Appendix A.3. VGG16

Table A3. Architecture of the CNN used based on VGG16.

Layer (Type)	Output Shape	Param
Input (Input Layer)	(None, 224, 224, 3)	0
Block1 Conv1 (Conv2D)	(None, 224, 224, 64)	1792
Block1 Conv2 (Conv2D)	(None, 224, 224, 64)	36,928
Block1 Pool (MaxPooling2D)	(None, 112, 112, 64)	0
Block2 conv1 (Conv2D)	(None, 112, 112, 128)	73,856
Block2 Conv2 (Conv2D)	(None, 112, 112, 128)	147,584
Block2 Pool (MaxPooling2D)	(None, 56, 56, 128)	0
Block3 Conv1 (Conv2D)	(None, 56, 56, 256)	295,168
Block3 Conv2 (Conv2D)	(None, 56, 56, 256)	590,080
Block3 Conv3 (Conv2D)	(None, 56, 56, 256)	590,080
Block3 Pool (MaxPooling2D)	(None, 28, 28, 256)	0
Block4 Conv1 (Conv2D)	(None, 28, 28, 512)	1,180,160
Block4 Conv2 (Conv2D)	(None, 28, 28, 512)	2,359,808
Block4 Conv3 (Conv2D)	(None, 28, 28, 512)	2,359,808
Block4 Pool (MaxPooling2D)	(None, 14, 14, 512)	0
Block5 Conv1 (Conv2D)	(None, 14, 14, 512)	2,359,808
Block5 Conv2 (Conv2D)	(None, 14, 14, 512)	2,359,808
Block5 Conv3 (Conv2D)	(None, 14, 14, 512)	2,359,808
Block5 Pool (MaxPooling2D)	(None, 7, 7, 512)	0
Flatten (Flatten)	(None, 2048)	0
Dense1 (Dense)	(None, 512)	262,656
Dropout1 (Dropout)	(None, 512)	0
Output (Dense)	(None, 3)	1539

- Total params: 14,978,883
- Trainable params: 14,978,883
- Non-trainable params: 0

Appendix A.4. Inception v3

Table A4. Architecture of the CNN used based on Inception v3.

Layer (Type)	Output Shape	Param
Inceptionv3 (Model)	(None, 2048)	21,802,784
Flatten (Flatten)	(None, 2048)	0
Dense1 (Dense)	(None, 512)	1,049,088
Dropout1 (Dropout)	(None, 512)	0
Output (Dense)	(None, 3)	1539

The inception model was imported with Keras with the following function: `keras.applications.InceptionV3()`.

The architecture is not displayed due to readability; the reader is referred to the original work of Szegedy et al. [37] for more details.

- Total params: 22,853,411
- Trainable params: 22,818,979
- Non-trainable params: 34,432

Appendix B. Investigating the Errors

Table A5. Errors explanation for Biarritz data. "Misclass." stands for misclassification during annotation, "Splash" corresponds to an intermediate storm regime between impact and overwash, "Vertical" corresponds to the presences of vertical features of runup.

Test					
Splash	Lighting	Misclass.	Hard to classify	Vertical	
2	1	5	3	5	
Validation					
Splash	Misclass.	Sand bags ?	Splash	Hard to classify	Vertical
1	2	1	2	2	4

Table A6. Errors explanation for Zarautz data. "Misclass." stands for misclassification during annotation, "Splash" corresponds to an intermediate storm regime between impact and overwash, "Vertical" corresponds to the presences of vertical features of runup. Finally, "SI" corresponds to an intermediate storm regime between swash and impact that was very close to the sea wall but did not impact.

Test						
Hard to classify	Lighting	Misclass.	SI	SI + Light	Splash	Vertical
10	30	7	22	2	14	1
Validation						
Hard to classify	Lighting	Misclass.	SI	SI + Light	Sandbag ?	Splash
5	27	3	26	2	4	4

References

1. Valchev, N.; Andreeva, N.; Eftimova, P.; Trifonova, E. Prototype of Early Warning System for Coastal Storm Hazard (Bulgarian Black Sea Coast). *CR Acad. Bulg. Sci.* **2014**, *67*, 977.
2. Van Dongeren, A.; Ciavola, P.; Martinez, G.; Viavattene, C.; Bogaard, T.; Ferreira, O.; Higgins, R.; McCall, R. Introduction to RISC-KIT: Resilience-increasing strategies for coasts. *Coast. Eng.* **2018**, *134*, 2–9. [CrossRef]
3. Sallenger, A.H.J. Storm Impact Scale for Barrier Islands. *J. Coast. Res.* **2000**, *16*, 890–895.
4. de Santiago, I.; Morichon, D.; Abadie, S.; Reniers, A.J.; Liria, P. A Comparative Study of Models to Predict Storm Impact on Beaches. *Nat. Hazards* **2017**, *87*, 843–865. [CrossRef]
5. Haigh, I.D.; Wadey, M.P.; Gallop, S.L.; Loehr, H.; Nicholls, R.J.; Horsburgh, K.; Brown, J.M.; Bradshaw, E. A user-friendly database of coastal flooding in the United Kingdom from1915 to 2014. *Sci. Data* **2015**, *2*, 1–13. [CrossRef]
6. Garnier, E.; Ciavola, P.; Spencer, T.; Ferreira, O.; Armaroli, C.; McIvor, A. Historical analysis of storm events: Case studies in France, England, Portugal and Italy. *Coast. Eng.* **2018**, *134*, 10–23. [CrossRef]
7. Abadie, S.; Beauvivre, M.; Egurrola, E.; Bouisset, C.; Degremont, I.; Arnoux, F. A Database of Recent Historical Storm Impact on the French Basque Coast. *J. Coast. Res.* **2018**, *85*, 721–725. [CrossRef]
8. André, C.; Monfort, D.; Bouzit, M.; Vinchon, C. Contribution of insurance data to cost assessment of coastal flood damage to residential buildings: Insights gained from Johanna (2008) and Xynthia (2010) storm events. *Nat. Hazards Earth Syst. Sci.* **2013**, *13*, 2003. [CrossRef]
9. Naulin, J.P.; Moncoulon, D.; Le Roy, S.; Pedreros, R.; Idier, D.; Oliveros, C. Estimation of insurance-related losses resulting from coastal flooding in France. *Nat. Hazards Earth Syst. Sci.* **2016**, *16*, 195–207. [CrossRef]
10. Ciavola, P.; Harley, M.; den Heijer, C. The RISC-KIT storm impact database: A new tool in support of DRR. *Coast. Eng.* **2018**, *134*, 24–32. [CrossRef]
11. Holman, R.A.; Stanley, J. The history and technical capabilities of Argus. *Coast. Eng.* **2007**, *54*, 477–491. [CrossRef]
12. Nieto, M.A.; Garau, B.; Balle, S.; Simarro, G.; Zarruk, G.A.; Ortiz, A.; Tintoré, J.; Álvarez-Ellacuría, A.; Gómez-Pujol, L.; Orfila, A. An open source, low cost video-based coastal monitoring system. *Earth Surf. Process. Landforms* **2010**, *35*, 1712–1719. [CrossRef]
13. Splinter, K.D.; Harley, M.D.; Turner, I.L. Remote sensing is changing our view of the coast: Insights from 40 years of monitoring at Narrabeen-Collaroy, Australia. *Remote Sens.* **2018**, *10*, 1744. [CrossRef]
14. Buscombe, D.; Carini, R.J. A data-driven approach to classifying wave breaking in infrared imagery. *Remote Sens.* **2019**, *11*, 859. [CrossRef]
15. Senechal, N.; Coco, G.; Bryan, K.R.; Holman, R.A. Wave runup during extreme storm conditions. *J. Geophys. Res. Ocean.* **2011**, *116*. [CrossRef]
16. Vousdoukas, M.I.; Wziatek, D.; Almeida, L.P. Coastal Vulnerability Assessment Based on Video Wave Run-up Observations at a Mesotidal, Steep-Sloped Beach. *Ocean. Dyn.* **2012**, *62*, 123–137. [CrossRef]
17. den Bieman, J.P.; de Ridder, M.P.; van Gent, M.R. Deep learning video analysis as measurement technique in physical models. *Coast. Eng.* **2020**, *158*, 103689. [CrossRef]
18. Stringari, C.E.; Harris, D.L.; Power, H.E. A Novel Machine Learning Algorithm for Tracking Remotely Sensed Waves in the Surf Zone. *Coast. Eng.* **2019**. [CrossRef]
19. Valentini, N.; Saponieri, A.; Molfetta, M.G.; Damiani, L. New algorithms for shoreline monitoring from coastal video systems. *Earth Sci. Inform.* **2017**, *10*, 495–506. [CrossRef]
20. Almar, R.; Cienfuegos, R.; Catalán, P.A.; Michallet, H.; Castelle, B.; Bonneton, P.; Marieu, V. A new breaking wave height direct estimator from video imagery. *Coast. Eng.* **2012**, *61*, 42–48. [CrossRef]
21. Andriolo, U.; Mendes, D.; Taborda, R. Breaking wave height estimation from Timex images: Two methods for coastal video monitoring systems. *Remote Sens.* **2020**, *12*, 204. [CrossRef]
22. Ondoa, G.A.; Almar, R.; Castelle, B.; Testut, L.; Leger, F.; Sohou, Z.; Bonou, F.; Bergsma, E.; Meyssignac, B.; Larson, M. Sea level at the coast from video-sensed waves: Comparison to tidal gauges and satellite altimetry. *J. Atmos. Ocean. Technol.* **2019**, *36*, 1591–1603. [CrossRef]
23. Holman, R.; Plant, N.; Holland, T. cBathy: A robust algorithm for estimating nearshore bathymetry. *J. Geophys. Res. Ocean.* **2013**, *118*, 2595–2609. [CrossRef]
24. Simarro, G.; Calvete, D.; Luque, P.; Orfila, A.; Ribas, F. UBathy: A new approach for bathymetric inversion from video imagery. *Remote Sens.* **2019**, *11*, 2722. [CrossRef]
25. Thuan, D.H.; Binh, L.T.; Viet, N.T.; Hanh, D.K.; Almar, R.; Marchesiello, P. Typhoon impact and recovery from continuous video monitoring: A case study from Nha Trang Beach, Vietnam. *J. Coast. Res.* **2016**, *75*, 263–267. [CrossRef]
26. Otsu, N. A threshold selection method from gray-level histograms. *IEEE Trans. Syst. Man Cybern.* **1979**, *9*, 62–66. [CrossRef]
27. Almar, R.; Blenkinsopp, C.; Almeida, L.P.; Cienfuegos, R.; Catalan, P.A. Wave runup video motion detection using the Radon Transform. *Coast. Eng.* **2017**, *130*, 46–51. [CrossRef]
28. Pan, S.J.; Yang, Q. A survey on transfer learning. *IEEE Trans. Knowl. Data Eng.* **2009**, *22*, 1345–1359. [CrossRef]
29. Valentini, N.; Balouin, Y. Assessment of a smartphone-based camera system for coastal image segmentation and sargassum monitoring. *J. Mar. Sci. Eng.* **2020**, *8*, 23. [CrossRef]
30. Morichon, D.; de Santiago, I.; Delpey, M.; Somdecoste, T.; Callens, A.; Liquet, B.; Liria, P.; Arnould, P. Assessment of Flooding Hazards at an Engineered Beach during Extreme Events: Biarritz, SW France. *J. Coast. Res.* **2018**, *85*, 801–805. [CrossRef]

31. Abadie, S.; Butel, R.; Mauriet, S.; Morichon, D.; Dupuis, H. Wave Climate and Longshore Drift on the South Aquitaine Coast. *Cont. Shelf Res.* **2006**, *26*, 1924–1939. [CrossRef]
32. De Santiago, I.; Morichon, D.; Abadie, S.; Castelle, B.; Liria, P.; Epelde, I. Video monitoring nearshore sandbar morphodynamics on a partially engineered embayed beach. *J. Coast. Res.* **2013**, *65*, 458–463. [CrossRef]
33. Bengio, Y.; Goodfellow, I.; Courville, A. *Deep Learning*; MIT Press: Cambridge, MA, USA, 2017; Volume 1.
34. LeCun, Y.; Bottou, L.; Bengio, Y.; Haffner, P. Gradient-based learning applied to document recognition. *Proc. IEEE* **1998**, *86*, 2278–2324. [CrossRef]
35. Krizhevsky, A.; Sutskever, I.; Hinton, G.E. Imagenet Classification with Deep Convolutional Neural Networks. *Adv. Neural Inf. Process. Syst.* **2012**, *25*, 1097–1105. [CrossRef]
36. Simonyan, K.; Zisserman, A. Very deep convolutional networks for large-scale image recognition. *arXiv* **2014**, arXiv:1409.1556.
37. Szegedy, C.; Vanhoucke, V.; Ioffe, S.; Shlens, J.; Wojna, Z. Rethinking the inception architecture for computer vision. In Proceedings of the IEEE Conference on Computer Vision and Pattern Recognition, Las Vegas, NV, USA, 27–30 June 2016; pp. 2818–2826.
38. Japkowicz, N.; Stephen, S. The class imbalance problem: A systematic study. *Intell. Data Anal.* **2002**, *6*, 429–449. [CrossRef]
39. Buda, M.; Maki, A.; Mazurowski, M.A. A Systematic Study of the Class Imbalance Problem in Convolutional Neural Networks. *Neural Netw.* **2018**, *106*, 249–259. [CrossRef]
40. Johnson, J.M.; Khoshgoftaar, T.M. Survey on deep learning with class imbalance. *J. Big Data* **2019**, *6*, 27. [CrossRef]
41. Deng, J.; Dong, W.; Socher, R.; Li, L.J.; Li, K.; Fei-Fei, L. Imagenet: A large-scale hierarchical image database. In Proceedings of the 2009 IEEE Conference on Computer Vision and Pattern Recognition, Miami Beach, FL, USA, 20–25 June 2009; pp. 248–255.
42. Uchida, S.; Ide, S.; Iwana, B.K.; Zhu, A. A further step to perfect accuracy by training CNN with larger data. In Proceedings of the 2016 15th International Conference on Frontiers in Handwriting Recognition (ICFHR), Shenzhen, China, 23–26 October 2016; pp. 405–410.
43. Snoek, J.; Larochelle, H.; Adams, R.P. Practical Bayesian Optimization of Machine Learning Algorithms. *arXiv* **2012**, arXiv:1206.2944.
44. Ganin, Y.; Lempitsky, V. Unsupervised domain adaptation by backpropagation. In Proceedings of the International Conference on Machine Learning, Lille, France, 6–11 July 2015; pp. 1180–1189.
45. Dodge, S.; Karam, L. A study and comparison of human and deep learning recognition performance under visual distortions. In Proceedings of the 2017 26th International Conference on Computer Communication and Networks (ICCCN), Vancouver, BC, Canada, 31 July–3 August 2017; pp. 1–7.
46. Simarro, G.; Bryan, K.R.; Guedes, R.M.; Sancho, A.; Guillen, J.; Coco, G. On the use of variance images for runup and shoreline detection. *Coast. Eng.* **2015**, *99*, 136–147. [CrossRef]
47. Andriolo, U.; Sánchez-García, E.; Taborda, R. Operational use of surfcam online streaming images for coastal morphodynamic studies. *Remote Sens.* **2019**, *11*, 78. [CrossRef]
48. Mole, M.A.; Mortlock, T.R.; Turner, I.L.; Goodwin, I.D.; Splinter, K.D.; Short, A.D. Capitalizing on the surfcam phenomenon: A pilot study in regional-scale shoreline and inshore wave monitoring utilizing existing camera infrastructure. *J. Coast. Res.* **2013**, 1433–1438. [CrossRef]
49. Andriolo, U. Nearshore wave transformation domains from video imagery. *J. Mar. Sci. Eng.* **2019**, *7*, 186. [CrossRef]
50. Poelhekke, L.; Jäger, W.; van Dongeren, A.; Plomaritis, T.; McCall, R.; Ferreira, Ó. Predicting Coastal Hazards for Sandy Coasts with a Bayesian Network *Coast. Eng.* **2016**, *118*, 21–34. [CrossRef]
51. Plomaritis, T.A.; Costas, S.; Ferreira, Ó. Use of a Bayesian Network for Coastal Hazards, Impact and Disaster Risk Reduction Assessment at a Coastal Barrier (Ria Formosa, Portugal). *Coast. Eng.* **2018**, *134*, 134–147. [CrossRef]

remote sensing MDPI

Article

Fog Season Risk Assessment for Maritime Transportation Systems Exploiting Himawari-8 Data: A Case Study in Bohai Sea, China

Pei Du [1], Zhe Zeng [1,*], Jingwei Zhang [2], Lu Liu [2], Jianchang Yang [3], Chuanping Qu [1], Li Jiang [1] and Shanwei Liu [1]

1 College of Oceanography and Space Informatics, China University of Petroleum, Qingdao 266580, China; s19160002@s.upc.edu.cn (P.D.); z20160077@s.upc.edu.cn (C.Q.); s17010066@s.upc.edu.cn (L.J.); 20110052@upc.edu.cn (S.L.)

2 Marine Geophysical Prospecting Branch, Bureau of Geophysical Prospecting Incorporated, China National Petroleum Corporation, Tianjin 300457, China; zhangjingwei@cnpc.com.cn (J.Z.); liulu_01@cnpc.com.cn (L.L.)

3 Offshore Oil Engineering Company Limited, Tianjin 300461, China; yangjc1@cooec.com.cn

* Correspondence: zengzhe@upc.edu.cn

Abstract: Sea fog is a disastrous marine phenomenon for ship navigation. Sea fog reduces visibility at sea and has a great impact on the safety of ship navigation, which may lead to catastrophic accidents. Geostationary orbit satellites such as Himawari-8 make it possible to monitor sea fog over large areas of the sea. In this paper, a framework for marine navigation risk evaluation in fog seasons is developed based on Himawari-8 satellite data, which includes: (1) a sea fog identification method for Himawari-8 satellite data based on multilayer perceptron; (2) a navigation risk evaluation model based on the CRITIC objective weighting method, which, along with the sea fog identification method, allows us to obtain historical sea fog data and marine environmental data, such as properties related to wind, waves, ocean currents, and water depth to evaluate navigation risks; and (3) a way to determine shipping routes based on the Delaunay triangulation method to carry out risk analyses of specific navigation areas. This paper uses global information system mapping technology to get navigation risk maps in different seasons in Bohai Sea and its surrounding waters. The proposed sea fog identification method is verified by CALIPSO vertical feature mask data, and the navigation risk evaluation model is verified by historical accident data. The probability of detection is 81.48% for sea fog identification, and the accident matching rate of the navigation risk evaluation model is 80% in fog seasons.

Keywords: sea fog identification; Himawari-8; navigation risk evaluation; CRITIC weighting method

1. Introduction

The visibility at sea will be greatly reduced in foggy weather [1], and ships sailing in fog tend to slow down and proceed carefully. However, collisions with other ships and rocks, and ship stranding, occur frequently every year in fog seasons [2]. Bohai Sea and Yellow Sea in China have a high incidence of sea fog [3], which has a great impact on shipping navigation and route planning. When using traditional, ground-based observational methods of sea fog (e.g., as obtained at meteorological stations), it may be hard to obtain the sea fog distribution of a whole sea area, which makes it difficult to assess navigation risks in fog seasons. In recent years, with the development of remote sensing technology, sea fog monitoring methods with large ranges and high frequency have gradually matured [4], which makes it possible to develop navigation risk assessment models in fog seasons. Indeed, it is of great significance for sailors to make correct navigational decisions when sailing in foggy weather.

There is little research in the field of navigation risk assessment that applies to remote sensing technology for sea fog detection, although many methods of detecting sea fog by remote sensing have been developed. For instance, the multiband threshold algorithm [5] and U-Net deep learning method [6] were used for sea fog detection based on Moderate Resolution Imaging Spectroradiometer (MODIS) satellite data. An unsupervised learning algorithm was used for sea fog detection based on Communication, Ocean, and Meteorological Satellite (COMS) data [7], while a dual-satellite method was proposed by [8], which combined Himawari-8 and FY-4A satellite data to detect sea fog at dawn. The authors of [9] proposed a decision tree approach that combined Himawari-8 and Geostationary Ocean Color Imager (GOCI) satellite data to detect sea fog. However, ship navigation risk assessments require long-term and daily sea fog data, which limits existing sea fog detection methods based on multi-source remote sensing satellites or satellites with a long revisit period. Himawari-8 is a geostationary orbit satellite launched by Japan in 2014, which can provide stable remote sensing images at intervals of 10 min [10], but few sea fog detection studies based on individual Himawari-8 satellite data have been conducted. The authors of [11] proposed a normalized difference snow index (NDSI) threshold algorithm for sea fog detection, but they noted that distinguishing between low clouds and sea fog could be further improved. Therefore, establishing a sea fog identification method based on Himawari-8 satellite data, as well as deriving a navigation risk estimation model based on remote sensing data, are great challenges for both research fields.

In navigation risk estimation research, weighting methods and analysis methods have been used to construct navigation risk assessment models. Some examples include the Fuzzy Analytic Hierarchy Process (Fuzzy-AHP) [12], formal safety assessment (FSA) [13], and Fuzzy reasoning [14], which were used to construct navigation risk estimation models. The authors of [15] analyzed navigation risks in the Arctic with a Bayesian network. According to these studies, due to a lack of objective data, experienced experts inevitably need to be used as a supplementary approach, leading to some subjective uncertainties in the estimations. Objective data (such as historical accident data) are considered to have more potential in providing unbiased risk information [16–18], such as being used as training data in a Bayesian network. Historical accident data and automatic identification system (AIS) data can be used instead of an experienced expert, which have further revealed potential links between different risk factors [19,20]. However, apart from training Bayesian networks, objective data are rarely used in other navigational risk estimation research. In addition, these studies had a good analytical capability for specific ships and accidents, but were unable to reflect the spatial distribution and variation of risks.

In order to enhance spatial analysis capability, the spatial multi-objective decision method was applied to the navigation risk estimation research. For instance, Fuzzy-AHP [21,22] and the grey relationship analysis (GRA) method were used to weigh evaluation criteria, and a navigation risk evaluation model based on spatial multi-decision technology was constructed [23]. Questionnaire data and expert experience data have also been used [21,22], which led to strong subjectivity, and the study by [22] regarding low cloud data was used to replace the sea fog. Therefore, to provide better navigation risk estimation in fog seasons, it is necessary to integrate sea fog detection technology into navigation risk estimation models, and calculate the weights of relevant criteria using objective data.

The main aims of this paper are to: (1) develop a sea fog identification method model based on Himawari-8 remote sensing imaging data, and obtain a large historical sea fog data set of Bohai Sea and its surrounding waters. Then, the main fog season(s) and fog area(s) are distinguished by statistical analyses; (2) determine the criteria weights from accident data using an objective weight method to avoid subjective uncertainties in the weight determination, and use data of sea fog, wind, waves, currents, and water depth to construct a navigation risk estimation model based on spatial multi-decision technology; and (3) assess the navigation risk during each of the four seasons, divide the shipping routes into different navigation areas, and analyze the risks in the specific navigation areas during each season.

2. Materials

2.1. Study Area

The study area in this paper includes Bohai Sea and its surrounding waters in China. As shown in Figure 1, the study area ranges from 117° E to 123° E, 35.5° N to 41° N. This region includes the main ports in five provinces around the Bohai Economic Rim, including Qingdao Port, Yantai Port, Dalian Port, Tianjin Port, Yingkou Port, etc. The study area is the main area of maritime trade in northern China, and also an important navigable area in China. The study area is adjacent to North Korea, South Korea, Japan, and other countries in eastern Asia. The navigable environment of Bohai Sea is relatively complex, and the average water depth of the sea area is less than 30 m [24]. Influenced by monsoons, the climate of Bohai Sea and Yellow Sea presents obvious seasonal and spatial characteristics. In winter, influenced by cold air, the wind is strong, and the waves and surges produced have a great influence on the navigability of ships [25]. In summer, as warm and humid air flows northward, sea fog is easily generated and has a great influence on ship navigability [26].

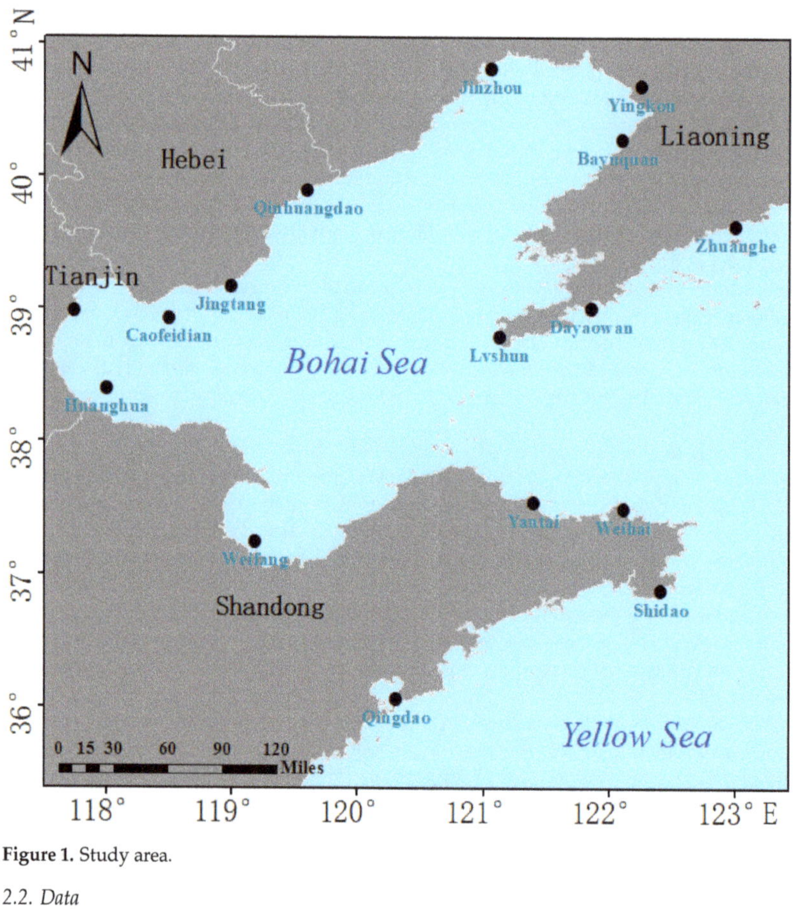

Figure 1. Study area.

2.2. Data

For our sea fog identification method, Imagery of Advanced Himawari Imager (AHI) sensor from Himawari-8 satellite and CALIPSO lidar level-2 vertical feature mask (VFM) data are used for sea fog sample construction, classifier training, and accuracy verification. AHI data around UTC 09:00 every day from 2015 to 2020 are used to obtain the daily sea fog information in the research area. These data are combined with wind and wave data from the ECMWF Reanalysis v5 model (ERA5), ocean currents data from the Hybrid Coordinate Ocean Model (HYCOM), and water depth data from General Bathymetric Chart of the Oceans (GRBCO) to statistically ascertain the occurrence of adverse sea conditions. Second, historical accident data of the sea area are used to calculate the weights of the criteria and to verify the risk evaluation model. The shipping route data are also used to divide the navigation section area so that risk analyses can be conducted on specific routes or parts thereof. The source, data introduction, related references, time, and use of the various data are shown in Table 1.

Table 1. Data information.

Name	Source	Data Description	Time	Purpose
AHI data	Japan Aerospace Exploration Agency (https://www.eorc.jaxa.jp/ptree (accessed on 1 July 2021))	Remote sensing images from the AHI sensor [10]	2015–2020	Sea fog identification and calculating the frequency of monthly sea fog
CALIPSO-VFM	National Aeronautics and Space Administration (https://search.earthdata.nasa.gov/search (accessed on 7 July 2021))	Classified data of clouds and aerosols from CALIPSO lidar level-2 product [27]	2018	Sea fog samples and verifying the accuracy
Wind	European Centre for Medium-Range Weather Forecasts (https://www.ecmwf.int/ (accessed on 15 June 2021))	10 m wind reanalysis data from ERA5 [28]	2015–2020	Calculating the frequency of strong wind
Sea wave	European Centre for Medium-Range Weather Forecasts (https://www.ecmwf.int/ (accessed on 15 June 2021))	Significant Wave Height reanalysis data from ERA5 [28]	2015–2020	Calculating the frequency of big waves
Ocean currents	National oceanic and atmospheric administration (https://www.ncei.noaa.gov/ (accessed on 16 June 2021))	Analyses currents data from HYCOM [29]	2016–2020	Calculating the frequency of big currents
Water depth	British Oceanographic Data Centre (https://www.bodc.ac.uk/ (accessed on 16 June 2021))	Global terrain model from GEBCO [30]	2019	Obtaining water depth data
Accident data	Maritme Safety Admininstration of the people's Republic of China (https://www.msa.gov.cn/ (accessed on 1 March 2021))	Accident data collected and disclosed by the authorities	2015–2020	Weighting factors and validating the risk model
Shipping routes	Maritme Safety Admininstration of the people's Republic of China (https://www.msa.gov.cn/ (accessed on 1 March 2021))	Officially recommended shipping route [31]	2019	Subdivision shipping route

3. Methods

In order to evaluate the navigation risk in Bohai Sea and its surrounding waters, first, a sample database based on AHI data and CALIPSO vertical feature mask (CALIPSO-VFM) data is established, and a sea fog identification method based on Multilayer Perceptron (MLP) method is developed. Second, the method is used to obtain a large historical sea fog data set of the study area, which is combined with other evaluation factors to calculate the occurrence frequency of severe sea conditions. Then, the criteria importance through intercriteria correlation (CRITIC) objective weighting method is used to determine the weights of the evaluation criteria, and obtain navigation risk maps of the area via global information system (GIS) mapping technology. In order to further analyze the risks, the Delaunay triangulation method is used to divide the sea area into different waterway regions, and the risk values for the specific waterway areas are calculated. Finally, two validation methods are used to validate the sea fog identification method and navigation risk estimation model, and a sensitivity analysis of the model is carried out. A method flow chart of this process is shown in Figure 2.

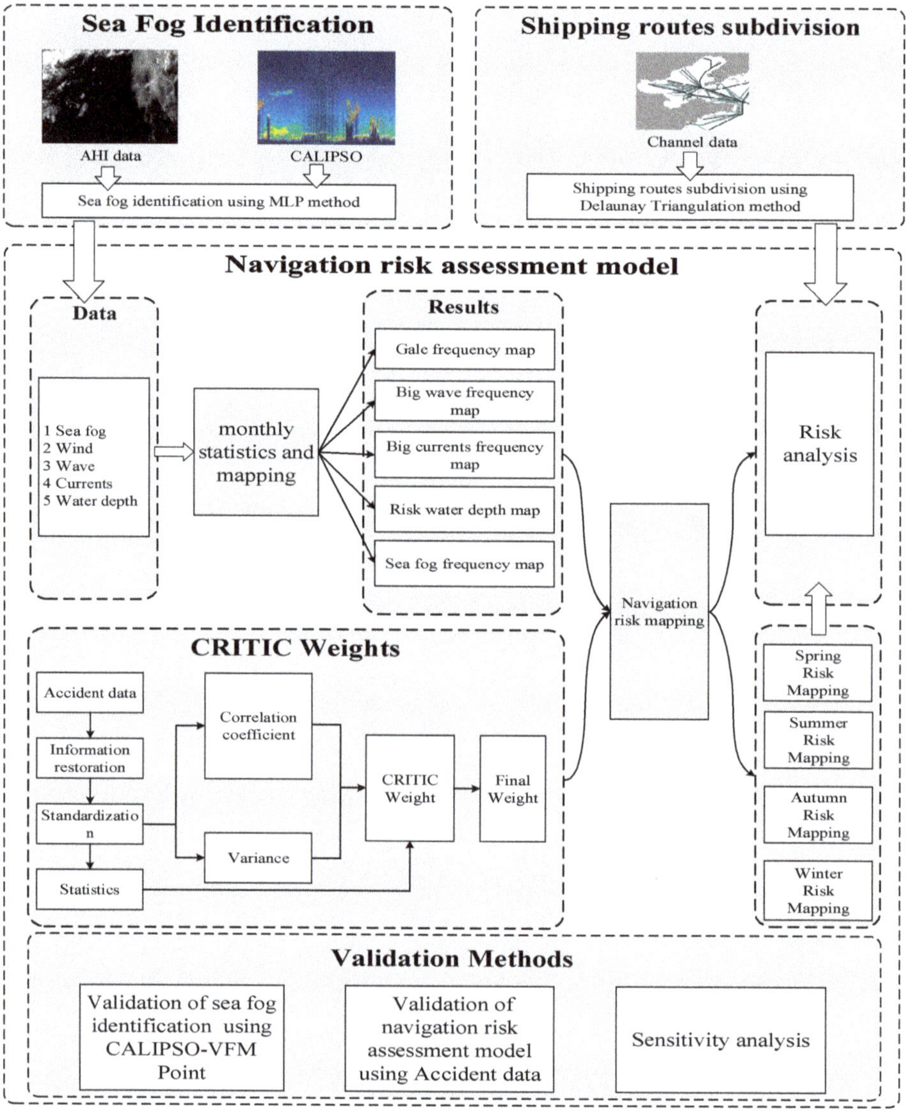

Figure 2. Flow chart of the navigation risk evaluation method.

3.1. Sea Fog Identification

In this paper, the MLP method is used to identify sea fog in the study area. First, sea fog sample points are selected from the CALIPSO-VFM and AHI data sets. Then, the spectral characteristics and fusion spectra of sea fog and non-sea-fog samples are analyzed to select features that benefit the classification. After that, sea fog and non-sea-fog binary classification models are constructed by the MLP method, and the optimal parameters are selected through training experiments. The specific processes are shown in Figure 3.

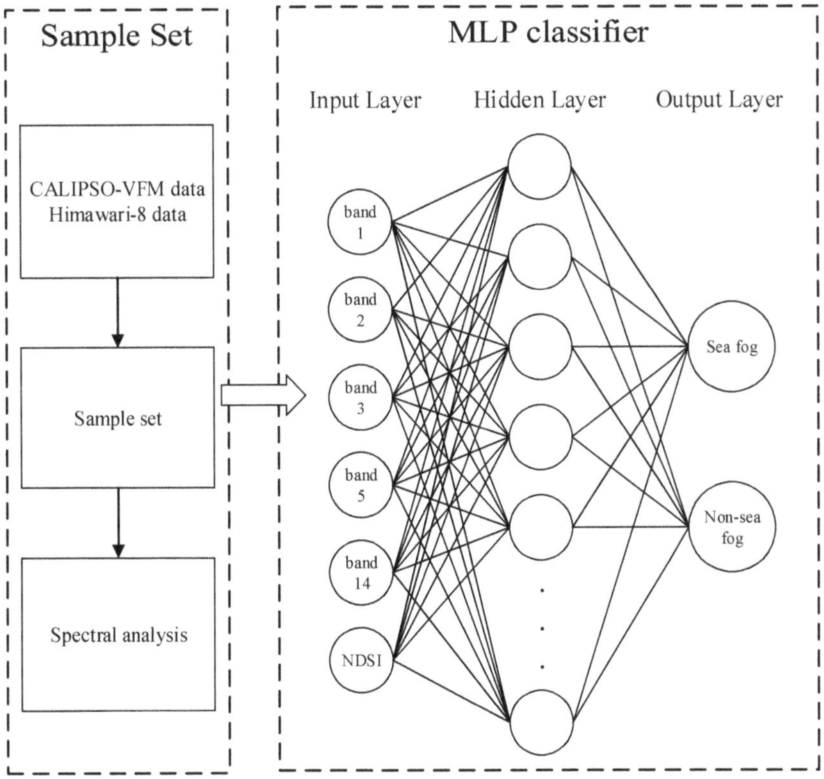

Figure 3. Flow chart of sea fog identification method.

3.1.1. Sample Selection of Sea Fog

It is difficult to distinguish between sea fog and low clouds and stratus clouds through visual interpretations [32]. Therefore, a sea fog sample database is established using CALIPSO-VFM data to assist the selection of sea fog samples from AHI images [5, 33]. Figure 4 shows the selection method, where Figure 4a is an AHI image on 15 June 2018 at 05:10 UTC, and Figure 4b is the classification result when using CALIPSO-VFM data from 15 June 2018 at 05:17 UTC. In the area in the red box in Figure 4b, the CALIPSO-VFM data are used to identify places where clouds are in contact with the sea surface, which are considered as sea fog points. From these, we extract the spectral values from the corresponding positions in the AHI images (e.g., the red dots in Figure 4a). Then, the spectral values at the sea surface, low clouds, and medium-high level clouds are obtained using the same approach. A total of 9141 samples of sea fog, 10,071 samples of the sea surface, 8733 samples of low clouds, and 10,771 samples of middle-high level clouds were selected and used in this paper.

3.1.2. Feature Selection

After obtaining the spectral features of each object, it is necessary to analyze the spectra to select features with better classification effect. The visible band of the Advanced Himawari Imager sensor onboard Himawari-8 consists of channels 1–3 (central wavelengths of 0.47 μm, 0.51 μm, and 0.64 μm, respectively); channels 4–6 are near-infrared bands (central wavelengths of 0.86 μm, 1.6 μm, and 2.3 μm, respectively), and channels 7–16 are infrared and far-infrared bands (central wavelengths of 3.9 μm, 6.2 μm, 6.9 μm, 7.3 μm, 8.6 μm, 9.6 μm, 10.4 μm, 11.2 μm, 12.4 μm, and 13.3 μm, respectively). The visible and near-infrared spectral characteristic curves of different objects are shown in Figure 5a. In the first three bands, the mid-high level clouds and the sea surface are significantly different from the sea fog and low clouds, but it is difficult to distinguish between low clouds and sea fog. With an increase of wavelength, the reflectance of all kinds of features gradually decrease, but the rate of change of sea fog in band 5 is significantly slower than that of other features, and the difference between the sea fog and the low clouds becomes obvious. According to [34], there are a large number of water droplets of 1 μm diameter in sea fog, which increase the reflectance measured in bands 4 and 5 and slow down the rate of change of reflectance in band 5; such behavior provides an obvious means to distinguish sea fog from other objects.

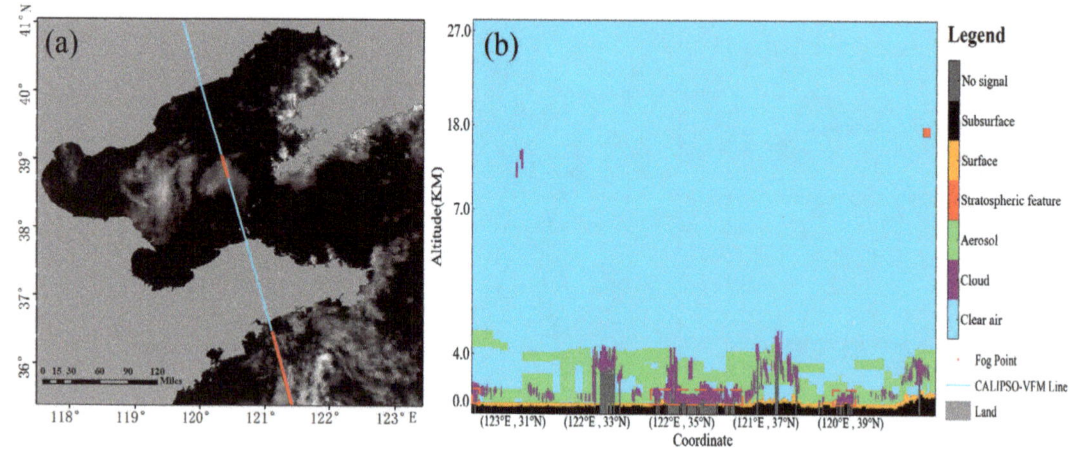

Figure 4. Sample selection using CALIPSO-VFM data, (**a**) AHI image on 15 June 2018 at 05:10 UTC. (**b**) Objects classification of CALIPSO-VFM data from 15 June 2018 at 05:17 UTC.

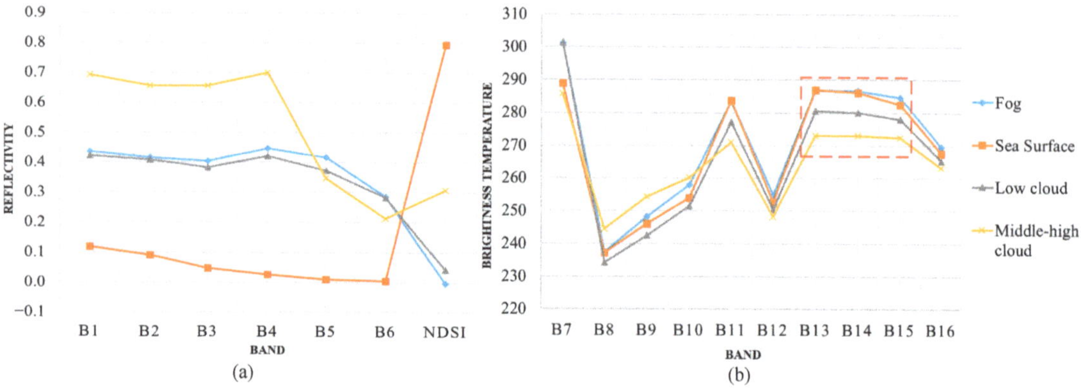

Figure 5. Spectral analysis, (**a**) the spectral reflectivity of bands 1–6 and NDSI for AHI data; (**b**) the brightness temperature of bands 7–16 for AHI data, the red box represents the regions of bands 13, 14, and 15 with obvious differences in brightness temperature of each objects.

To identify more features, NDSI [11] is constructed with data from bands 2 and 5 (i.e., 0.51 μm and 1.6 μm), calculated using Equation (1) as:

$$\text{NDSI} = \frac{B_{0.51\mu\text{m}} - B_{1.6\mu\text{m}}}{B_{0.51\mu\text{m}} + B_{1.6\mu\text{m}}}, \tag{1}$$

where $B_{0.51\mu m}$ and $B_{1.6\mu m}$ represent bands with central wavelengths of 0.51 μm and 1.6 μm, respectively, in the AHI data. It can be seen from Figure 5a that NDSI provides a significant distinction between sea fog and low clouds. There are also distinct intervals between different objects.

Figure 5b shows the characteristic curves of the different infrared bands. The brightness temperatures of sea fog and the sea surface are very similar, and cannot be used to effectively distinguish between them, even though the average height of low clouds is higher than that of sea fog, and the brightness temperature of the former is slightly lower. The brightness temperatures of middle-high clouds are the lowest, and this difference is more obvious in bands 13, 14, and 15 (see, e.g., the red dotted box in Figure 5b). Therefore, these three bands can be selected to further distinguish low clouds and sea fog.

Based on the spectral analysis, we selected bands 1, 2, 3, 5, and 14 and NDSI as classification features to construct the sea fog sample set.

3.1.3. Sea Fog Recognition Method Based on MLP

As a mature classification method, Multilayer Perceptron (MLP) method has been applied to various remote sensing image classification tasks [35,36]. In this paper, sea fog identification is taken as a binary-classification task to train the MLP classifier. The low clouds, sea surface, and medium-high level clouds are unified into a non-sea-fog category. The objective function of the MLP binary classification method can be described as shown in Equation (2) [37]:

$$\min_{w,b}\frac{1}{2}||h_{w,b}(f(wx+b))-y||^2, \tag{2}$$

where w and b represent the weight and bias of each layers, y represents the sample labels, x represents the sample values, $h_{w,b}$ represents the output of the output layer, and f represents the activation function. The kernel objective in this paper is the sigmoid function, which can be described as given in Equation (3):

$$f(x)=\frac{1}{1+\exp(-x)}, \tag{3}$$

Through training the MLP classifier, each parameter of the layers is obtained. The main training process consists of four steps:

- Step 1: Normalize the sample values, which are scaled between [0, 1];
- Step 2: Divide the training set and testing set, and train the MLP classifier using the former;
- Step 3: Cross validation to determine the optimal parameters; and
- Step 4: Complete MLP training and verify the accuracy of the classifier.

The partition ratio of training set to test set is 7:3, and the hidden layer sizes in this study is 100 × 100, and the random state is 10, the optimizer solver is "adam". Through training, a sea fog identification model based on AHI data is obtained, which is used to obtain a large amount of historical sea fog information in the research area. The verification method and accuracy results of sea fog identification are described in Sections 3.3.1 and 4.1.

3.2. Navigation Risk Assessment Model

After creating a suitable sea fog data source, the navigation risk assessment model can be constructed according to various environmental elements in the study area. This section mainly includes four parts: first, the accident data are statistically analyzed to select the environmental factors that have the greatest impact on the navigation of ships in the study area. Second, the data of each evaluation factor are preprocessed. Third, we introduce a detailed process of determining the weights of each evaluation factor by the CRITIC objective weighting method, and then construct the navigation risk assessment model. Finally, we use the Delaunay triangulation method to divide the sea area into different navigational regions, and calculate the navigation risks in the specific navigational areas for further risk analysis.

3.2.1. Criteria for Navigation Risk

The identification of relevant criteria is a very critical step in the construction of an assessment model. The navigation risks mentioned in this paper refer to the loss of people or property and the possible impact of the navigation environment, which may consist of harsh conditions, resulting in the improper operation of the ship in the course of navigation, and which may cause ship collisions (i.e., the collision of two or more ships), collisions with stationary objects (i.e., with a reef, coast, or building), grounding (where the bottom of the ship touches the bottom of the sea due to insufficient water depth), or a sunken ship (sinking of ship) [38]. In order to obtain the criteria that may lead to identifying navigation risks more objectively, we use accident data from Maritime Safety Administration of the People's Republic of China in study areas to determine criteria that have a greater impact on ship navigation. In this paper, environmental risks are mainly used for assessment, so the accidents caused by purely human factors need to be eliminated. After screening, a total of 39 accident records were included. The distribution of accident data is shown in Figure 6, where different color points represent accidents in different seasons.

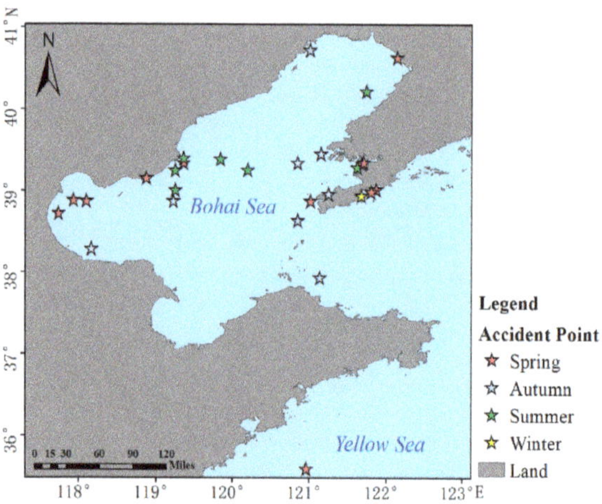

Figure 6. Distribution of accident points.

The accident assessment section of each accident report details the main and secondary causes of the accident. As such, we used the accident assessment report to count the number of accidents caused by the different assessment factors. When selecting accident records, certain rules should be observed. For example, if there is a sentence whose meaning is similar to "heavy fog in the sea, poor visibility, crew neglected to look out, resulting in ship collision" in the accident cause assessment, it is considered that the accident is related to sea fog and "1" is added to the number of accidents caused by sea fog in the statistics. If there is a sentence whose meaning is similar to "the waters of the accident were affected by strong winds, and the ship failed to evacuate the sea in time, resulting in the collision of the ship", it is considered that strong winds had an impact on the accident, and we added "1" to the number of accidents caused by strong winds to the statistics. When the accident is caused by a combination of various factors, such as "due to the superposition of strong wind, cold air, and storm surge in the accident water area" recorded in the report, it is considered that strong winds and big waves had an impact on the accident, and "1" is added to the statistics of each category. The statistical results of the final criteria involved in the historical accidents are shown in Table 2.

Table 2. Statistics of accident-related factors.

	Fog	Wind	Wave	Currents	Water Depth	Rainfall	Total
Number	13	17	18	7	9	4	68
Proportion	19.2%	25.0%	26.5%	10.3%	13.2%	5.8%	100%
Ranking	3	2	1	5	4	6	

Finally, according to the statistical information and data support, the top five criteria were selected, which were waves, wind, sea fog, ocean currents, and water depth. These five criteria are meteorological and hydrological factors which are especially relevant in meteorological navigation. Specifically, gale weather has an impact on both sailing and berthing ships. Sailing ships are affected by gales when sailing, which will reduce the controllability of the ships. If berthed ships are not fully anchored, the hull may shift under greater wind pressure, resulting in grounding or a collision. When big waves occur, they are usually accompanied by strong winds. It is difficult for ships to adapt to the peaks and troughs of sea waves, and when ships with lower tonnages are hit by big waves, their external structure will be subjected to greater external forces, which may cause damage to the ship, which is very dangerous during navigation. Currents will also affect ship navigation. In narrow sea areas, the speed and direction of ocean currents are unstable, which makes it more difficult to effectively use seamanship. When sailing against the currents, the energy consumption of a ship increases, and the difficulty of navigation also increases [39].

Sea fog is the main reason affecting the visibility at sea. Many ship collisions, as well as touching and grounding accidents, are caused by sea fog. Generally, ships will avoid sailing in fog. However, when sea fog arises, it may be difficult for sailing ships to leave the fog area promptly. When sea fog occurs, the sea waves are generally small but limited by the field of vision. It is difficult for the crew to fully observe the surrounding navigation environment, and the probability of accidents increases greatly [40]. Water depth is a factor that must also be considered when planning passage through a channel. Before sailing, the captain should plan the route reasonably according to the ship's draught and abundant water depth. Generally speaking, the nearshore sea is shallow, so when nearshore navigation is affected by other factors, grounding and collisions may easily occur. In addition, the water depth is related to sediment in the sea area, which may cause a change of water depth due to the movement of the sea water. If the relevant properties of the navigation area are not known in advance, accidents can easily occur.

In general, the five kinds of navigation environment criteria identified by the accident data statistics have a great impact on ship navigation, and are also the five main environmental factors about which the maritime industry are very concerned. At present, wind, waves, currents, and terrain factors can be supported by the existing product level data, but it is still difficult to obtain sea fog data in large areas. Therefore, this paper uses a remote sensing identification method to obtain the daily sea fog occurrence in Bohai Sea and its surrounding waters over a five-year period, using the method described in Section 3.1.3.

3.2.2. Data Processing

Five evaluation factors with a great impact on ship navigation were selected in Section 3.2.1. However, before conducing the risk assessments, the risk factor value of each criterion should be obtained in each of the four seasons. First, the spatial resolution of the gridded data are unified to $0.125° \times 0.125°$. Second, the risk values of each factor under severe sea conditions are calculated from the gridded wind, wave, ocean currents, and sea fog data from 2016 to 2020. Then, the risk grid values of each factor are standardized. When setting the thresholds for the different evaluation criteria, we follow the methods employed in previous research [21,41].

First, the sea fog criterion calculates the occurrence frequency of sea fog in each grid in the different seasons following Equation (4):

$$F_i^j = \frac{\sum f_{i,k}}{d^j}, \tag{4}$$

where F_i^j represents the occurrence frequency of sea fog in grid i in season j, $f_{i,k}$ represents the occurrence of sea fog in grid i on day k, $f_{i,k} \in [0,\ 1]$, 0 represents no sea fog, 1 represents sea fog, and d^j represents the total number of days in season j.

The wind criterion, for wind speeds greater than 11.7 m/s in each grid, in different seasons is calculated via Equation (5):

$$G_i^j = \frac{\sum g_{i,k}}{d^j}, \tag{5}$$

where G_i^j represents the frequency of wind speeds greater than 11.7 m/s in grid i in season j, $g_{i,k}$ represents wind conditions for grid i on day k, $g_{i,k} \in [0,\ 1]$, 0 represents wind speeds slower than 11.7 m/s, 1 represents winds speeds greater or equal to 11.7 m/s, and d^j represents total days in season j.

Calculation of the wave and currents criteria are same as for the wind criterion, where the thresholds for the wave and currents criteria are 2 m and 1 m/s, respectively.

The water depth is a static criterion, where each grid risk value of water depth is calculated using Equation (6):

$$D_i = \begin{cases} 0 & d_i > 30\ \text{m} \\ 1 - \frac{d_i}{30} & 0\ \text{m} < d_i < 30\ \text{m} \\ 1 & d_i < 0\ \text{m} \end{cases}, \tag{6}$$

where D_i represents the risk value of water depth in grid i, and d_i represents the original water depth value.

After calculating these criteria, it is necessary to standardize the grid values to 0–1. The standardization formula used in this paper is shown in Equation (7):

$$x' = \frac{x - x_{min}}{x_{max} - x_{min}}, \tag{7}$$

where x' represents the standardized value, x represents the original risk value, x_{min} represents the minimum value, and x_{max} represents the maximum value.

3.2.3. Weighting the Criteria Using CRITIC

Weights can reflect the importance of each evaluation criterion, and the method of weighting is an important step in building a spatial, multi-objective decision model. The traditional way to obtain weighting data is from an expert's knowledge or a questionnaire, which inevitably involve some subjective uncertainties. With the aim of eliminating subjectivity as much as possible, we collected accident data from 2015 to 2021 in Bohai Sea and Yellow Sea, and reproduced the values of each evaluation factor during the accident using the CRITIC weighting method to get the criteria weights. The CRITIC method determines weights according to the discretization of each criterion and the relationships between them [42], and is widely used in weight determination problems [43,44]. In this paper, the CRITIC method is used to get the criteria weights from historical accident data. The specific weight determination steps are as follows.

- Step 1. The accident value of each criterion is standardized using Equation (7).
- Step 2. The standardized data are organized into an evaluation matrix. The columns represent different criteria, and the rows represent the standardized value of the same criteria, as shown in Equation (8):

$$\begin{bmatrix} x'_{11} & x'_{12} & \cdots & x'_{1k} \\ x'_{21} & x'_{22} & & x'_{2k} \\ \vdots & \vdots & \ddots & \vdots \\ x'_{n1} & x'_{n2} & \cdots & x'_{nk} \end{bmatrix}, \quad (8)$$

where k represents the total number of criteria, n represents the total number of accident records, and x' represents the standardized value.
- Step 3. Calculate the correlation coefficients, and a correlation coefficient matrix is obtained, expressed as Equation (9). Then, the conflict coefficient of each evaluation factor is calculated by Equation (10):

$$\begin{bmatrix} r_{11} & r_{12} & \cdots & r_{1k} \\ r_{21} & r_{22} & & r_{2k} \\ \vdots & \vdots & \ddots & \vdots \\ r_{k1} & r_{k2} & \cdots & r_{kk} \end{bmatrix}, \quad (9)$$

$$R_z = \sum_{m=1}^{k} (1 - r_{z,m}) z \in [1, \ldots, k], \quad (10)$$

where R_z represents the conflict coefficient of criterion z, and $r_{z,m}$ represents the correlation coefficients between criteria z and m. Note that a bigger correlation between criterion z with other criteria produces a smaller R_z.
- Step 4. Calculate the sample standard deviation of each criterion using Equation (11):

$$S_z = \sqrt{\frac{\sum_{l=1}^{n} \left(x'_l - \overline{x'} \right)^2}{n-1}}, \quad (11)$$

where S_z represents the sample standard deviation of criterion z, and $\overline{x'}$ is the mean of the standardized values.
- Step 5. The amount of information for criterion z is calculated with the conflict coefficient R_z obtained in Step 3 via Equation (12):

$$C_z = S_z \cdot R_z, \quad (12)$$

where C_z represents the amount of information of criterion z.
- Step 6. Normalize the amount of information and get the CRITIC weights for each criterion following Equation (13):

$$w_z^c = \frac{C_z}{\sum_{m=1}^{k} C_m}, \quad (13)$$

where w_z^c represents the CRITIC weight of criterion z.

Through the above six steps, the CRITIC method can be used to obtain the different criteria weights. When the correlations between a criterion and other criteria are large and the variances are small, the CRITIC weight of this criterion will be relatively small. Thus, this method can make a tradeoff between the correlations and the standard deviations of the different factors, and hence give more reasonable weight values. The final weight of each criterion is combined with the accident statistics (see Table 2) using Equation (14):

$$w_z = \frac{w_z^c \times Q_z}{\sum_{m=1}^{k} w_m^c \times Q_m}, \tag{14}$$

where w_z represents the final weight of criterion z, and Q_z represents the normalized statistics of criterion z, obtained from Table 2.

The correlation coefficients of the five criteria are shown in Table 3. The high correlation coefficient between the wind and wave criteria indicates a certain correlation between wind and waves in accidents. The weak negative correlation between sea fog and wind reflects that it is not easy to generate sea fog in gale weather. The correlations between ocean currents with the other factors are small reflecting that the ocean currents are less affected by other environmental factors. The weighting results are shown in Table 4 in descending order: sea fog, waves, wind, depth, and currents.

Table 3. Correlation table of CRITIC-determined weights.

	Wind	Wave	Sea Fog	Currents	Water Depth
Wind	1	0.76	−0.15	0.23	0.19
Wave	0.76	1	−0.17	0.04	0.12
Sea fog	−0.15	−0.17	1	−0.01	−0.12
Currents	0.23	0.04	−0.01	1	−0.21
Depth	0.19	0.12	−0.12	−0.21	1

Table 4. Weights.

Name	Wind	Wave	Sea Fog	Currents	Water Depth
R_z	2.97	3.24	4.46	3.95	4.02
CRITIC weights	0.14	0.17	0.36	0.17	0.14
Final weights	0.19	0.23	0.37	0.09	0.10
Rank	3	2	1	5	4

The navigational risk value is calculated through Equation (15):

$$R_i = \sum_{z=1}^{k} w_z \times x'_{i,z}, \tag{15}$$

where R_i represents the navigational risk value in grid i, and $x'_{i,z}$ represents the standardized value of criterion z in grid i. The calculated navigational risk values are standardized using Equation (7). After standardization, the final navigational risk assessment values are obtained, which are divided into four levels: [0, 0.25), [0.25, 0.5), [0.5, 0.75), and [0.75, 1], representing "safe", "a little safer", "a little dangerous", and "dangerous", respectively.

3.2.4. Shipping Routes Subdivision

To provide a suitable risk analysis of specific shipping lanes in Bohai Sea and its surrounding waters, 107 main shipping routes officially recommended by the China Maritime Safety Administration in the study area [31] are considered in this paper (as shown in Figure 7). Next, the shipping routes are manually simplified (as shown in Figure 8a). Then, the Delaunay triangulation method is used to divide the navigation area into the main shipping routes. The main steps of this process are described as follows.

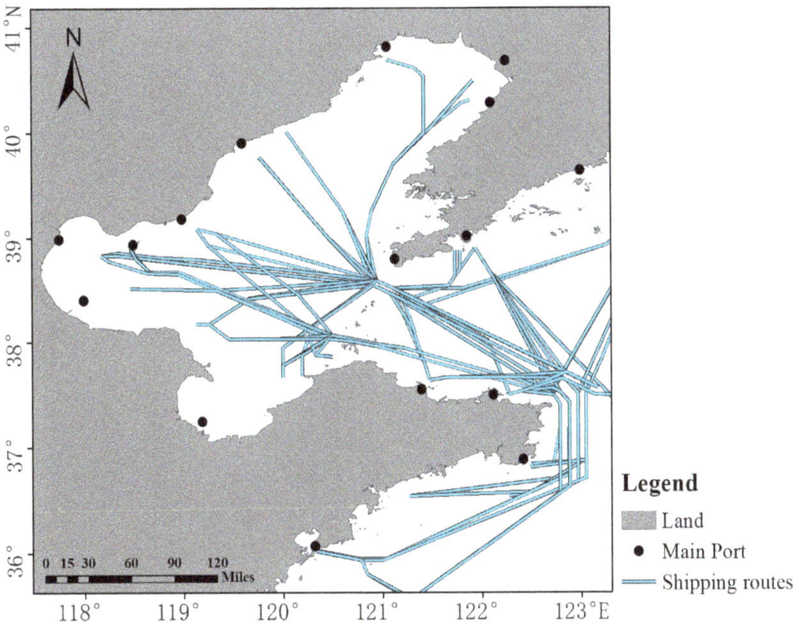

Figure 7. Shipping route distribution.

- Step 1: Merge and simplify the dense shipping routes;
- Step 2: Generate equally spaced route points along the simplified shipping routes;
- Step 3: Use the Delaunay triangulation method to generate a regional Tyson polygon network with route points as the source;
- Step 4: Merge polygons from the same source to form a route subdivision;
- Step 5: Trim the boundaries manually.

A subdivided region is obtained after trimming, where the sea area is divided into different regions according to the shipping routes, as shown in Figure 8a–c. The black line segments in Figure 8a show the simplified shipping routes, and the green points are the route points generated at equal intervals along the routes. Figure 8b shows the Tyson polygon network generated by the route points, and Figure 8c shows the regional channel network after merging and trimming.

By dividing the sea areas into different routes, a risk analysis can be clearly conducted on each specific route area. The mean risk value represents each route's region risk for every area, which is calculated as Equation (16):

$$\overline{R_d} = \frac{R_{sum}^d}{A_d}, \tag{16}$$

where $\overline{R_d}$ represents the risk value of region d, R_{sum}^d represents the total risk value of region d, and A_d represents the area of region d. By classifying the risks in each region of each route, the risks therein can be further analyzed. Section 4.4 describes the analysis results.

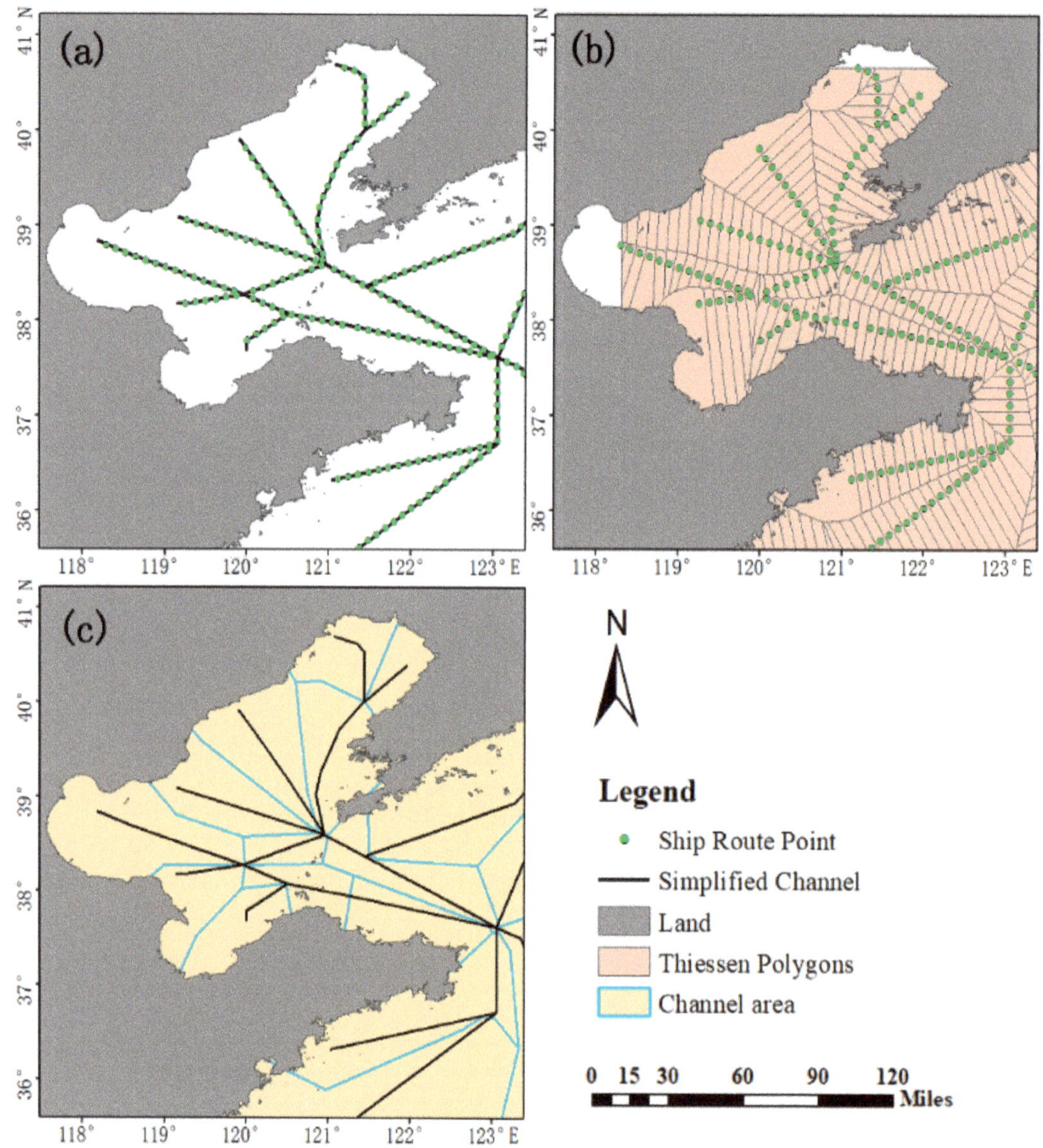

Figure 8. Channel area division process, (**a**) the simplified shipping routes and points; (**b**) the regional Tyson polygon using Delaunay triangulation method with route points as the source; (**c**) shipping route areas after merging and trimming.

3.3. Validation Methods

3.3.1. Sea Fog Identification Validation Using the CALIPSO-VFM Points

In this paper, CALIPSO-VFM data are used for sea fog verification. The CALIPSO-VFM data of Bohai Sea and Yellow Sea from January to September in 2018 are selected to identify different types of ground objects along the trajectory lines, which are treated as different object points. The observing times of the AHI data and CALIPSO-VFM data are controlled within 10 min. After obtaining the sea fog inversion results using the MLP method, the confusion matrix is counted, and the spatial distributions of the track points are obtained, as shown in Figure 9. Four types of ground objects are clearly indicated here, but only sea fog and non-sea-fog are required for the analysis made in this paper. The detailed classification made here allows us to further analyze the algorithm's accuracy. This paper compares and analyzes the results obtained with the proposed MLP method with those obtained by the NDSI threshold method proposed by [11], and the specific results are shown in Section 4.1.1.

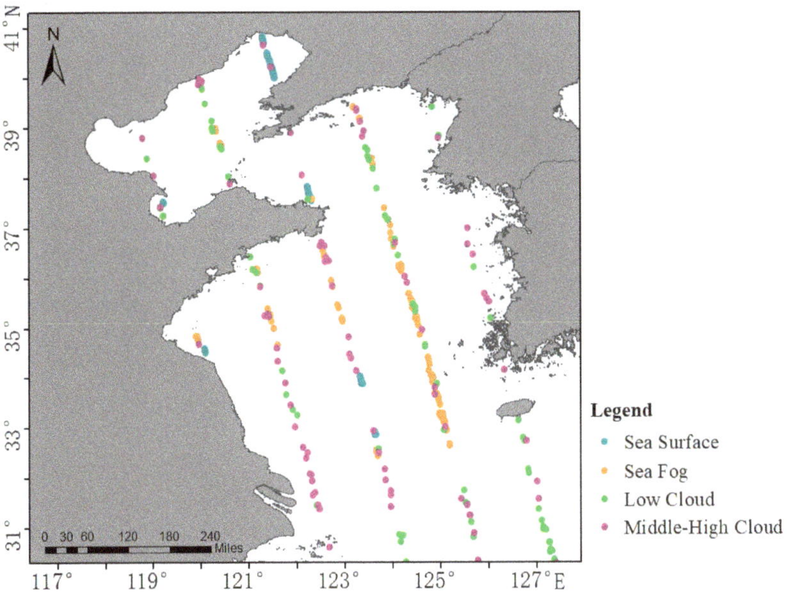

Figure 9. Distribution of CALIPSO-VFM verification points around UTC 05:00 from January to September in 2018.

3.3.2. Navigation Risk Evaluation Validation Using the Accident Points

This paper uses accident data to verify the accuracy of the evaluation model. The collected accident reports are treated as accident points, which are divided into four subsamples based on the four seasons (e.g., spring, summer, autumn, and winter). The specific distribution of the accident points is shown in Figure 6. When matching the accident points with the navigation risk evaluation results in the same season, if an accident point occurs in a high-risk area (i.e., a risk value greater than or equal to 0.5), the match is successful; if it occurs in a low-risk area (i.e., a risk value less than 0.5), the match is not successful. The matching degree of the model with the accidents was verified by counting the matching rate, and the results are shown in Section 4.2.

3.3.3. Sensitivity Analysis

A sensitivity analysis can be used to analyze the stability of a model and observe the stability of the evaluation criteria weights. The one-at-a-time method is a widely used sensitivity analysis method for spatial multi-objective decision technology [45,46], where the sensitivity increase/decrease of the spatial evaluation results is calculated by way of the aggregate mean, which has been applied to the risk evaluation model in this paper.

The main steps of the method are as follows. First, for criteria a, the weight value is increased and decreased using 10% equal intervals. The scaling range is [−50%, 50%], and the new weight of criteria a is obtained using Equation (17):

$$\overline{w_a}(cr) = (1 + cr) * w_a \,, \tag{17}$$

where $\overline{w_a}(cr)$ represents the new weight of criteria *a*, *cr* is the increased/decreased value, and w_a represents the original weight. To ensure that the summed weights equal 1, Equation (18) is calculated for the remaining criteria:

$$\overline{w}_h(cr) = w_h \times \frac{1-\overline{w_a}}{1-w_a}, \tag{18}$$

where $\overline{w}_h(cr)$ represents the weights adjusted for the remaining criteria, $(h \neq a)$, and w_h represents the original weight of criterion *h*.

Then, the new risk evaluation values are calculated using the adjusted weights, as shown in Equation (19):

$$\mathrm{R}(w_a, cr) = \overline{w_a} \times x_a' + \sum_{a \neq h}^{n} \overline{w_h} \times x_h', \tag{19}$$

where $\mathrm{R}(w_a, cr)$ represents the new risk evaluation value for each grid, x_a' represents the standardized value of criterion *a* for each grid, and x_h' represents the standardized value of criterion *h* for each grid.

The mean of the absolute change rate (MACR) of the whole region is calculated to obtain the change rate of criterion *a* with changes *cr*, following Equation (20):

$$\mathrm{MACR}(w_a, cr) = \sum_{j=1}^{S} \frac{1}{S} \left| \frac{R_j(w_a, \mathrm{cr}) - R_j^0}{R_j^0} \right| \times 100\%, \tag{20}$$

where $\mathrm{MACR}(w_a, cr)$ represents the MACR of the risk evaluation in the whole region, $R_j(w_a, \mathrm{cr})$ represents the risk evaluation value of the *j*-th grid after weight adjustment, and R_j^0 represents the original risk evaluation value, *S* represents the total number of grids.

If the adjustment amplitude of a criterion weight is far less than the change amplitude of the model result, it indicates that the proposed model has good stability. The results of the sensitivity analysis are shown in Section 4.3.

4. Results

4.1. Sea Fog Identification

4.1.1. Identification Precision

The MLP sea fog identification method was applied using Python programming, and the algorithm was verified using the CALIPSO-VFM classification points; the results it produced were then compared with those obtained with the NDSI threshold method proposed by [11]. The accuracy evaluation indices used in this paper are based on the work of [7], including the probability of detection (POD), probability of false detection (PFD), probability of missing detection (PMD), and equitable threat score (ETS). The calculation formulas of the four indicators are respectively shown in Equations (21)–(24):

$$\mathrm{POD} = \frac{H}{H+M}, \tag{21}$$

$$\mathrm{PFD} = \frac{F}{F+C}, \tag{22}$$

$$\mathrm{PMD} = \frac{M}{H+M}, \tag{23}$$

$$\mathrm{ETS} = \frac{H-(H+F)(H+M)/N}{H+M+F-(H+F)(H+M)/N} \tag{24}$$

where *H* represents the number of verification data points related to sea fog, *C* represents the number of verification data points that are all non-sea fog related, *F* represents the number of non-sea-fog data points that the algorithm incorrectly identifies as sea fog, and M represents that the verification data that are sea fog, but the algorithm incorrectly identifies as non-sea-fog. N represents the total number of points.

The confusion matrix of the results from the two algorithms (i.e., the MLP sea fog identification algorithm and NDSI threshold method) is shown in Table 5, and the precision comparison results are shown in Table 6.

Table 5. Confusion matrix of sea fog classification.

		Sea Fog	**Non-Sea Fog**
		H	F
Sea fog	MLP NDSI	88 96	17 75
		M	C
Non-sea-fog	MLP NDSI	20 12	206 148

Table 6. Accuracy of sea fog classification.

Methods	**POD**	**PMD**	**PFD**	**ETS**
MLP	81.48%	7.62%	18.51%	59.22%
NDSI threshold	88.89%	33.63%	11.11%	31.61%

It can be seen from the results that the POD of both MLP sea fog identification algorithm and NDSI threshold method are more than 80%; indeed, the NDSI threshold algorithm identifies eight more sea fog points relative to the MLP method. However, the NDSI threshold algorithm identifies more non-sea-fog points as sea fog, resulting in lower accuracy indices tested by the algorithm. These non-sea-fog points are mainly low clouds, indicating that the NDSI threshold algorithm is not able to effectively distinguish between sea fog and low clouds. The MLP sea fog classification method is relatively stable for the various accuracy indices. Although the POD value of the MLP method is smaller than that of the NDSI threshold algorithm, the former algorithm is more able to distinguish between sea fog and non-sea fog points than the NDSI threshold algorithm, with an ETS of 59.22%. Therefore, the MLP sea fog classification method constructed in this paper performs better than the NDSI threshold algorithm on the whole.

4.1.2. Sea Fog Statistics

We applied the constructed sea fog identification algorithm to AHI data to get the sea fog information of 1442 days in the study area from 2016 to 2019, and calculated the sea fog occurrence of each grid in the four seasons. The location of Bohai Sea is about 37° N, and according to the climate of the region, we set February to April as spring, May to July as summer, August to October as autumn, and November to January of the next year as winter. Following these classifications, the fog occurrence of the four seasons in the study area is shown in Figure 10.

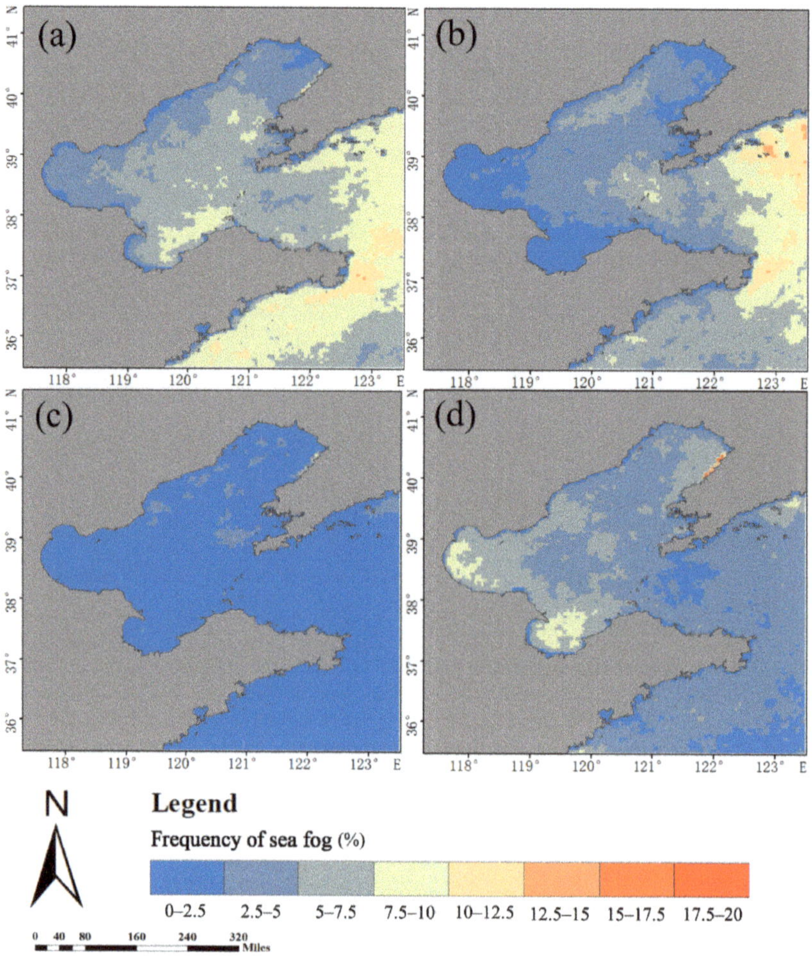

Figure 10. Frequency distribution of sea fog in (**a**) spring; (**b**) summer; (**c**) autumn; (**d**) winter.

The sea fog in the study areas begins from the winter, and goes until the next autumn. The occurrence frequencies of sea fog in spring and summer are higher, while those in autumn and winter are lower; thus, the seasonal occurrence of sea fog shows obvious differences. From winter to the next autumn, the high incidence area of sea fog shows a trend of developing from south to north, and from west to north.

From the perspective of spatial distribution, in winter, the higher frequency areas of sea fog are in the west of Bohai Sea, and with an eastward movement tendency with time. In spring, higher frequency areas of sea fog are in the Yellow Sea and south of Bohai Sea, with a northward movement tendency with time. The two tendencies gather in the south of Bohai Bay in summer, when the frequency of sea fog reaches its highest value in the whole year.

On the other side, the frequency of sea fog in spring and summer is greater than that in autumn and winter, and in winter, strong winds are frequent, which have a greater impact on navigation than sea fog [25]. Therefore, this paper considers the spring and summer as the main fog seasons (February to July). This result will be useful for navigational risk analyses.

4.2. Navigation Risk Evaluation Validation

The navigation risk evaluation model was coded in the R language and ArcGIS software. The matching results between the evaluation results and accident points are shown in Figure 11a–d, the latter of which respectively shows the results from in spring, summer, autumn, and winter. The specific matching results are shown in Table 7.

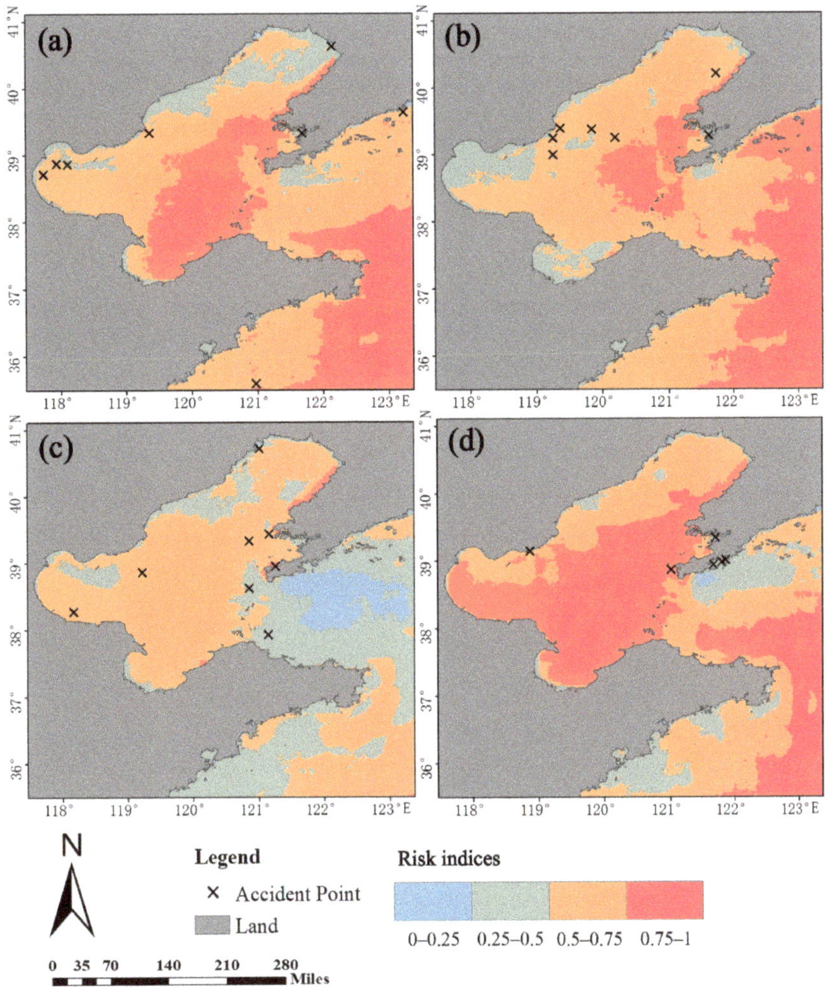

Figure 11. Navigation risk assessment and accident points matching results in (**a**) spring; (**b**) summer; (**c**) autumn; (**d**) winter.

Table 7. Accident matching table.

	Spring	Summer	Autumn	Winter	Summary
Success	6	6	6	2	20
All accident numbers	8	7	8	6	29
Matching rate	75%	85%	75%	33%	69.9%

The matching degree of the navigation risk evaluation model is 69.9%. In summer, the matching degree reaches 85%, where only one accident point is not successfully matched. In winter, the matching rate is low, only 33%, and four accident points are not successfully matched. Of the four unsuccessfully matched accident points in winter, three are concentrated on the eastern side of Bohai Sea, and occur around a similar time (within three days) and in a similar area. The concentrated accidents have high randomness, leading to a low matching degree in winter. Except for winter, the matching degrees of accidents in the other three seasons are good, where only 1–2 points are not matched successfully. In the fog season (spring and summer), the matching degree of the proposed model reaches 80%; thus, the evaluation performance is great in the fog season.

It should be noted that the small amount of accident information is distributed randomly in space and time. The overall matching degree of the model is 69.9%, and the matching degree during the fog season is 80%. It can be considered that the model has a good matching ability to accidents in the case of limited accident data, especially for fog seasons.

4.3. Sensitivity Analysis

A sensitivity analysis can be used to analyze the fitting degree between model weights and criteria values. Figure 12 shows the sensitivity analysis results of the one-to-a-time method.

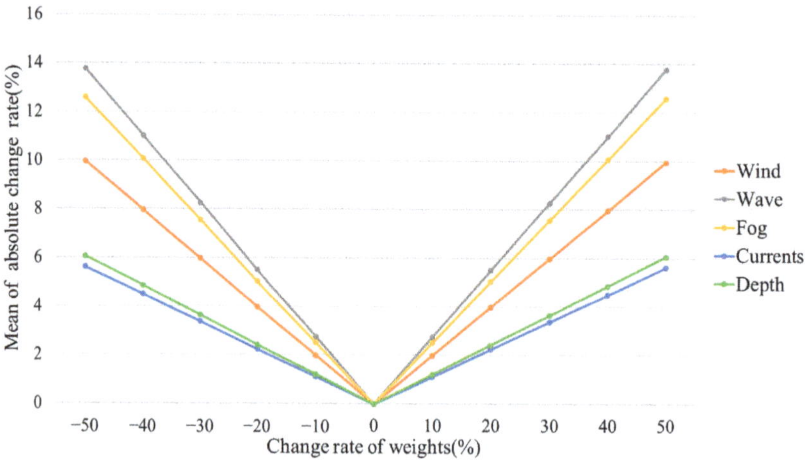

Figure 12. Sensitivity analysis results.

It can be seen from the results that the most sensitive criterion is waves, and the least sensitive criteria are currents. The descending order of sensitivity is waves, sea fog, wind, water depth, and currents. The maximum change rate is 14% when the change amplitudes of the criteria weights reach a maximum of 50%, and the corresponding change ratio is 0.28. The change rates of the evaluation results are much smaller than those of the criteria weights, which means the weights obtained by the CRITIC objective weight method are in good agreement with the criteria values, and the proposed navigation risk evaluation model has good stability.

4.4. Navigation Risk Analysis

According to the results of the navigation risk assessment in the four seasons given in Figure 11, winter has the greatest navigation risk in the study area throughout the year. In winter, accompanied by large winds and waves, sea fog weather also occurs occasionally, and the navigation risk in the sea area is high. The descending order of risk levels in the other seasons is spring, summer, and autumn.

In winter, the navigation risk in the southern part of the study area is greater than in the northern part. In spring, the high-risk navigation area is concentrated in the central part of Bohai Sea, extending from south to north. The high-risk area in spring has a trend of developing to the outer side of Bohai Sea. In summer, the high-navigation risk area retreats to outside of the Bohai Strait, and the risk value in the northern Yellow Sea increases. In autumn, the risk value in the sea area decreases to a minimum, although the western part of Bohai Sea has a relatively high value.

For further analysis, the evaluation results were calculated for every shipping region, as shown in Figure 13. In winter and spring, high-risk shipping regions are in the south, where the risks are high on the routes to Laizhou Port, Yantai Port, Harbor Port, Shidao Port, and Qingdao Port. In summer and autumn, high-risk shipping regions are in the north, and the risks are higher on the routes to Yinkou Port, Dalian Port, and Dandong Port. Other high-risk routes are from Qingdao Port to Yantai Port in spring and summer, and from Tianjin Port to Yantai Port and Shidao Port in spring and summer.

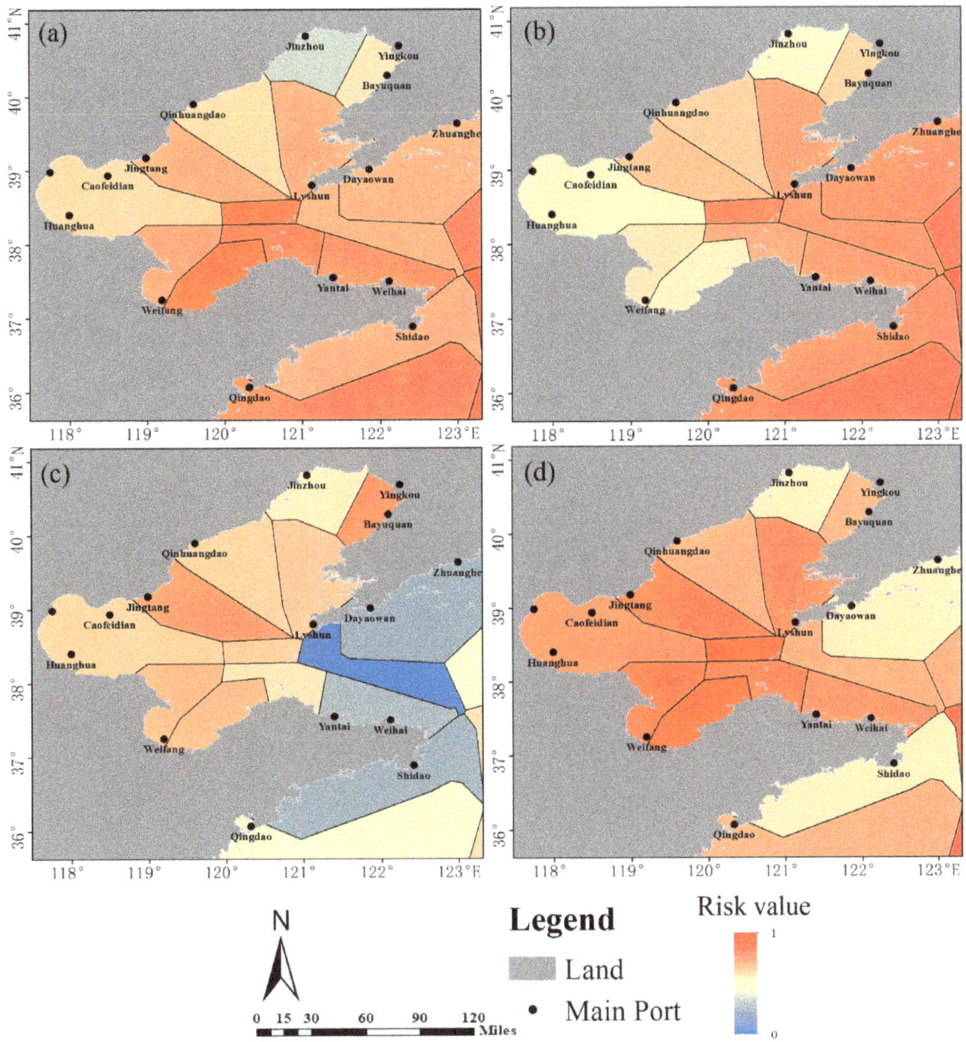

Figure 13. Risk distribution in the waterway areas in (**a**) spring; (**b**) summer; (**c**) autumn; (**d**) winter.

5. Discussion

In the navigation risk assessment framework proposed in this paper, we utilized a remote sensing sea fog identification method to obtain a large amount of accurate historical sea fog data in Bohai Sea and its surrounding waters. In the experiment, we found that the statistical results of sea fog are disturbed by low clouds, which could bring some abnormal statistic values (for instance, some areas with more than 200 days of sea fog in a year). In order to improve the classification accuracy, we used CALIPSO-FVM data to construct the sea fog sample set, and selected the features that contribute to the classification by spectral analysis. Compared with the NDSI threshold method in [11], the POD of proposed method decreased by 7.41%, but the ETS increased by 27.61%, which has a better distinguishing performance on sea fog and low clouds. It is very useful to have the historical frequency statistics of sea fog. The spatial distribution and seasonal differences of sea fog frequency in Bohai Sea and its surrounding waters are very obvious (Figure 10), the anomaly statistics disappear, and the results can be applied to the navigation risk assessment model. However, the proposed method still has some limitations. The small sample set size and unbalanced sample proportion are two factors that restrict the classification performance. The number of accurate sea fog samples obtained from CALIPSO-VFM data in Bohai Sea and Yellow Sea is limited, resulting in an unbalanced sample proportion (the sample ratio of sea fog to non-sea fog data is 1:3). It is necessary to expand the sample set size further. Observation data from weather ships or ground stations may be a viable option, which will be performed in a future study.

In order to eliminate the subjectivity in the navigation risk assessment methods, we collected the ship accident data of the study area for many years, and used the objective weighting method to determine the weight of criteria, from the accident matching and sensitivity analysis results, it has achieved a good performance in Bohai Sea and its surrounding waters. There are still some limitations that could be ameliorated further. Firstly, the spatial resolution and data performance of datasets used in this paper are quite different (for instance, the resolution of wind data is $0.25° \times 0.25°$, but the resolution of the currents is $0.125° \times 0.125°$). Some data need to be unified by interpolation, as it is expected that high-quality data could get the better risk-mapping results. Secondly, we significantly consider the navigation environment factors in the model, and some thresholds about ship factors follow the methods employed in previous research [21,41]. In the analysis about the ship, even the human factors could be added in the model and the applicability of the model will get a better performance.

In addition, with the development of geostationary orbit satellite and satellite constellation observation technology, ocean state observation data like wind, sea surface temperature, sea fog, etc. can be obtained in near real time. The historical or monitoring data of sea fog derived from Himawari-8 data have been integrated into a risk model of the proposed method; however, real-time sea fog observation data will be expected to be applied in risk assessment when the temporal resolution is further improved by satellite constellation. Using the real-time sea state observation data to evaluate and forecast navigation risks will greatly enhance the risk warning capability of the model, but it is a great challenge to develop a navigation risk forecast system. The dynamic Bayesian network could be a solution: for example, Paper [47] used dynamic Bayesian network to predict the risk of current and future time. Although this study only considers some risk factors of the ship itself and does not make use of the observed data of the space marine environment, it may be a feasible method for the prediction of navigation risk. Secondly, if the forecast data of the sea state (like wind, currents, etc.) could be obtained, the spatial multi- criteria decision-making method like this paper proposed can be used to evaluate risk on the basis of the forecast data, which will improve the spatial expression ability of the model. In general, the development of a spatial navigation risk prediction system is a great challenge, but it is worth performing it in future work.

6. Conclusions

This paper proposed a navigation risk evaluation method for Bohai Sea and its surrounding waters in the fog season. The MLP method was used to build a sea fog identification method, and with this method, a sample of sea fog historical information in the study area, obtained from AHI data, was obtained. The CRITIC method was used to weight the various criteria, which included sea fog, wind, waves, currents, and water depth. Then, a navigation risk evaluation was conducted using GIS mapping technology. For a better risk analysis of specific shipping lanes in the study area, the Delaunay triangulation method was used to divide the navigation area, and specific risk evaluation results of the navigation area were obtained. The POD of the sea fog identification method was 81.48%, the EST was 59.22%, and the accident matching rate of the navigation risk evaluation model was 80% in the fog season. The stability of the model was verified by a sensitivity analysis.

This study can provide a reference for ship navigation, ship route planning, and risk control in the four seasons in Bohai Sea and its surrounding waters. At the same time, the division of the waterway area can provide some planning advice for the various ports therein and the China Maritime Safety Administration, and may provide some help for maritime cargo transport and maritime rescue.

Author Contributions: Z.Z., P.D. conceived and designed the research, and performed the experiments. P.D., C.Q., L.J. wrote and edited the paper. J.Z., L.L., J.Y., S.L. analyzed the result and gave advice for the preparation and revision of the paper. Z.Z. and S.L. supervised the project. All authors have read and agreed to the published version of the manuscript.

Funding: This research was supported by the National Key R&D program 2017YFC1405600, and Natural Science Foundation of Shandong Province 2019GHY112034.

Institutional Review Board Statement: Not applicable.

Informed Consent Statement: Not applicable.

Data Availability Statement: The data used in this paper are all public data sets, which can be obtained free of charge through websites and books. The data sources are detailed in Table 2.

Acknowledgments: We thank the European Centre for Medium-Range Weather Forecasts, and Japan Aerospace Exploration Agency for Public available data. We thank the Maritime Safety Administration of the people's Republic of China for public ship accident reports. Finally, thanks to the verification data support of sea fog by Xiaofei Guo, and anonymous reviewers for constructive and helpful suggestions on the manuscript.

Conflicts of Interest: The authors declare no conflict of interest.

References

1. Xian, J.; Han, Y.; Huang, S.; Sun, D.; Zheng, J.; Han, F.; Zhou, A.; Yang, S.; Xu, W.; Song, Q.; et al. Novel Lidar algorithm for horizontal visibility measurement and sea fog monitoring. *Opt. Express* **2018**, *26*, 34853–34863. [CrossRef] [PubMed]
2. Gao, Y.; Jiang, G. Research on Influencing Factors and Countermeasures of Fog Navigation in Weihai Harbour. In Proceedings of the 5th International Conference on Education, Management, Arts, Economics and Social Science, Sanya, China, 10–11 November 2018.
3. Dorman, C.E.; Mejia, J.; Koračin, D.; McEvoy, D. World marine fog analysis based on 58-years of ship observations. *Int. J. Clim.* **2019**, *40*, 145–168. [CrossRef]
4. Fu, G.; Guo, J.; Pendergrass, A.; Li, P. An analysis and modeling study of a sea fog event over the Yellow and Bohai Seas. *J. Ocean Univ. China* **2008**, *7*, 27–34. [CrossRef]
5. Wu, D.; Lu, B.; Zhang, T.; Yan, F. A method of detecting sea fogs using CALIOP data and its application to improve MODIS-based sea fog detection. *J. Quant. Spectrosc. Radiat. Transf.* **2014**, *153*, 88–94. [CrossRef]
6. Chunyang, Z.; Jianhua, W.; Shanwei, L.; Hui, S.; Yanfang, X. Sea fog detection using U-net deep learning model based on MODIS data. In Proceedings of the 2019 10th Workshop on Hyperspectral Imaging and Signal Processing: Evolution in Remote Sensing (WHISPERS), Amsterdam, The Netherlands, 24–26 September 2019.
7. Shin, D.; Kim, J.-H. A New Application of Unsupervised Learning to Nighttime Sea Fog Detection. *Asia Pac. J. Atmos. Sci.* **2018**, *54*, 527–544. [CrossRef]

8. Yang, J.-H.; Yoo, J.-M.; Choi, Y.-S. Advanced Dual-Satellite Method for Detection of Low Stratus and Fog near Japan at Dawn from FY-4A and Himawari-8. *Remote Sens.* **2021**, *13*, 1042. [CrossRef]
9. Kim, D.; Park, M.-S.; Park, Y.-J.; Kim, W. Geostationary Ocean Color Imager (GOCI) Marine Fog Detection in Combination with Himawari-8 Based on the Decision Tree. *Remote Sens.* **2020**, *12*, 149. [CrossRef]
10. Bessho, K.; Date, K.; Hayashi, M.; Ikeda, A.; Imai, T.; Inoue, H.; Kumagai, Y.; Miyakawa, T.; Murata, H.; Ohno, T.; et al. An Introduction to Himawari-8/9—Japan's New-Generation Geostationary Meteorological Satellites. *J. Meteorol. Soc. Jpn. Ser. II* **2016**, *94*, 151–183. [CrossRef]
11. Ryu, H.; Hong, S. Sea Fog Detection Based on Normalized Difference Snow Index Using Advanced Himawari Imager Observations. *Remote Sens.* **2020**, *12*, 1521. [CrossRef]
12. Sahin, B.; Kum, S. Risk Assessment of Arctic Navigation by using Improved Fuzzy-AHP Approach. *Int. J. Marit. Eng.* **2015**, *157*, 241–250.
13. Hu, S.; Fang, Q.; Xia, H.; Xi, Y. Formal safety assessment based on relative risks model in ship navigation. *Reliab. Eng. Syst. Saf.* **2007**, *92*, 369–377. [CrossRef]
14. Balmat, J.-F.; Lafont, F.; Maifret, R.; Pessel, N. A decision-making system to maritime risk assessment. *Ocean Eng.* **2011**, *38*, 171–176. [CrossRef]
15. Zhang, C.; Zhang, D.; Zhang, M.; Lang, X.; Mao, W. An integrated risk assessment model for safe Arctic navigation. *Transp. Res. Part A Policy Pract.* **2020**, *142*, 101–114. [CrossRef]
16. Zhang, D.; Yan, X.; Zhang, J.; Yang, Z.; Wang, J. Use of fuzzy rule-based evidential reasoning approach in the navigational risk assessment of inland waterway transportation systems. *Saf. Sci.* **2016**, *82*, 352–360. [CrossRef]
17. Zhang, G.; Thai, V.V. Expert elicitation and Bayesian Network modeling for shipping accidents: A literature review. *Saf. Sci.* **2016**, *87*, 53–62. [CrossRef]
18. Lim, G.J.; Cho, J.; Bora, S.; Biobaku, T.; Parsaei, H. Models and computational algorithms for maritime risk analysis: A review. *Ann. Oper. Res.* **2018**, *271*, 765–786. [CrossRef]
19. Zhang, D.; Yan, X.P.; Yang, Z.L.; Wall, A.; Wang, J. Incorporation of formal safety assessment and Bayesian network in navigational risk estimation of the Yangtze River. *Reliab. Eng. Syst. Saf.* **2013**, *118*, 93–105. [CrossRef]
20. Jiang, M.; Lu, J.; Yang, Z.; Li, J. Risk analysis of maritime accidents along the main route of the Maritime Silk Road: A Bayesian network approach. *Marit. Policy Manag.* **2020**, *47*, 815–832. [CrossRef]
21. Wang, J.; Li, M.; Liu, Y.; Zhang, H.; Zou, W.; Cheng, L. Safety assessment of shipping routes in the South China Sea based on the fuzzy analytic hierarchy process. *Saf. Sci.* **2014**, *62*, 46–57. [CrossRef]
22. Zhou, X.; Cheng, L.; Li, M. Assessing and mapping maritime transportation risk based on spatial fuzzy multi-criteria decision making: A case study in the South China sea. *Ocean Eng.* **2020**, *208*, 107403. [CrossRef]
23. Zhang, S.; Jing, Z.; Li, W.; Wang, L.; Liu, D.; Wang, T. Navigation risk assessment method based on flow conditions: A case study of the river reach between the Three Gorges Dam and the Gezhouba Dam. *Ocean Eng.* **2019**, *175*, 71–79. [CrossRef]
24. Yang, S.; Song, B.; Ye, S.; Xu, Q.; He, L.; Zhao, G.; Li, J.; Laws, E.A. Regional-Scale distributions of pollen and spore assemblages in alluvium around the Bohai Sea: An essential step toward understanding marine palynological sources in China. *Mar. Geol.* **2019**, *415*, 105968. [CrossRef]
25. Lv, X.; Yuan, D.; Ma, X.; Tao, J. Wave characteristics analysis in Bohai Sea based on ECMWF wind field. *Ocean Eng.* **2014**, *91*, 159–171. [CrossRef]
26. Tian, M.; Wu, B.; Huang, H.; Zhang, H.; Zhang, W.; Wang, Z. Impact of water vapor transfer on a Circum-Bohai-Sea heavy fog: Observation and numerical simulation. *Atmos. Res.* **2019**, *229*, 1–22. [CrossRef]
27. Winker, D.M.; Vaughan, M.A.; Omar, A.; Hu, Y.; Powell, K.A.; Liu, Z.; Hunt, W.H.; Young, S. Overview of the CALIPSO Mission and CALIOP Data Processing Algorithms. *J. Atmos. Ocean. Technol.* **2009**, *26*, 2310–2323. [CrossRef]
28. Hersbach, H.; Bell, B.; Berrisford, P.; Hirahara, S.; Horanyi, A.; Muñoz-Sabater, J.; Nicolas, J.; Peubey, C.; Radu, R.; Schepers, D.; et al. The ERA5 global reanalysis. *Q. J. R. Meteorol. Soc.* **2020**, *146*, 1999–2049. [CrossRef]
29. Chassignet, E.P.; Hurlburt, H.E.; Smedstad, O.M.; Halliwell, G.; Hogan, P.J.; Wallcraft, A.J.; Baraille, R.; Bleck, R. The HYCOM (HYbrid Coordinate Ocean Model) data assimilative system. *J. Mar. Syst.* **2007**, *65*, 60–83. [CrossRef]
30. Mayer, L.; Jakobsson, M.; Allen, G.; Dorschel, B.; Falconer, R.; Ferrini, V.; Lamarche, G.; Snaith, H.; Weatherall, P. The Nippon Foundation—GEBCO Seabed 2030 Project: The Quest to See the World's Oceans Completely Mapped by 2030. *Geosciences* **2018**, *8*, 63. [CrossRef]
31. Maritime Safety Administration of the People's Republic of China. *Sailing Direction On Chinese Coast: North Area*; China Communications Press Limited Liability Company: Beijing, China, 2018; pp. 27–198.
32. Nowak, D.; Ruffieux, D.; Agnew, J.L.; Vuilleumier, L. Detection of Fog and Low Cloud Boundaries with Ground-Based Remote Sensing Systems. *J. Atmos. Ocean. Technol.* **2008**, *25*, 1357–1368. [CrossRef]
33. Xiao, Y.-F.; Zhang, J.; Qin, P. An Algorithm for Daytime Sea Fog Detection over the Greenland Sea Based on MODIS and CALIOP Data. *J. Coast. Res.* **2019**, *90*, 95–103. [CrossRef]
34. Hao, Z.; Pan, D.; Gong, F.; Chen, J. Sea fog characteristics based on MODIS data and streamer model. In Proceedings of the Remote Sensing of Clouds and the Atmosphere XIV, Berlin, Germany, 31 August–3 September 2009.
35. Sharifzadeh, F.; Akbarizadeh, G.; Kavian, Y.S. Ship Classification in SAR Images Using a New Hybrid CNN–MLP Classifier. *J. Indian Soc. Remote Sens.* **2018**, *47*, 551–562. [CrossRef]

36. Costache, R.; Pham, Q.B.; Corodescu-Roșca, E.; Cîmpianu, C.; Hong, H.; Linh, N.T.T.; Fai, C.M.; Ahmed, A.N.; Vojtek, M.; Pandhiani, S.M.; et al. Using GIS, Remote Sensing, and Machine Learning to Highlight the Correlation between the Land-Use/Land-Cover Changes and Flash-Flood Potential. *Remote Sens.* **2020**, *12*, 1422. [CrossRef]
37. Zhang, C.; Pan, X.; Li, H.; Gardiner, A.; Sargent, I.; Hare, J.; Atkinson, P.M. A hybrid MLP-CNN classifier for very fine resolution remotely sensed image classification. *ISPRS J. Photogramm. Remote Sens.* **2017**, *140*, 133–144. [CrossRef]
38. Formela, K.; Neumann, T.; Weintrit, A. Overview of Definitions of Maritime Safety, Safety at Sea, Navigational Safety and Safety in General. *TransNav Int. J. Mar. Navig. Saf. Sea Transp.* **2019**, *13*, 285–290. [CrossRef]
39. Chen, C. Case study on wave-current interaction and its effects on ship navigation. *J. Hydrodyn.* **2018**, *30*, 411–419. [CrossRef]
40. Heo, K.-Y.; Park, S.; Ha, K.-J.; Shim, J.-S. Algorithm for sea fog monitoring with the use of information technologies. *J. Meteorol. Appl.* **2014**, *21*, 350–359. [CrossRef]
41. Pacuraru, F.; Domnisoru, L. Numerical investigation of shallow water effect on a barge ship resistance. *IOP Conf. Ser. Mater. Sci. Eng.* **2017**, *227*, 012088. [CrossRef]
42. Diakoulaki, D.; Mavrotas, G.; Papayannakis, L. Determining objective weights in multiple criteria problems: The critic method. *Comput. Oper. Res.* **1995**, *22*, 763–770. [CrossRef]
43. Žižović, M.; Miljković, B.; Marinković, D. Objective methods for determining criteria weight coefficients: A modification of the CRITIC method. *Decis. Mak. Appl. Manag. Eng.* **2020**, *3*, 149–161. [CrossRef]
44. Mukhametzyanov, I. Specific character of objective methods for determining weights of criteria in MCDM problems: Entropy, CRITIC and SD. *Decis. Mak. Appl. Manag. Eng.* **2021**, *4*, 76–105. [CrossRef]
45. Bai, J. Estimating Multiple Breaks One at a Time. *Econ. Theory* **1997**, *13*, 315–352. [CrossRef]
46. Xu, E.; Zhang, H. Spatially-Explicit sensitivity analysis for land suitability evaluation. *Appl. Geogr.* **2013**, *45*, 1–9. [CrossRef]
47. Khan, B.; Khan, F.; Veitch, B. A Dynamic Bayesian Network model for ship-ice collision risk in the Arctic waters. *Saf. Sci.* **2020**, *130*, 104858. [CrossRef]

remote sensing

MDPI

Article

The InflateSAR Campaign: Developing Refugee Vessel Detection Capabilities with Polarimetric SAR

Peter Lanz [1,2,*], Armando Marino [3], Morgan David Simpson [3], Thomas Brinkhoff [2], Frank Köster [1,4] and Matthias Möller [5]

1 Department of Computing Science, Carl von Ossietzky University of Oldenburg, Ammerländer Heerstraße 114-118, 26129 Oldenburg, Germany; frank.koester@dlr.de
2 Institute for Applied Photogrammetry and Geoinformatics, Jade University Oldenburg, Ofener Str. 16/19, 26121 Oldenburg, Germany; thomas.brinkhoff@jade-hs.de
3 Department of Biological and Environmental Sciences, University of Stirling, Stirling FK9 4LA, UK; armando.marino@stir.ac.uk (A.M.); m.d.simpson@stir.ac.uk (M.D.S.)
4 Institute for AI Safety and Security, German Aerospace Center (DLR), Rathausallee 12, 53757 Sankt Augustin, Germany
5 Faculty for Humanities and Cultural Sciences, Otto-Friedrich-University of Bamberg, Am Kranen, 96045 Bamberg, Germany; matthias.moeller@uni-bamberg.de
* Correspondence: peter.lanz@uni-oldenburg.de

Abstract: In the efforts to mitigate the ongoing humanitarian crisis at the European sea borders, this work builds detection capabilities to help find refugee boats in distress. For this paper, we collected dual-pol and quad-pol synthetic aperture radar (SAR) data over a 12 m rubber inflatable in a test-bed lake near Berlin, Germany. To consider a real scenario, we prepared the vessel so that its backscattering emulated that of a vessel fully occupied with people. Further, we collected SAR imagery over the ocean with different sea states, categorized by incidence angle and by polarization. These were used to emulate the conditions for a vessel located in ocean waters. This setup enabled us to test nine well-known vessel-detection systems (VDS), to explore the capabilities of new detection algorithms and to benchmark different combinations of detectors (detector fusion) with respect to different sensor and scene parameters (e.g., the polarization, wind speed, wind direction and boat orientation). This analysis culminated in designing a system that is specifically tailored to accommodate different situations and sea states.

Keywords: polarimetry (PolSAR); constant false alarm rate (CFAR) detector; polarimetric detector; sub-look detector; vessel-detection system (VDS); ship detection; synthetic aperture radar (SAR); disaster mitigation, refugee boat

1. Introduction

The ongoing humanitarian crisis at Europe's southern maritime border has made it the world's deadliest border. For many years, migrants have been risking their lives to cross the Mediterranean Sea under very poor conditions on routes that are frequently controlled by human traffickers. Up-to-date and historic data and statistics about the maritime migration into Europe and its toll on human life can be found on the web services of the Missing Migrants Project [1], the International Organization for Migration [2] and the United Nations Refugee Agency [3].

Search and rescue organisations on the ground attempt to mitigate the impact of the continuously high numbers of maritime disasters, and one of the main hindrances is the difficulty of quickly locating migrant boats in distress. These operations use different strategies for maritime surveillance, such as vessels, airplanes and drones, and satellite-based search is another useful tool for this task. For such a time-critical application in disaster mitigation, synthetic aperture radar (SAR) is appropriate and practicable due to its all-weather and its 24 hour imaging capabilities. Moreover, the detection of ships and other maritime objects with SAR has a long research history (e.g., [4–11]).

With the rapid evolution in processing power, the pattern recognition community also added an important contribution to ship detection. However, many of the traditional methods, such as region selections, scale-invariant feature transform (SIFT), support vector machine (SVM) and histogram of oriented gradients (HOG), are typically associated with high computing costs. With deep learning, the time performance increases because it uses GPU acceleration. Different approaches using convolutional neural networks (CNN) began to evolve rapidly, and we see different deep-learning models for object detection (e.g., [12,13]) and image-classification tasks.

The region-based convolutional neural network (RCNN; [14]) and the sliding window were followed by Fast-RCNN [15] and Faster-RCNN [16,17]. Improvements and reduced complexity were achieved by using the softmax function instead of SVMs by using multiresolution convolutional features and by dividing the large detection area into region of interest (ROI) images. Ref. [18] used this approach for the detection of ships.

Ref. [19] presented a grid-CNN (G-CNN), which is a combination of a backbone-CNN (B-CNN) and a detection-CNN (D-CNN) to further improve and speed up automatic ship detection. Other approaches, such as YOLO [20] and YOLOv2, attempt to embrace the whole image during the training and testing period [21]. Ref. [22] used a deep neural network (DNN) for SAR image classification with the sliced Wasserstein distance (SWD) to provide a better solution to the optimization problem.

In our research, we attempt to adapt well-known detectors and develop detection capabilities with SAR for 12 m long rubber inflatables. This specific type of sea vehicle is predominantly used to cross the central Mediterranean Sea. In most cases, they are sent to the ocean packed with more than 100 passengers. Shorter versions of such boats (10 or 8 m long), but identical in construction, are used on other routes, for example in the Aegean Sea.

Our previous work on the development of automatic detectors for such migrant rubber boats was based on SAR data that we collected from an original 12×3.5 m inflatable [23]. Those data enabled us to examine the radar pattern of our special maritime vessel under different combinations of scene parameters (e.g., the incidence angle and resolution) and sensor parameters (e.g., the boat orientation) and to evaluate them for their impact on the detectability of the inflatable. We tested and compared a variety of existing automatic vessel detectors [24]. Our results are, we think, noteworthy and legitimate and offer significant insights regarding the behaviour of the inflatable when hit by microwaves and the interactions with the surrounding water surface. With this, we can detect the empty inflatable on a lake.

For this research, we collected data to create a more realistic scenario by simulating passengers on the vessel and combining this with SAR imagery from the ocean. This enabled us to analyse the SAR vessel-detection capabilities for migrant rubber inflatables for a variety of sensor settings and scene parameters, such as the wave height and wave direction. We present a polarimetric analysis of the backscattering behaviour of the fully occupied inflatable and benchmark a variety of well-known detectors with respect to several influencing parameters. To improve the detection quality, we experiment with new algorithms with adaptations of the Intensity Dual-Polarization Ratio Anomaly Detector (iDPolRAD) [25] and with the combination of detectors.

2. Materials and Methods

2.1. The Human Scattering Experiment

In this initial experiment, we evaluated which material was best suited to emulate the presence of humans on a rubber inflatable and was sufficiently durable to remain in the vessel on a lake for several months. During the whole campaign, the material properties that influence the scattering behaviour (roughness, shape, water content and dielectric property) must remain constant. With a mobile/handheld radar sensor, we compared the radar backscatter behaviour of a number of material with that of humans. The experiments involved:

- Four volunteers sitting close to each other on the ground of an empty room. We chose the sitting posture because that most closely resembles the real situation in the migrant inflatable. The data include different arrangements: four people in a row perpendicular to the sensor line of sight (LoS) ('H4×1'), two rows of two people behind each other ('H2×2') and all four people in one column behind each other parallel to the LoS ('H1×4').
- Water-soaked clay pebbles, packed in 30 × 40 cm air-tight plastic bags. The bags themselves are invisible to microwaves and the soaked clay pebbles, as they are roundish objects smaller than the wavelength and with a similar water content to the human body and, thus, should appear similar to the uppermost body parts (heads and shoulders). We took data from two bags perpendicular to the LoS ('C2×1'), two bags parallel to the LoS ('C1×2'), two bags sitting on top of each other ('C1×1×2') and two bags stacked with one large bag (30 × 60 cm) standing behind them ('C1×1×2+1').
- Steel wool clumped to random 20 cm diameter balls to imitate the top layer of passengers in a boat. The acquisitions involved six balls in two rows ('S2×3') and two balls plus four 5 × 10 × 60 cm (h,w,l) steel wool layers not clumped but stretched out in the front ('S2+4').

The dielectric constant ϵ' is an important factor during the interaction with electromagnetic waves. For a given shape, the lower the dielectric constant, the higher the backscatter intensity from volume scattering [26]. A very low dielectric constant, such as air or PVC (all dry) renders them almost invisible to microwaves and increases the penetration depth. The dielectric property is a function of the temperature and the microwave frequency and of course the exact composition of the particular material. Dry clay pebbles would be almost invisible to microwaves due to their very low ϵ' (Table 1). For that reason, we water-soaked the clay pebbles and used air-tight bags to preserve a similar water content to that of the human body (about 80%) throughout the acquisition campaign.

Table 1. The dielectric properties for a list of materials involved in our detection scenario at 20 °C (* at 37 °C) [27]. The value for sea water, given the typical salinity of the Mediterranean Sea, is about 38 g per kilo [28].

Material	Dielectric Constant ϵ'	Loss Tangent tan δ
Air	1	depends on weather
Blood *	58	0.27
Fat *	5.5	0.21
Muscle *	49	0.33
Nylon	2.4	0.0083
Polyethylene	2.25	-
Water, fresh [29]	80	-
Sea water [30,31]	70	-
Sea ice [26]	4	0.5
Sandy soil (dry)	2.55	0.0062
Clay bricks	3.7-4.5	-
Metals	infinite	-
Plywood	2.5	

The loss tangent is the relation between the dielectric constant ϵ' and the loss factor ϵ'' and describes the conductivity of a medium:

$$\tan(\delta) = \frac{\epsilon''}{\epsilon'} \tag{1}$$

The dielectric permittivity ϵ_c is defined as a complex number:

$$\epsilon_c = \epsilon' - i\epsilon'' \quad (2)$$

consisting of the dielectric constant (real) and the loss factor (imaginary). The latter describes the electromagnetic loss of a medium. A low electromagnetic loss generally results in an additional contribution from volume scattering [26].

Fresh water has an ϵ' similar to blood or muscle tissue, which justifies/explains our use of water-soaked clay pebbles. Steel wool has, as with all magnetic matter, a very high dielectric constant, and its ϵ_c is infinite. Due to that and due to its shape, we expect it to behave similarly to human bodies and to have a strong volume backscattering signal.

Figure 1 compares the intensity images of all three test cases. The test objects are placed at the same distance from the sensor (y-axis) whereas the x-axis represents the number of pulses. In this case, 65 consecutive pulses were emitted and recorded. The radar return from H2×2 is speckled in time due to movement, and the signal is more spread out since human bodies occupy a wider space along the LoS than the comparably small test objects (Figure 1, left). The wall in the background is more clearly visible in the scenarios with clay pebbles and steel wool since the objects used are much smaller than four people, occupy much less space and produce less occlusion (Figure 1, centre and right). To evaluate how many sacks are needed, we summed up the intensity values of all pulses of the different materials and the humans within each scenario. The comparison of these backscattering intensities revealed that approximately 90 bags of soaked clay would emulate 80 occupants.

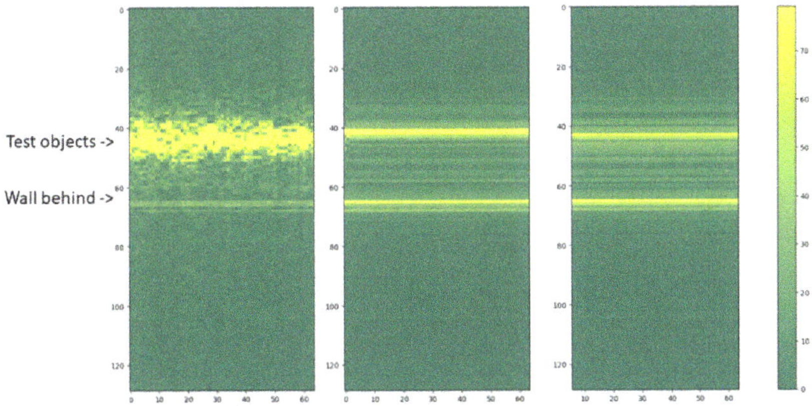

Figure 1. Example results (in dB) of the human backscattering experiments: H2×2 (**left**), C2×2×2 (**centre**) and S2×3 (**right**). The y-axes show the distance from the sensor in decimetres.

2.2. Data Campaign and Data Collection

We set up the data campaign on a small lake in northern Germany in spring 2022. The test object was a 12 × 3 m rubber (1.2 mm PVC) inflatable with a wooden floor and no metal parts at all.

Due to its dielectric properties, the boat itself is expected to be hardly visible using microwaves (see Table 1; [23]). Finally, we placed 90 bags of soaked clay pebbles in a way that makes them cover the whole area of the vessel, including the inflated volumes that passengers use to sit on (Figure 2).

(a) Top view

(b) Side view

Figure 2. SAR data acquisition setup in the test-bed lake.

We expected the bags to leave a similar radar signature compared to a group of 80 people sitting in the boat. First, the bags were expected to simulate the volumetric structure created by many people with their shoulders and heads placed close to each other. Second, the water in the volume-like clay pebbles, preserved over the whole time by air-tight bags, should resemble the water content of the human body. Our approach to simulate the backscattering of a human body in different frequencies is supported by similar studies. For example, Ref. [32] used wet sand for Ultra Wide Band (UWB) radar. In our experiment, we expect a stronger intensity signature compared to an empty boat (see [24]).

Thus, prepared, the boat sat on the lake perpendicular to the LoS (in a N-S direction) for the first two months. In a second phase, we oriented it at 45° to the LoS and acquired data for another two months. These two geometries should represent two diametrically different scenarios when it comes to scattering mechanisms. The boat oriented 90° relative to the LoS should generate a strong radar backscattering since it exposes the maximum amount of surfaces facing the LoS and being capable of scattering the EM-waves directly back through double reflections [23]. The boat oriented at 45°, on the other hand, is expected to scatter some of the multiple reflections away from the sensor. This is because the volume scattering expected by the human-equivalent and from the vessel facets (for example in case they are wet) is not fully isotropic.

The data collected comprise full-pol data from Cosmo-SkyMed Second Generation (CSG), dual-pol Stripmap data in several combinations from TerraSAR-X (TSX) and very high-resolution VV-pol data from ICEYE (Table 2).

Table 2. Overview of the data collected for the boat on the lake with 80 simulated people.

Mission	Mode	Average Pixel Size (m^2)	Polarization	Incidence Angle	Datasets
TerraSAR-X	Stripmap	4.4	Dual-pol: HH VV, HV HH, VH VV	Low, medium and high	46
Cosmo-SkyMed	Spotlight	2.7	Quad-Pol	Medium and high	4
ICEYE	High-res. Spotlight	0.6	Single-Pol: VV	Low and medium	4

For the CSG and the TSX data, we attempted to cover different combinations of sensor and scene setting, such as incidence angle, dual-pol polarization and the orientation of the boat. This enabled us to analyse the influence of these parameters on the scattering behaviour of the inflatable and on the performance of the detectors.

Table 3 quantifies the available TSX datasets. The categories for the incidence angles are defined as smaller than 28° for 'low', from 28° to 43° for 'medium' and greater than 43° for 'high'. The figures list the datasets where an unambiguous identification of the boat on the lake was possible; figures in brackets represent unsuccessful identifications. The reason for the latter was mainly strong Bragg scattering due to low incidence angles. We speak of an unambiguous identification when the target-to-clutter ratio (TCR) of at least one of the polarization channels is larger than three times the clutter standard deviation σ in dB:

$$TCR \geq 3 * \sigma_{clutter} \tag{3}$$

where TCR is defined as the ratio between the maximum backscattering of the rubber inflatable and the mean of the backscattering of the local lake clutter. In dB, the ratio has a difference. It is useful to express σ^0 in the decibel (dB) scale since it can be statistically modelled with a Gaussian distribution. In linear scale, the fit of a generalized gamma distribution would be much more appropriate and (3) would have to be adapted accordingly.

Table 3. TSX data parameter matrix of the 'full' inflatable on the lake.

	90 Degrees			45 Degrees		
	Low	**Medium**	**High**	**Low**	**Medium**	**High**
HH VV	1 (1)	4	3	3 (2)	5	5
HV HH	1	2	2	1 (1)	3	2
VH VV	1	2	1	1 (1)	2	2

The high variability of the ocean surface due to wind speed, wind direction, swell and other factors has grave implications for its scattering behaviour. The backscattering intensity changes with the water wave height, frequency, form and orientation relative to the LoS of the radar. Moreover, the combination of scattering mechanisms changes with the existence and density of wave features, such as foam or breaking waves [33–37]. Consequently, higher sea states are expected to hamper the detection task.

Figure 3a visualizes an example for the increased radar backscattering intensity from 2.1 m waves. On the right, the corresponding detection probability map from the intensity-based cell averaging-constant false alarm rate (CA-CFAR) detector shows how the waves can raise the false positive rate.

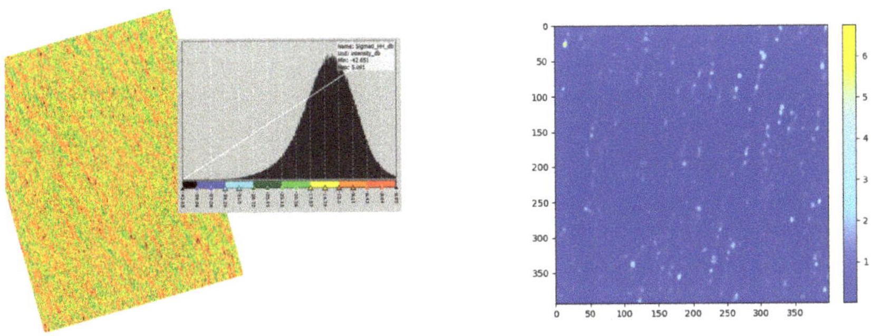

(**a**) (**b**)

Figure 3. Radar signature of 2.1 m waves in the up/down direction, HH-pol and TSX Stripmap (left) and its implications for the detection task (right) (@ DLR 2014). (**a**) Radar signature (σ^0 in dB); (**b**) CA-CFAR probability map. Axis labels are in meters.

To analyse and quantify the influence of the significant wave height H_s on the performance of the detectors, we gathered a collection of SAR data from the ocean. These ocean data comprise different H_s for a selection of combinations of wave directions, incidence angles and dual-pol combinations. We arranged our collection classifying the wave direction (relative to the LoS) in two groups:

- Cross-wind waves: these are waves that move perpendicular to the LoS. They move in the direction of the satellite azimuth, which is close to N-S.

- Up/down-wind waves: here, the waves move in the range direction.

The wave direction is—particularly for low wind speed/low waves—an important impact factor for the backscattering behaviour of the water surface (e.g., [38]). The LoS was calculated using the heading of the satellite defined by the azimuth angle, which is 10.2° for TSX. That means a heading of 349.8° for ascending orbits and 190.1° for descending orbits. We accepted an average deviation of 10° between the wave direction and the respective directions for up/down wind and cross wind.

We determined the wave direction and H_s of the TSX-archive datasets with the help of in situ wave-sensor buoys from the Copernicus Marine Environment Monitoring Service In Situ Thematic Assembly Centre (CMEMS In Situ TAC). In fact, we only used SAR data near in situ buoys where the wave height and wave direction data were available. To verify the data, we double checked the plausibility with data from nearby buoys of that network and with estimations from a number of Copernicus wave models covering different ocean areas:

- North Sea: Atlantic–European North-West Shelf-Wave Physics Reanalysis.
- Mediterranean Sea and West Gibraltar region: Mediterranean Sea Waves Reanalysis.
- Arctic Ocean: Arctic Ocean Wave Hindcast.
- All other maritime regions not covered by a high-resolution wave model: Global Ocean Waves Reanalysis WAVERYS.

From these models, we extracted and used parameters, such as the *sea surface wave significant height (VHM0 or Hm0 (m)), significant height of waves Hs on the water body (VGHS (m))* and *average height highest wave (VAVH)* for the wave height, *wave principal direction at spectral peak (VPED), direction from which the waves are coming relative to true north (VDIR), direction from which the wind is coming relative to true north (WDIR)* and *the wave-induced horizontal Eulerian mean current (HCDT)*, to determine the main direction of the waves. The latter is useful since the Eulerian mean current is always in the direction of the waves [39].

We followed the Beaufort scale (BFT) to define our wave height categories in Table 4, and each category corresponds to a specific sea state defined in that scale. This collection enables us to examine the impact of a variety of sea states on the radar backscattering and on the detection task. As Table 4 shows, the collection of ocean data was subdivided according to the polarization, the incidence angle (low, medium and high) and the principal wave direction relative to the LoS (up/down and cross).

Table 4. Collection of TSX dual-pol ocean data covering different wave directions, wave heights, polarimetric channel combinations and incidence angles.

Polarization	HH VV				HV HH		VH VV	
Incidence Angle	Low	Low	Medium	High	Medium	High	Medium	High
Wave Direction	Cross	Up/Down	Up/Down	Up/Down	Up/Down	Up/Down	Up/Down	Up/Down
Wave Height (m)								
0.4–0.8 (BFT3)	✓	✓	✓	✓	✓	✓	✓	✓
0.8–1.5 (BFT4)	✓	✓	✓	✓	✓		✓	✓
1.5–2.5 (BFT5)		✓			✓	✓	✓	✓
2.5–3.5 (BFT6)			✓	✓				
3.5–4.5 (BFT7)						✓		
4.5–6.5 (BFT8)	✓							

Figure 4 provides an overview of our collection of TSX archive SAR data of different wave heights and the water surface mean backscattering energy by polarization. The acquisitions from the lake test bed are included as well ($H_s = 0$). It shows a general trend of stronger backscattering from the water surface with increasing H_s. Moreover, small incidence angles tend to give very high backscattering intensities due to larger Bragg scattering. In co-pol datasets, VV shows a slightly higher mean intensity value throughout the data collection (HH: −15.5 dB and VV: −14.4 dB). At medium and high incidence angles, the return from the water surface is very low—especially for cross-pol channels.

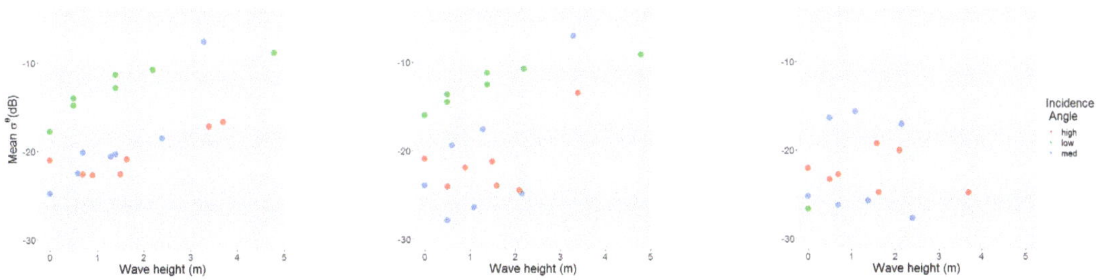

(**a**) HH-pol. (**b**) VV-pol. (c) Cross-pol (HV or VH).

Figure 4. TSX dual-pol data collection of different wave heights and their mean backscattering intensity; one diagram for each polarization.

To facilitate the testing of vessel detector algorithms at different sea states, we fused the simulated 'full' inflatable with the datasets showing the ocean. All important sensor parameters (resolution/acquisition mode, incidence angle and polarization), were considered, and we replaced pixels in the images from the ocean with the pixels of the boat. The pixels of the boat on the lake were chosen using an intensity-based approach as described in Formula (3). Figure 5 shows a ground truth map with four different signatures of the inflatable (yellow) inserted at random positions into the ocean data (purple). This binary mask with yellow pixels represent the 'positive true' case in the following analysis. Again, we took extra care to only replace pixels where the satellite parameters were similar.

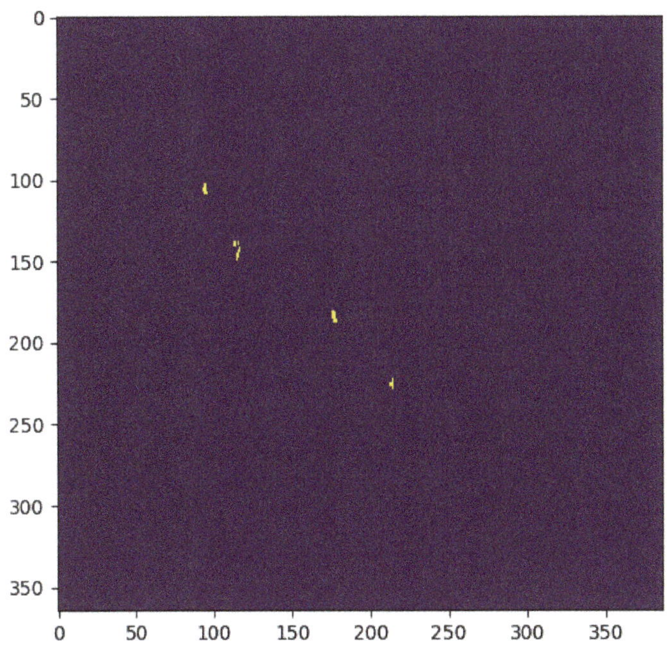

Figure 5. A 350 × 350 m ground truth map after insertion of the positive true pixels (yellow) of four different acquisitions from the rubber inflatable. Purple pixels represent the ocean surface.

2.3. Polarimetric Analysis of the Inflatable

Decomposition theorems (DT) provide a physical interpretation of the signals scattered from a target by considering them as a combination of several components. To better understand the scattering mechanisms at the inflatable, we used quad-pol CSG data for a selection of well-known coherent and incoherent polarimetric decompositions as listed in Table 5. In this context, the term *coherent* indicates the decomposition of the scattering matrix [S] (Pauli, Cameron), including the phase information, whereas incoherent decompositions (Yamaguchi, Cloude–Pottier) use the averaging coherency [C] or covariance [(]T] matrix [40].

The most commonly known coherent model-based decomposition is the Pauli decomposition where the scattering matrix [S] is the sum of the complex quantities of odd-bounce (e.g., single bounce), even-bounce (e.g., double bounce) and 45° rotated double bounce (see Table 5). The Cloude–Pottier decomposition or Entropy/Anisotropy/Alpha decomposition (H/A/$\underline{\alpha}$) [41,42] proposes the existence of three orthogonal scattering mechanisms. It belongs to the category of eigenvector-based TDs and is especially useful for partial targets. To retrieve its parameters, it is necessary to form the covariance matrix in the Pauli basis.

Working on the assumption of a monostatic system and reciprocity, the Pauli vector is $k_p = \frac{1}{\sqrt{2}}[HH + VV, HH - VV, 2 * HV]^T$. This matrix is then called coherency matrix [T]. In the quad-pol reciprocal scenario, [T] gives three independent scattering mechanisms along its main axis: surface scattering $T_{11} = \langle \underline{k_1} \underline{k_1^{*T}} \rangle$, double-bounce scattering $T_{22} = \langle \underline{k_2} \underline{k_2^{*T}} \rangle$ and volume scattering $T_{33} = \langle \underline{k_3} \underline{k_3^{*T}} \rangle$. The three roll-invariant parameters H, A and $\underline{\alpha}$ can be used to describe the quantitative proportion between the three scattering mechanisms. For our detection task, H was found to be particularly useful.

Table 5. Polarimetric decompositions and parameters. The Yamaguchi parameters had their orientation removed.

Parameter	Decomposition	Note
Alpha	Cloude–Pottier	
Entropy	Cloude–Pottier	
Single Bounce	Yamaguchi Y4R	
Double Bounce	Yamaguchi Y4R	
Volume Scattering	Yamaguchi Y4R	
Helix Scattering	Yamaguchi Y4R	
Symmetry	Yamaguchi Y4R	Huynen Target Generator A0
Irregularity/Double Bounce	Yamaguchi Y4R	Huynen Target Generator B0-B
Non-symmetry	Yamaguchi Y4R	Huynen Target Generator B0+B
Even bounce	Pauli	|HH-VV|
Even bounce 45° oriented	Pauli	|HV|
Odd bounce	Pauli	|HH+VV|
Trihedral	Cameron	
Dipole	Cameron	
Narrow Diplane	Cameron	
Diplane	Cameron	
Left Helix	Cameron	
Right Helix	Cameron	
Cylinder	Cameron	
1/4 Wave Device	Cameron	

The model-based Yamaguchi decomposition [43] added the helix parameter to the Freeman and Durden decomposition [44], adding an asymmetric component to the scattering model. Helix scattering is fully determined by the imaginary part of the T_{23} element of the coherency matrix. This element is expected to be strong when there is no reflection symmetry, such as in forests or 45° oriented built-up areas.

It is expected to be weak for flat surfaces and parallel built-up areas. Polarimetric reflection symmetry means there is no correlation between the co-pol and the cross-pol channels [45]:

$$\langle S_{HH} S_{HV}^* \rangle = \langle S_{HV} S_{VV}^* \rangle = 0 \tag{4}$$

The Cameron decomposition [46] uses the Pauli spin matrices to decompose and to classify [S] in several steps. The scattering [S] matrix is described as the complex sum of these basis matrices. Each of them is associated with one of the elementary scattering mechanisms: single scattering from plane surfaces, diplane scattering from corner reflectors oriented 45° and non-symmetric components. The Cameron decomposition is a multistage procedure.

First, the degree of reciprocity (in Cameron's formula the angle θ) is calculated in order to examine whether the reciprocal components dominate or not by separating the symmetric and non-symmetric parts. For the scattering matrix of reciprocal objects, the elements S_{12} and S_{21} are equal. If the reciprocal part is dominant, the reciprocal term is decomposed into symmetric and asymmetric scatterers. Representative [S] matrices for asymmetric left and right helices are:

- Left helix: $S_{hl} = \frac{1}{2}\begin{bmatrix} 1 & i \\ i & -1 \end{bmatrix}$
- Right helix: $S_{hr} = \frac{1}{2}\begin{bmatrix} 1 & -i \\ -i & -1 \end{bmatrix}$

If the [S] matrix exhibits asymmetry, its share of left helix and right helix is calculated. In a final step, if the matrix is symmetric, it is compared to a list of symmetric scatterers.

- Trihedral: $\frac{1}{\sqrt{2}}\begin{bmatrix} 1 & 0 \\ 0 & 1 \end{bmatrix}$
- Cylinder: $\frac{1}{\sqrt{5}}\begin{bmatrix} 2 & 0 \\ 0 & 1 \end{bmatrix}$
- Diplane: $\frac{1}{\sqrt{2}}\begin{bmatrix} 1 & 0 \\ 0 & -1 \end{bmatrix}$
- Narrow diplane: $\frac{1}{\sqrt{5}}\begin{bmatrix} 2 & 0 \\ 0 & -1 \end{bmatrix}$
- Dipole: $\begin{bmatrix} 1 & 0 \\ 0 & 0 \end{bmatrix}$
- 1/4 wave device: $\frac{1}{\sqrt{2}}\begin{bmatrix} 1 & 0 \\ 0 & i \end{bmatrix}$

where, according to Cameron, a symmetric scatterer has an axis of symmetry in the plane of the radar line of sight (LoS).

The Cameron decomposition is a coherent DT, such as, for example, the Krogager decomposition [47]. This group of DTs are particularly useful in the case of one dominant scattering mechanism. Cameron's cylinder and narrow diplane scattering can be composed of trihedral, dihedral and dipole scattering. Together with the quarter wave device, they can be called fundamental scattering mechanisms [48].

2.4. Detector Comparison and Detector Fusion

We tested and benchmarked nine different vessel-detection systems (VDS; Table 6) algorithms using receiver operating characteristic (ROC) curves. ROC curves are plots of the probability of detection (P_d) versus the probability of false alarm (P_{fa}). To quantify and compare the results from the ROC curves, we estimated the area under the curve (AUC) and grouped them for different wave heights and detectors.

ROC curves are computed by varying the threshold between a minimum and a maximum value depending on the histogram of the detector output. The AUC is a value for the classifier's performance: an AUC of 1 is the best achievable result and means 100% detection with no false alarms. An AUC of 0.5 is obtained for detectors making random choices. All VDS were implemented using a sliding window of a specific size (CUT) with a guard window around it and a training window around the guard window.

Table 6. Overview of the vessel-detection algorithms being tested and their parameterization.

Detector	Cells Under Test (CUT) Window Size	Guard Window Size	Train Window Size
Polarimetric Symmetry Detector (PolSym)	1	2	5
Polarimetric Notch Filter (PNF)	5	12	36
Polarimetric Entropy Detector (PolEntropy)	2	-	10
Polarimetric Match Filter (PMF)	5	12	36
Intensity Depolarization Ratio Anomaly Detector (iDPolRAD)	1	12	36
Surface Intensity Depolarization Ratio Anomaly Detector (SiDPolRAD)	1	12	36
Sub-look Correlation Detector (SubCorr)	1	-	36
Polarimetric Whitening Filter (PWF)	5	12	36
Cell Averaging Constant False Alarm Rate (CA-CFAR)	1	24	36

During a preliminary phase of benchmarking and detector tuning, we selected, for each detector, the parameterization that translates into the best performance (Table 6). PolEntropy and the SubCorr worked better without a guarding window, so we removed it for those two. The iDPolRAD is actually intended for use as a volume detector and designed to work with the cross–polarization over co–polarization ratio [25].

Since we applied it to all available combinations of dual-pol data, including HH VV, we refer to it here as PolRatio1 when used with cross-pol data and PolRatio3 when used with co-pol data. The same is true for its surface and double-bounce scattering detecting variant SiDPolRAD [24], which we rename as PolRatio2 when used with cross-pol data and PolRatio4 when used with co-pol data. For VH VV and HV HH, the two detectors are defined as:

$$PolRatio1: \frac{\langle|cross\text{-}pol|^2\rangle_{test} - \langle|cross\text{-}pol|^2\rangle_{train}}{\langle|co\text{-}pol|^2\rangle_{train}} * \langle|cross\text{-}pol|^2\rangle_{test} \quad (5)$$

$$PolRatio2: \frac{\langle|co\text{-}pol|^2\rangle_{test} - \langle|co\text{-}pol|^2\rangle_{train}}{\langle|cross\text{-}pol|^2\rangle_{train}} * \langle|co\text{-}pol|^2\rangle_{test} \quad (6)$$

Applying those two to the HH VV polarization combination, we express them as:

$$PolRatio3: \frac{\langle|HH|^2\rangle_{test} - \langle|HH|^2\rangle_{train}}{\langle|VV|^2\rangle_{train}} * \langle|HH|^2\rangle_{test} \quad (7)$$

$$PolRatio4: \frac{\langle|VV|^2\rangle_{test} - \langle|VV|^2\rangle_{train}}{\langle|HH|^2\rangle_{train}} * \langle|VV|^2\rangle_{test} \quad (8)$$

A possible physical interpretation for PolRatio3 is 'dihedral scattering' since the HH channel is stronger compared to the VV channel when we have a horizontal dihedral. The interpretation for PolRatio4 is 'Bragg scattering' since VV is stronger than HH in the case of Bragg scattering.

At this point, we were able to identify the most promising detectors for each dual-pol combination of polarimetric channels against the background of different wave heights. To improve the detection capabilities for our vessel, we tested different combinations of the top ranking detectors.

Each detector produces a detection probability map. This map is then classified by applying a threshold to generate the detection mask. This mask is a binary image showing the detector decision at a specific threshold. Each pixel receives either a zero value for 'no boat' or a one value for 'boat detected'. To fuse binary detection masks coming from different detectors, we tested different combinations of the logical operators 'OR' and 'AND' (Figure 6). Further, we tested detection algorithms that use the intensity and the ratio between the T_{11} (surface scattering), T_{12} (compound H– and V–dipoles, [49]) and T_{22} (double-bounce scattering) elements of the covariance matrix. The double-bounce scattering can come from the side of the inflatable and the passengers inside, which, together with the water surface, form dihedral structures. The term T_{22} can be expanded as:

$$T_{22} = \frac{1}{2}\langle|S_{HH} - S_{VV}|^2\rangle = \frac{1}{2}\langle|S_{HH}|^2\rangle + \frac{1}{2}\langle|S_{VV}|^2\rangle - \langle Re\{S_{HH}S_{VV}^*\}\rangle \quad (9)$$

In the dihedral scattering, the most significant feature is that the co-polarized components HH and VV are in opposite phase. Therefore, T_{22} increases to the maximum when the last part of Equation 9 becomes smaller than zero.

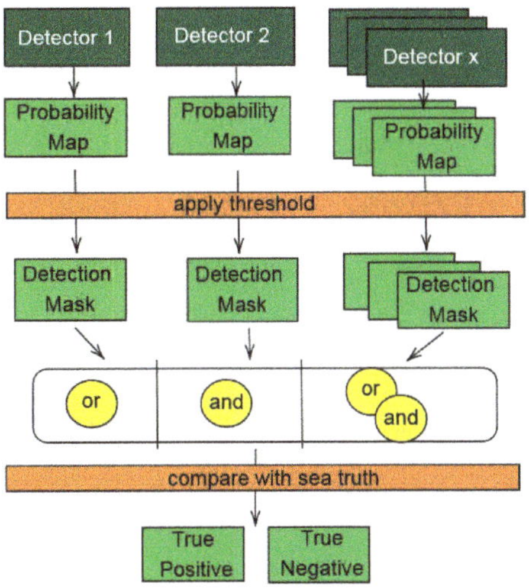

Figure 6. Detector melange flow diagram.

3. Results

3.1. Qualitative Inspection of High Resolution Data

At this point, we present three high-resolution spotlight images from ICEYE. These single-VV-pol images provide insight into the location and spatial variation of scattering mechanisms producing the radar backscattering of the rubber inflatable.

Figure 7 shows the boat with different sensor parameters at a sub-metre resolution of about 0.8 m. All three allow for a good visual identification and a precise size estimation. Strong scattering responses (bright pixels) are almost uniformly spread throughout the vessel. At a low incidence angle (Figure 7a), the mean response from the water surface is at about 10 dB and, compared to higher incidence angles, is increased by Bragg scattering. The vessel response is stronger as well; however, that increase is less strong. The wave structure of the water seems to be disturbed by the boat in the southeast–northwest direction, where there are also areas of lower backscattering.

This theory is backed by recordings from the Deutsche Wetterdienst (DWD) of the wind direction being 330° at the time of acquisition. At medium incidence angles, the mean radar response from the water surface (Figure 7b,c) is around 16 dB, and the TCR is higher. The vessel orientation relative to the LoS is clearly visible.

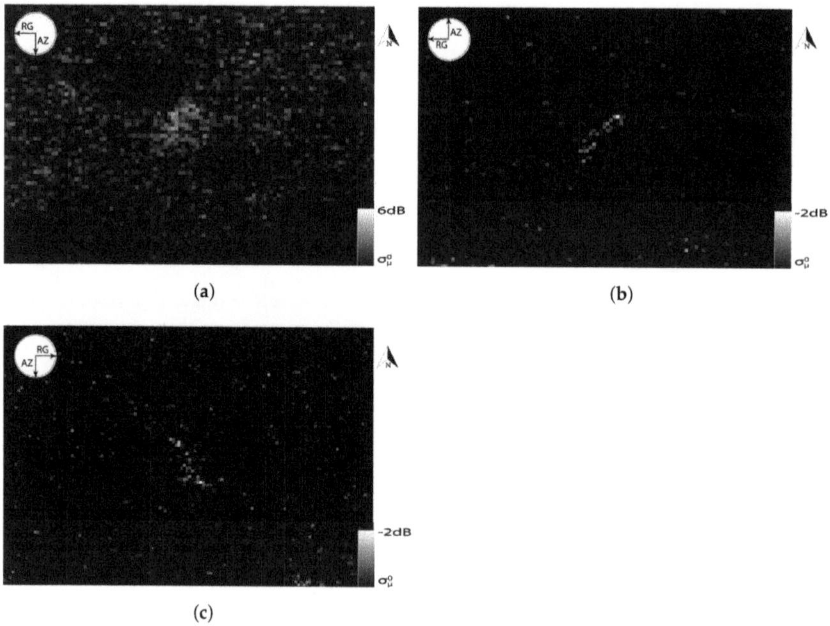

Figure 7. High-resolution ICEYE VV-pol of the vessel at different wave heights and their mean backscattering intensity. μ = mean of the water surface backscattering, σ = standard deviation of the water surface backscattering (@ ESA 2022). (**a**) Low incidence angle, 45° inclined boat. (**b**) Medium incidence angle, 45° inclined boat. (**c**) Medium incidence angle, orthogonal boat.

3.2. Polarimetric Scattering Analysis

A pixel-wise Pauli RGB image analysis (Figure 8) shows that the boat triggers different scattering mechanisms. For the filtering of [C] and [T] matrices, we used a rather small 3×3 window to preserve the spatial resolution as much as possible.

Table 7 lists the contributions of scattering mechanisms according to the Pauli decomposition. Double-bounce and single-bounce scattering clearly dominate. Volume scattering plays a minor role across the range of different acquisition parameters and seems to be stronger at low incidence angles.

Table 7. Pauli scattering mechanisms (dimensionless, comparative quantities, normalized to 0->1).

	Double Bounce	Volume	Single Bounce
low inc. angle, inclined vessel	0.41	0.25	0.41
low inc. angle, orthogonal vessel	0.33	0.17	0.33
medium inc. angle, orthogonal vessel	0.68	0.26	0.73

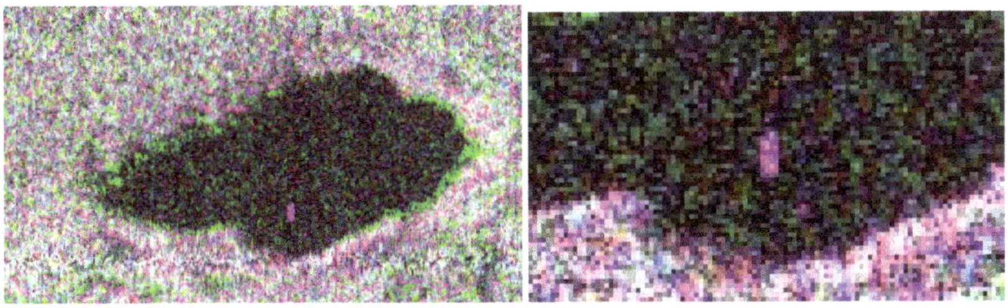

(**a**) (**b**)

Figure 8. Pauli R (double bounce) G (volume scattering) B (single bounce) composites for the orthogonal vessel at a medium incidence angle (@ ASI 2022). (**a**) Pauli RGB from the lake test bed. (**b**) Pauli RGB from a section of the lake with the rubber inflatable in the centre.

The Cloude–Pottier decomposition delivers low entropy at low incidence angles and medium entropy for medium incidence angles (Table 8, Figure 9). The mean alpha angle varies throughout the three acquisitions, indicating a dominance of surface scattering and double-bounce scattering as well as minor contributions from volume scattering.

Table 8. The results of the Cloude–Pottier decomposition of the inflatable (dimensionless, comparative quantities, normalized to 0->1).

	H	Mean Alpha
Low inc. angle, inclined vessel	0.46	0.32
Low inc. angle, orthogonal vessel	0.46	0.65
Medium inc. angle, orthogonal vessel	0.58	0.54

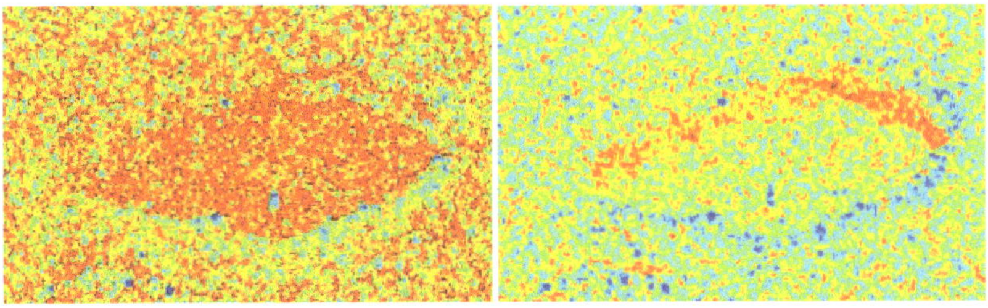

(**a**) Entropy (**b**) Mean alpha

Figure 9. Cloude–Pottier decomposition results in a scale from red (high) to blue (low) for the orthogonal vessel at a medium incidence angle (@ ASI 2022).

The Yamaguchi Y4R decomposition reveals the dominance of single-bounce scattering in the two acquisitions when the vessel was oriented orthogonally (Table 9, Figure 10). Double-bounce scattering varies to a great extent. Volume and helix scattering do not contribute significantly in any section of the inflatable.

Table 9. Yamaguchi Y4R decomposition (dimensionless, comparative quantities, normalized to 0->1).

	Helix	Single Bounce	Volume	Double Bounce
Low inc. angle, inclined vessel	0.00	0.07	0.01	0.49
Low inc. angle, orthogonal vessel	0.00	0.48	0.01	0.00
Medium inc. angle, orthogonal vessel	0.00	0.80	0.01	0.39

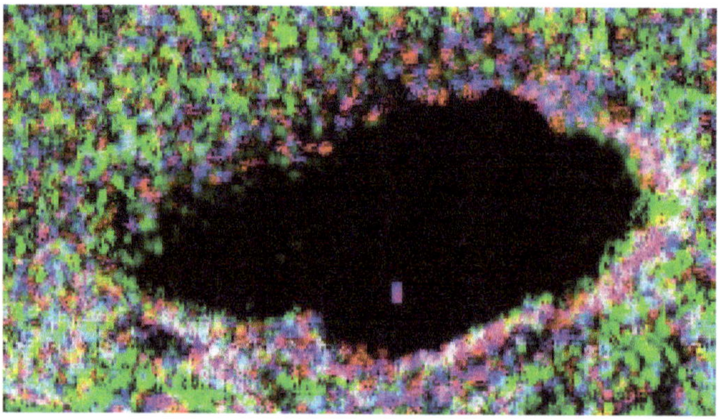

Figure 10. Yamaguchi Y4R RGB-composite showing symmetric (red), irregular/double-bounce (green) and non-symmetric (blue) scattering of the orthogonal vessel at a medium incidence angle (@ ASI 2022).

According to the Cameron decomposition (Table 10), the inflatable is dominated by scattering from dipoles with minor contributions from narrow diplanes and scattering from cylinders. It appears that $\frac{1}{4}$ wave device scattering contributes more at low incidence angles. Trihedral scattering plays a minor role. Asymmetric helix scattering only contributes to a minor extent. There is no direct and simple explanation for the occurrence of these mechanisms; however, they could be related to multiple interactions (in the resolution cell) between the pellet sacks and the water underneath.

Table 10. Cameron decomposition (dimensionless, comparative quantities, normalized to 0->1).

	Trihedral	Dipole	Narrow Diplane	Diplane	Cylinder	1/4 Wave Device	Left Helix	Right Helix
low inc. angle, inclined vessel	0.08	0.47	0.17	0.07	0.26	0.18	0.03	0.01
low inc. angle, orthogonal vessel	0.15	0.53	0.22	0.06	0.11	0.21	0.02	0.01
medium inc. angle, orthogonal vessel	0.06	0.57	0.15	0.06	0.23	0.10	0.00	0.00

All four decompositions show a dominance of surface scattering and double-bounce scattering. Pauli and Cloude–Pottier show that the entropy is at a medium level since, with double-bounce and single-bounce scattering, there are two dominant mechanisms. The result of the Yamaguchi decomposition show an unequal distribution of single-bounce and double-bounce scattering and no double-bounce scattering in one acquisition. This agrees with the Cameron decomposition in showing very little asymmetric helix scattering.

3.3. Detector Testing

The first results are a comprehensive comparison of the performance of nine detectors for three different categories of incidence angles and three different polarization channel combinations. Here, we used the simulated data that combine acquisitions over the sea and the lake. Figure 5 shows an example of a sea truth mask after embedding the boat into the ocean background.

We tested the detectors with two different boat orientations and using all the combinations of these sensor parameters with different wave directions and wave heights, according to the availability of ocean data as listed in Table 4).

For dual co-pol data, most detectors performed worse with increasing H_s and decreasing incidence angles. Very interestingly, the PolEntropy had increasing, or at least stable, AUCs for higher sea states. The fact that it searches for high entropy shows that the entropy of the water surface declines with increasing H_s. This is mostly due to noise floor issues where low sea states produce high entropy backscattering that is close to the noise floor. A high incidence angle can reduce the backscattering even below the noise floor. That is why the PolEntropy detector is the only detector with very low performance at high incidence angles and for calm water surfaces (Figure 11g,h).

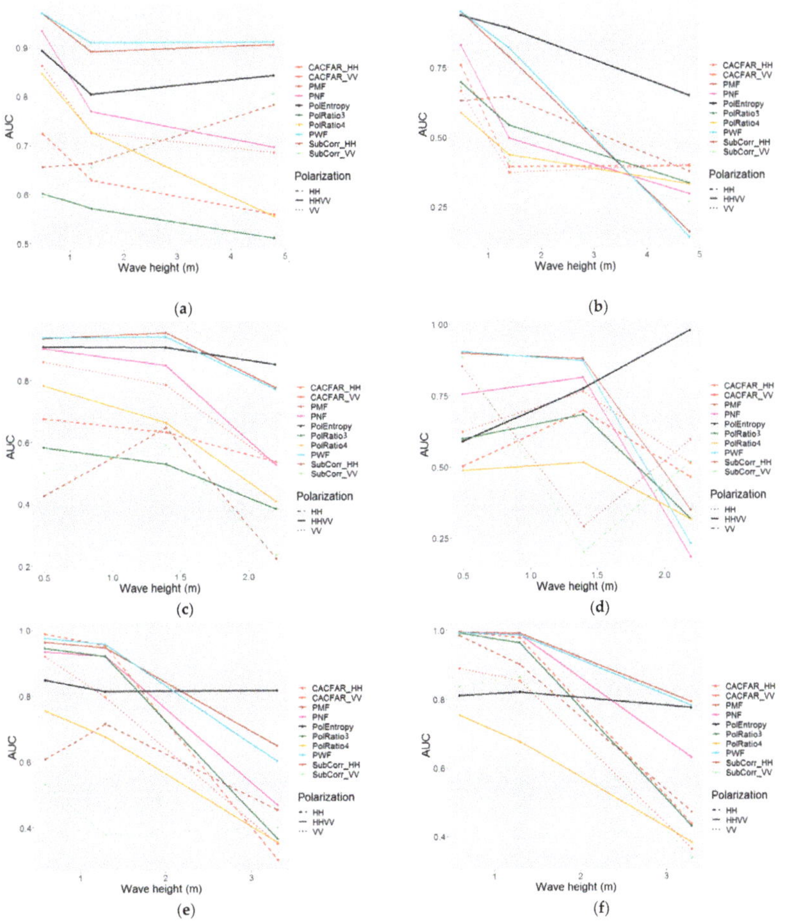

Figure 11. *Cont.*

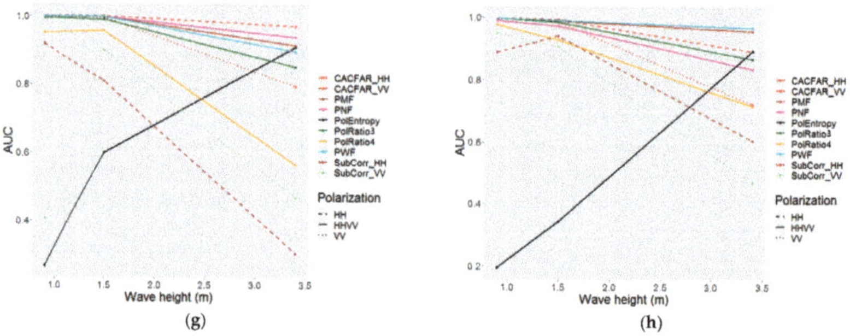

Figure 11. Comparison of the detector AUCs for different wave heights for HH VV. (**a**) Wind: cross, incidence angle: low and boat orientation: 45°. (**b**) Wind: cross, incidence angle: low and boat orientation: 90°. (**c**) Wind: up/down, incidence angle: low and boat orientation: 45°. (**d**) Wind: up/down, incidence angle: low and boat orientation: 90°. (**e**) Wind: up/down, incidence angle: medium and boat orientation: 45°. (**f**) Wind: up/down, incidence angle: medium and boat orientation: 90°. (**g**) Wind: up/down, incidence angle: high and boat orientation: 45°. (**h**) Wind: up/down, incidence angle: high and boat orientation: 90°.

The detector performance for HV HH polarization (Figure 12) reveals good and stable performances for most of the detectors at medium incidence angles. The entropy here seems to be less reliable for higher sea states than it is for co-pol data.

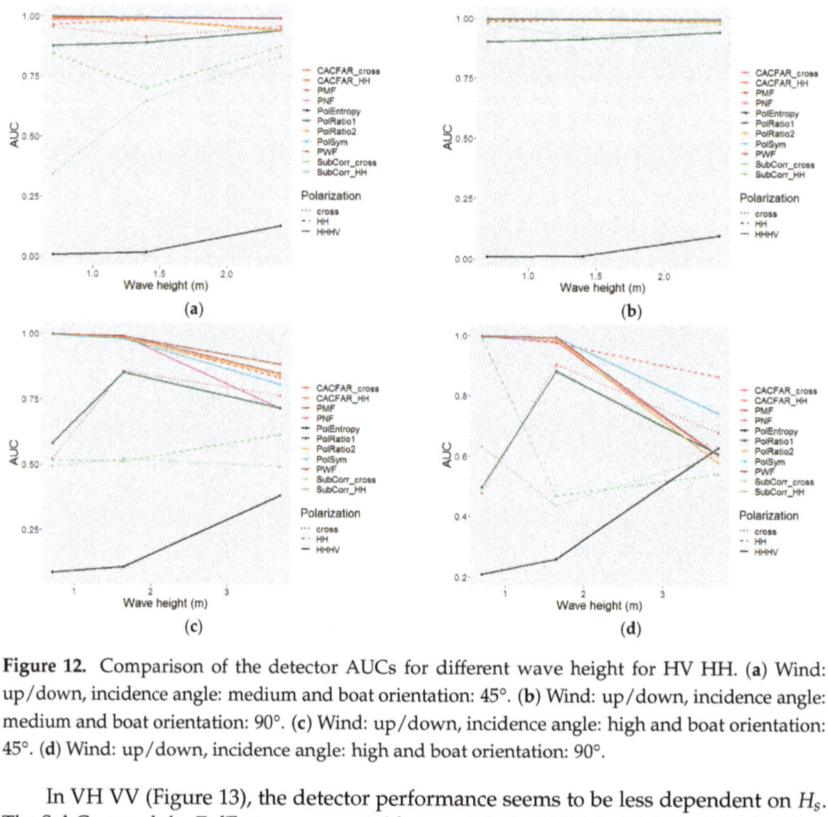

Figure 12. Comparison of the detector AUCs for different wave height for HV HH. (**a**) Wind: up/down, incidence angle: medium and boat orientation: 45°. (**b**) Wind: up/down, incidence angle: medium and boat orientation: 90°. (**c**) Wind: up/down, incidence angle: high and boat orientation: 45°. (**d**) Wind: up/down, incidence angle: high and boat orientation: 90°.

In VH VV (Figure 13), the detector performance seems to be less dependent on H_s. The SubCorr and the PolEntropy are not able to match the PWF, the PMF and the PolSym.

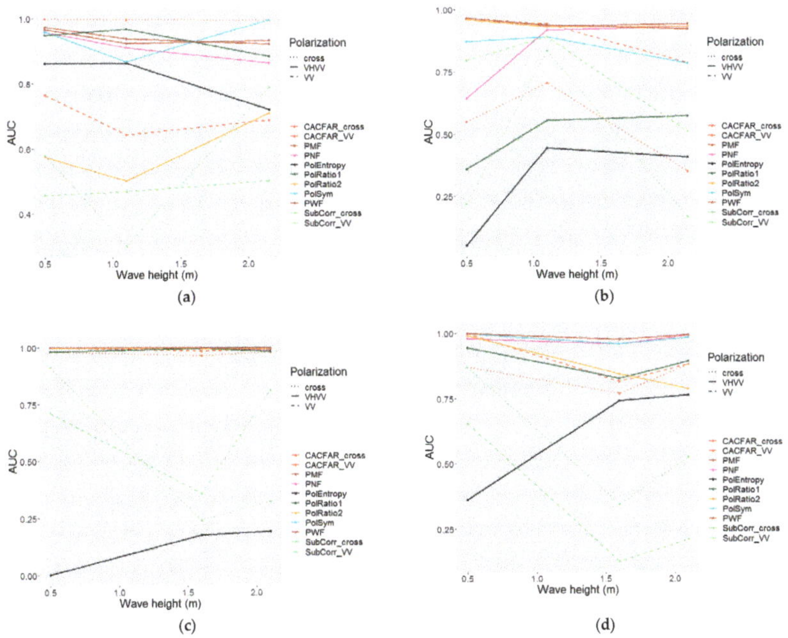

Figure 13. Comparison of the detector AUCs for different wave heights for VH VV. (**a**) Wind: up/down, incidence angle: medium and boat orientation: 45°. (**b**) Wind: up/down, incidence angle: medium and boat orientation: 90°. (**c**) Wind: up/down, incidence angle: high and boat orientation: 45°. (**d**) Wind: up/down, incidence angle: high and boat orientation: 90°.

Summing up the behaviour of the entropy of the water surface for different polarizations shows that:

- HH VV: the water surface has lower entropy values than the boat except for low sea states and/or high incidence angles.
- HV HH: medium angles: the entropy of the water stays higher than that of the boat; high angles: the entropy of water stays higher than that of the boat.
- VH VV: the entropy of water is lower than that of the boat.

In the following section, one table compares the mean AUCs of all detectors, the next explores the detector performance at higher sea states, and the last visualizes the influence of different boat orientations. They also give an indication of the overall influence of the incidence angle and the polarization.

Table 11 compares the overall performance of all detector algorithms. The best detectors, on average, are the PWF, the PMF and the PolSym. Low incidence angles are not a good choice when it comes to maritime object detection, since increased backscattering from the water surface decreases the TCR. Medium and high incidence angles are more useful for our task. The comparison of combinations of polarizations is strongly biased by different sets of wave heights; for example, the category for medium incidence angles in HV HH does not include any data with high waves.

Table 11. The AUCs of the vessel-detection algorithms with different sensor parameters. The background colours indicate different AUCs from white (low AUCs) to dark green (very high AUCs).

Polarization		**HH VV**		**HV HH**		**VH VV**		
Incidence Angle	**Low**	**Medium**	**High**	**Medium**	**High**	**Medium**	**High**	**Avg**
PMF	0.787	0.888	0.976	0.996	0.906	0.945	0.995	**0.928**
PWF	0.779	0.882	0.975	0.997	0.912	0.941	0.995	**0.926**
PNF	0.67	0.822	0.956	0.994	0.881	0.871	0.986	**0.883**
PolEntropy	0.834	0.813	0.534	0.046	0.279	0.559	0.373	**0.491**
PolRatio1/3	0.529	0.768	0.947	0.913	0.689	0.714	0.939	**0.786**
PolRatio2/4	0.554	0.599	0.849	0.983	0.9	0.768	0.936	**0.799**
SubCorr_HH	0.565	0.688	0.744	0.9	0.608			**0.701**
SubCorr_VV	0.528	0.557	0.684			0.604	0.481	**0.571**
SubCorr_cross				0.785	0.531	0.483	0.537	**0.584**
CACFAR_HH	0.6	0.775	0.975	0.98	0.943			**0.854**
CACFAR_VV	0.628	0.695	0.915			0.798	0.946	**0.797**
CACFAR_cross				0.972	0.701	0.766	0.909	**0.837**
PolSym				0.999	0.92	0.895	0.991	**0.951**
avg	**0.647**	**0.749**	**0.856**	**0.869**	**0.752**	**0.759**	**0.826**	

The same is true for high incidence angles in VH VV. Nevertheless, we see that all three combinations of polarization enable detectors to reach high AUCs.

As Table 12 shows, only results for data with a SPAN greater than −17 dB, it shows that the PMF, the PWF and the PolSym adapt more readily to high sea states. The entropy is more suitable for challenging situations, such as low incidence angles or high seas (high SPAN) at HH VV. For cross-pol, the PolSym seems to be a good choice, and the CACFAR_HH delivers very good results at high incidence angles. Again, all three combinations enable the detectors to deliver comparably good performances on average. When the SPAN reaches over −5 dB, all detectors except PMF, PWF and entropy fall to a level equivalent to random choice (AUC 0.5).

Table 12. The AUCs of the vessel-detection algorithms with high sea states (SPAN > −17 dB). The background colours indicate different AUCs from white (low AUCs) to dark green (very high AUCs).

Polarization		**HH VV**		**HV HH**		**VH VV**		
Incidence Angle	**Low**	**Medium**	**High**	**Medium**	**High**	**Medium**	**High**	**Avg**
PMF	0.787	0.843	0.935		0.728	0.945		**0.848**
PWF	0.779	0.83	0.932		0.745	0.941		**0.846**
PNF	0.67	0.75	0.887		0.662	0.871		**0.768**
PolEntropy	0.834	0.804	0.9		0.506	0.559		**0.721**
PolRatio1/3	0.529	0.669	0.858		0.664	0.714		**0.687**
PolRatio2/4	0.554	0.521	0.64		0.714	0.768		**0.639**
SubCorr_HH	0.565	0.634	0.455		0.58			**0.558**
SubCorr_VV	0.528	0.494	0.468			0.604		**0.523**
SubCorr_cross					0.551	0.483		**0.517**
CACFAR_HH	0.6	0.666	0.931		0.85			**0.762**
CACFAR_VV	0.628	0.59	0.758			0.798		**0.694**
CACFAR_cross					0.723	0.766		**0.745**
PolSym					0.776	0.895		**0.835**
avg	**0.647**	**0.68**	**0.776**		**0.682**	**0.759**		

Table 13 shows which detectors and sensor parameters deliver better AUCs for the boat when oriented at 45° (blue) or at 90° (orange) relative to the LoS. In general, the two different experimental setups deliver very comparable results, which backs the theory that the inflatable itself is effectively invisible to microwaves and does not scatter at all. This shows, moreover, that the most dominant scatterers are the passengers whose scattering behaviour is less dependent on the orientation. For low incidence angles in HH VV, the inclined boat is more easily detectable for most detectors. At medium angles, an orthogonal vessel seems to be slightly favourable. At high angles, the boat orientation has no impact at all.

Table 13. The AUCs of the vessel-detection algorithms with different orientations of the rubber vessel. The blue background colours indicate better AUCs for the 45° inclined vessel, the orange colours show higher AUCs for the vessel oriented at 90°.

Polarization	HH VV			HV HH		VH VV		
Incidence Angle	**Low**	**Medium**	**High**	**Medium**	**High**	**Medium**	**High**	**Avg**
PMF	0.23	−0.07	−0.01	0	0.08	0	0.01	**0.03**
PWF	0.25	−0.08	−0.02	0	0.09	0	0.01	**0.04**
PNF	0.22	−0.10	0.05	−0.01	0.04	0.08	0.02	**0.04**
PolEntropy	0.06	0.02	0.12	0.01	−0.17	0.51	−0.50	**0.01**
PolRatio1/3	0	−0.05	0	−0.02	0.05	0.44	0.1	**0.07**
PolRatio2/4	0.22	−0.01	−0.05	−0.02	0.09	−0.35	0.12	**0**
SubCorr_HH	0	−0.19	−0.13	−0.18	−0.12			**−0.13**
SubCorr_VV	0.07	−0.24	−0.18			−0.26	0.3	**−0.06**
SubCorr_cross			−0.35	−0.06	0.15	−0.09	**−0.09**	
CACFAR_HH	0.05	−0.06	0.03	−0.03	−0.01			**0**
CACFAR_VV	0.22	−0.01	0.03			−0.20	0.09	**0.03**
CACFAR_cross				−0.06	0.03	0.46	0.14	**0.14**
PolSym				0	0.02	0.09	0.02	**0.03**
avg	**0.13**	**−0.08**	**−0.02**	**−0.06**	**0**	**0.08**	**0.02**	**0.01**

Detectors that use using the cross-pol element (CACFAR_cross, PolRatio1) work slightly better with an inclined boat. The PolRatio1 is, when applied to VH VV or HV HH, called iDPolRAD. The SubCorr is the only detector for which an orthogonal vessel is preferable.

3.4. Detector Fusion

Combining detectors can increase the detection performance. We attempted to find the best combination of detectors for dual-pol cross-pol (VH VV and HV HH) and dual-pol co-pol (HHVV) data separately. In the cross-pol case, we tested the combination of PolRatio1 and PolRatio2. We used the logical operator 'OR' and called it PolRatioOR. This detector reaches an average AUC of 0.91.

Thus, this 'surface or volume anomaly detector' approaches the best-performing algorithms PolSym and PWF (Figure 14). The PMF reaches similar results, such as those of the PWF, and since their mechanisms are quite similar, we used only the PWF for this comparison. To further increase its performance, we tested various combinations of the PolRatioOR, the PolSym, the PWF and the CACFAR but without any further increase in the AUC.

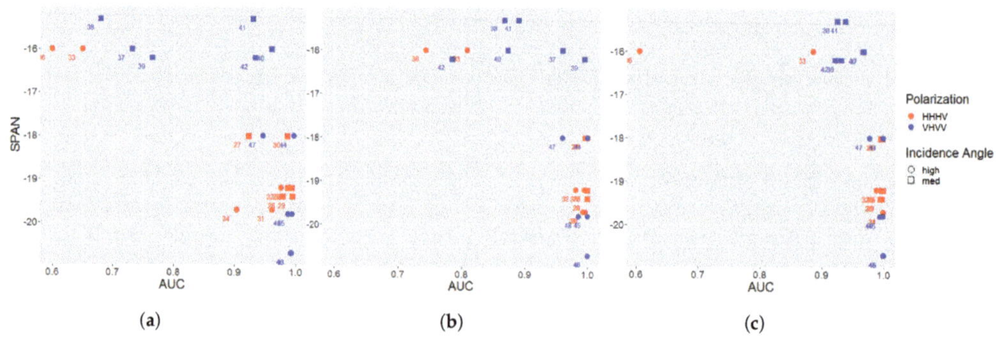

Figure 14. Comparison of the mean AUCs for different wave heights at VH VV and HV HH for PolRatioOR, PolSym and PWF. (**a**) PolRatioOR ($\overline{AUC}$: 0.91). (**b**) PolSym ($\overline{AUC}$: 0.951). (**c**) PWF ($\overline{AUC}$: 0.961).

The ROC curves for the HV HH dual cross-pol data reveal that wave heights larger than three metres significantly hamper the detection of the inflatable (Figure 15). If we only look at acquisitions taken at a medium incidence angle, a P_d of 90% is possible with a maximum P_{fa} of 0.11%. These results cover wave heights of up to 2.4 metres or BFT6.

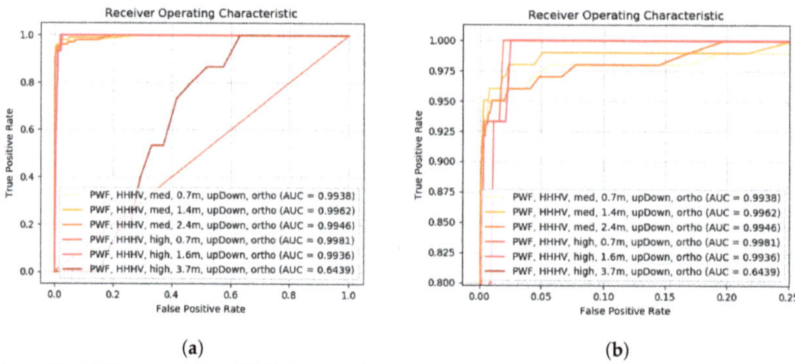

Figure 15. ROC curves of the PWF detector for HV HH data with different sea states and the vessel oriented orthogonally to the LoS. (**a**) ROC of PWF overview; (**b**) ROC of PWF zoomed.

The tests with VH VV dual cross-pol imagery give the impression that the PWF delivers better results at high incidence angles (Figure 16). Since the results are partly incoherent, it is difficult to express a recommendation regarding the wave height. The best results for the 90% detection rate were found for a wave height of 0.5 metres with a P_{fa} of 0.12% and for 2.1 metres with a P_{fa} of 1.07%.

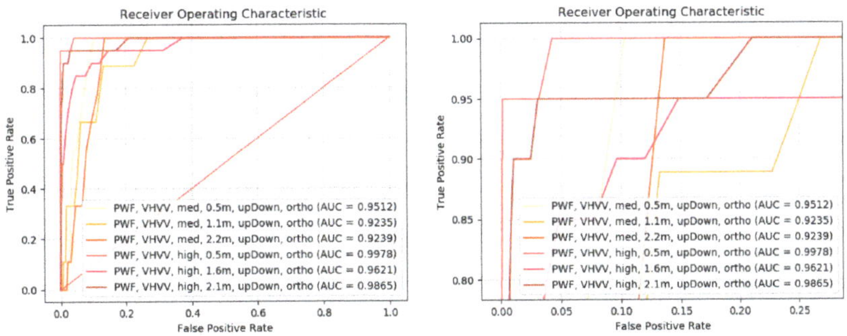

(a) (b)

Figure 16. ROC curves of the PWF detector for VH VV data with different sea states and the vessel oriented orthogonally to the LoS. (**a**) ROC of PWF overview. (**b**) ROC of PWF zoomed.

In the co-pol case (HH VV), we tested the combinations of a new double-bounce detector ('T_{22}'), PolEntropy, PWF and CACFAR. Again, we left out the PMF due to its very similar performance results. The T_{22} yielded an average AUC of 0.914 for HH VV data, thus, making it the best detector—closely ahead of PMF and PWF, each with an AUC of 0.909.

For our co-pol data (HH VV), the combination using the 'AND' operator for T_{22} and PolEntropy, which we called HT22AND, reached the highest AUC (Figure 17). In particular, the results for higher sea states/higher SPANs are promising. This effect can be attributed to the PolEntropy algorithm.

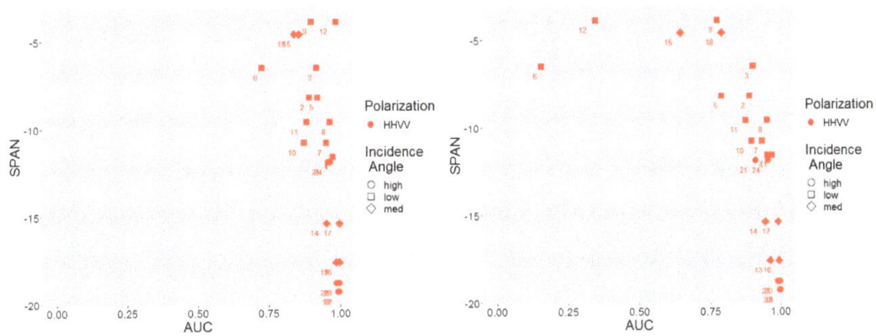

(a) (b)

Figure 17. Comparison of the mean AUCs for different wave heights at HH VV for HT22AND and PWF. (**a**) HT22AND ($\overline{AUC}$: 0.936). (**b**) PMF ($\overline{AUC}$: 0.884).

Figure 18a helps to describe the predictive strength of the HT22AND detector. We see very poor detection results at low incidence angles (P_{fa} 20% at a P_d of 80%). The same is true for wave heights in the range of three metres, with comparably high false alarm rates. If we look at medium and high incidence angles and wave heights below 1.5 m, the HT22AND reaches a P_d of 90% with a much better P_{fa} of, at most, 0.59% (Figure 18b).

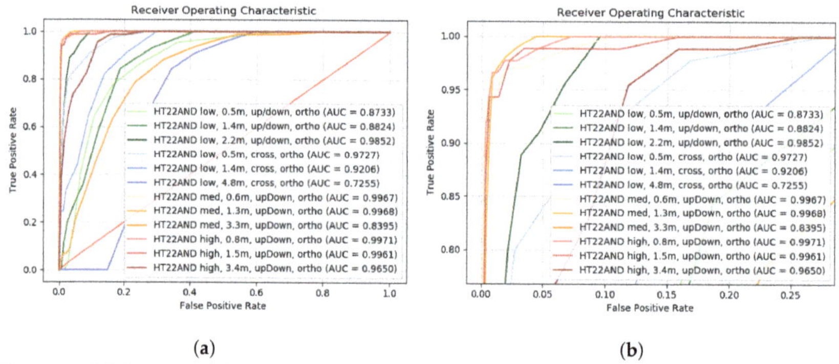

Figure 18. ROC curves of the HT22AND detector for HH VV data with different sea states and the vessel oriented orthogonally to the LoS. (**a**) HT22AND overview. (**b**) HT22AND zoomed.

3.5. Estimation of the Detection Quality

These detection results always include some inherent degree of uncertainty. A metric that gives an estimation of the detection quality would be of great help for the sea rescue teams on the ground. Two appropriate parameters for such a metric would be the estimated vessel size and the TCR. These help with the exclusion of large and strong scatterers, such as large metallic ships. Long high waves could have a TCR similar to the inflatable but could be excluded from the positive true list by virtue of their sheer size. Figure 19a shows that the TCR ranges between two and six times the standard deviation of the clutter of the lake. From the TCR, expressed in terms of the standard deviation of the background clutter, a contrast parameter could be calculated (as in [50], called 'significance').

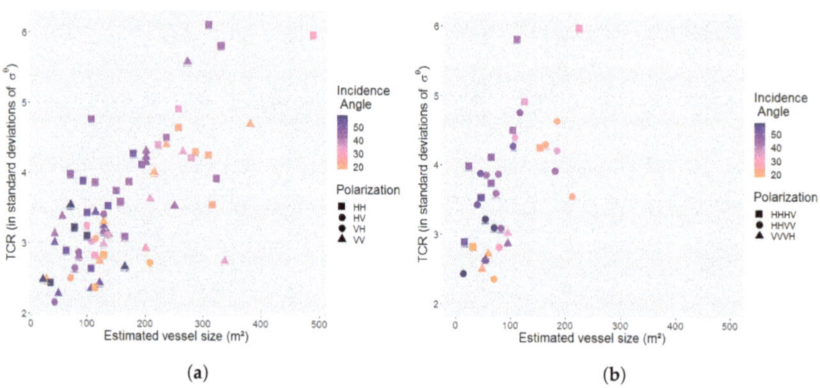

Figure 19. Detection quality estimation parameters TCR and estimated vessel size on the lake test bed. (**a**) Polarization channels shown individually. (**b**) Mean values of the two respective coherent dual-pol channels.

The vessel size estimations in our data range from 30 to 500 m^2 (the true size is 42 m^2). Low incidence angles seem to increase the overestimation. If only high incidence angles were used, the variation in estimations reduced to between 30 and 180 m^2. Figure 19a further reveals a linear correlation between the TCR and the vessel size estimation. Underlined symbols stand for an orthogonal orientation of the boat, the others mean the boat is inclined at 45° and the vessel orientation appears to have no significant impact on the detection quality.

The size estimation can be improved by using the mean value of the two dual-pol channels and by the application of a resizing formula from our previous work [23]. Now, the improved size estimations range between 20 and 250 m^2; for high incidence angles, they stay below 100 m² (Figure 19b). Future work on this could include an analysis of the behaviour of the TCR against the background of different sea states.

4. Discussion

In our approach, we combined maritime SAR data with a simulated fully occupied inflatable by pasting the vessel pixels into the ocean background. That enabled us to create test data that very closely resemble the real situation. The main assumptions that we made in this operation were that:

- The resolution was high enough to resolve the boat pixels separately from the ocean to a large extent.
- The interactions between the vessel and the water surface (flattening of the water surface below the vessel and those caused by the wind shadow on the lee side) were the same in our simulation and in the real situation.
- The wetness of the boat (salty spray on the inflatable and water inside the boat were the same in our simulation and in the real situation.

As a minor effect, we neglected the fact that the rolling of the boat on the ocean (yaw, pitch and roll) induced by waves might change the backscattering of the vessel. Further, the definition of positive true was achieved using the intensity values. That approach biases all results as it helps the intensity-based detectors to perform better compared to the others.

Another noteworthy difference is that the refugee vessel may be moving, even if very slowly. This movement of the vessel triggers certain interactions with the water surface. Since our database has only SAR imagery of a stationary vessel, we could not use and test the movement as a detection mechanism. Therefore, methods, such as along track interferometry (ATI; [51]), which attempt to discriminate non-moving from moving contributions by estimating a velocity-dependent attribute, could be used in the future to improve the effectiveness of migrant boat detection.

Future work could focus on testing the influence of spatial resolution on the different fusion strategies that we proposed. Further, different SAR frequencies (e.g., L-band and C-band) should be added to the test dataset. As mentioned above, the ground truth masks that we used may be improved by exploiting more detectors. Finally, the detectors presented here would gain more confidence if we were able to test them against data with real ground truth obtained from refugee boats.

5. Conclusions

In this work, we attempted to identify the best-performing detectors for a simulated fully occupied refugee inflatable in different weather conditions. For this task, we explored different X-band SAR data types, including TerraSAR-X (TSX), ICEYE and Cosmo-SkyMed Second Generation.

In a preliminary step, we conducted experiments with a ground SAR system to identify bags of wet clay pebbles to simulate humans in SAR. The data acquisition campaign setting was to use these bags to simulate the backscattering of 80 passengers in two different boat orientations (relative to the sensor line of sight) on a lake test bed. Additionally, we created a collection of dual co-pol (HH VV) and dual cross-pol TSX Stripmap data (HV HH and VH VV) from the open ocean with a variety of incidence angles, wave heights and wave directions. We combined the two collections to create a detector testing environment as close as possible to the real situation.

This enabled us to test existing detectors and to develop new variants that were specifically tailored for our rubber inflatable. To further increase the detection capabilities, we experimented with different combinations of detectors.

For dual cross-pol data, we tested a new combination of the volume-detecting Intensity Dual-Polarization Ratio Anomaly Detector (iDPolRAD) and its surface-detecting variant. It reached an overall area under the curve (AUC) of 0.91. For dual co-pol data, the combination of our volume-scattering detector with the polarimetric entropy detector (HT22AND) delivered the best results with an overall AUC of 0.94. The latter is especially interesting, since it seems to be less compromised by high sea states, and this is mainly caused by the polarimetric entropy detector. A deeper analysis of the best detectors resulted in a better estimation of their capabilities:

- With dual cross-pol channel combinations, the polarimetric whitening filter (PWF) was the best-performing detector.
- With HV HH, the PWF reached, at medium incidence angles, a detection rate of 90% with only 0.12% false detections. Our data support the assumption that the PWF can be used up to a maximum wave height of about 2.4 metres.
- With VH VV data, the PWF had a false detection rate of 1.07% at a 90% detection rate and up to 2.1 m wave height.
- For dual co-pol data, the HT22AND was the best detection algorithm with a false alarm rate of 0.59% at a detection rate of 90%. This is true for wave heights of up to 1.5 m and medium or high incidence angles.

The polarimetric match filter (PMF) delivered equivalent results to those of the PWF throughout all our test data. Although the false alarm rates presented here are not yet good enough to support an efficient detection system for high sea states, the study builds a basis for further research and development. For example, the false alarm rate could be lowered with the help of spatial filters that exclude single pixels as well as very large and very bright false detections.

Author Contributions: Conceptualization, data curation, formal analysis, visualization and methodology: P.L., A.M. and M.D.S.; Funding acquisition: P.L., A.M., T.B., F.K. and M.M.; Investigation: P.L. and A.M.; Project administration: P.L., T.B. and F.K.; Resources, A.M., T.B., F.K. and M.M.; Software and validation: P.L. and A.M.; Supervision, A.M., T.B., F.K. and M.M.; Writing—original draft, P.L.; Writing—review and editing, A.M. All authors have read and agreed to the published version of the manuscript.

Funding: This document is a result of the PhD program Safe Automation of Maritime Systems (SAMS) funded by the state of Lower Saxony and the internal research fund of the Jade University Oldenburg.

Informed Consent Statement: Informed consent was obtained from all subjects involved in the study.

Data Availability Statement: We used the following publicly archived datasets for in situ weather data: The Copernicus Marine Environment Monitoring Service at https://marine.copernicus.eu/, accessed on 30 March 2023. The Copernicus Marine Environment Monitoring Service In Situ Thematic Assembly Centre at https://marine.copernicus.eu/about/producers/insitu-tac, accessed on 30 March 2023.

Acknowledgments: We would like to thank all the people who have supported the theoretical and experimental/practical work described in this paper. We would also like to thank the German Aerospace Center DLR, the Agenzia Spaziale Italiana ASI and ICEYE as they provided friendly cooperation and the satellite data. Special thanks goes to the Wokule community who made the satellite data campaign a wonderful experience.

Conflicts of Interest: The authors declare no conflict of interest.

References

1. IOM. Missing Migrants Project. Available online: https://missingmigrants.iom.int/region/mediterranean (accessed on 23 March 2023).
2. IOM. Displacement Tracking Matrix. Available online: https://dtm.iom.int/europe/arrivals (accessed on 23 March 2023).
3. UNHCR. Mediterranean Situation. Available online: https://data.unhcr.org/en/situations/mediterranean (accessed on 23 March 2023).
4. Novak, L.M.; Sechtin, M.B.; Burl, M.C. *Algorithms for Optimal Processing of Polarimetric Radar Data*; Polarimetric Technology Handbook, GACIAC HB 92–01, 139–206, Chicago; Massachusetts Institute of Technology Lexington Lincoln Lab: Lexington, MA, USA, 1989; p. 99.
5. Crisp, D.J.; Redding, N.J. Ship Detection in Synthetic Aperture Radar Imagery. In Proceedings of the 12th Australasian Remote Sensing and Photogrammetry Conference, Fremantle, Australia, 18–22 October 2004; p. 10.

6. Liu, C.; Vachon, P.W.; Geling, G.W. Improved Ship Detection Using Polarimetric SAR Data. In Proceedings of the IGARSS 2004, 2004 IEEE International Geoscience and Remote Sensing Symposium, Anchorage, AK, USA, 20–24 September 2004; Volume 3, pp. 1800–1803. [CrossRef]
7. Yeremy, M.; Campbell, J.W.M.; Mattar, K.; Potter, T. Ocean Surveillance with Polarimetric SAR. *Can. J. Remote Sens.* **2001**, *27*, 328–344. [CrossRef]
8. Touzi, R.; Charbonneau, F.; Hawkins, R.K.; Murnaghan, K.; Kavoun, X. Ship-Sea Contrast Optimization When Using Polarimetric SARs. In Proceedings of the IGARSS 2001, Scanning the Present and Resolving the Future, Proceedings, IEEE 2001 International Geoscience and Remote Sensing Symposium (Cat. No. 01CH37217), Sydney, Australia, 9–13 July 2001; Volume 1, pp. 426–428.
9. Marino, A. A Notch Filter for Ship Detection with Polarimetric SAR Data. *IEEE J. Sel. Top. Appl. Earth Obs. Remote Sens.* **2013**, *6*, 1219–1232. [CrossRef]
10. Marino, A.; Sugimoto, M.; Nunziata, F.; Hajnsek, I.; Migliaccio, M.; Ouchi, K. Comparison of Ship Detectors Using Polarimetric Alos Data: Tokyo Bay. In Proceedings of the 2013 IEEE International Geoscience and Remote Sensing Symposium (IGARSS), Melbourne, Australia, 21–26 July 2013; pp. 2345–2348.
11. Nunziata, F.; Migliaccio, M.; Brown, C.E. Reflection Symmetry for Polarimetric Observation of Man-Made Metallic Targets at Sea. *IEEE J. Ocean. Eng.* **2012**, *37*, 384–394. [CrossRef]
12. Kang, M.; Ji, K.; Leng, X.; Lin, Z. Contextual Region-Based Convolutional Neural Network with Multilayer Fusion for SAR Ship Detection. *Remote Sens.* **2017**, *9*, 860. [CrossRef]
13. Zhao, J.; Guo, W.; Zhang, Z.; Yu, W. A Coupled Convolutional Neural Network for Small and Densely Clustered Ship Detection in SAR Images. *Sci. China Inf. Sci.* **2019**, *62*, 42301. [CrossRef]
14. Girshick, R.; Donahue, J.; Darrell, T.; Malik, J. Rich Feature Hierarchies for Accurate Object Detection and Semantic Segmentation. In Proceedings of the 2014 IEEE Conference on Computer Vision and Pattern Recognition, Columbus, OH, USA, 23–28 June 2014; pp. 580–587. [CrossRef]
15. Girshick, R. Fast R-Cnn. In Proceedings of the IEEE International Conference on Computer Vision, Santiago, Chile, 7–13 December 2015; pp. 1440–1448.
16. Ren, S.; He, K.; Girshick, R.; Sun, J. Faster R-CNN: Towards Real-Time Object Detection with Region Proposal Networks. In *Advances in Neural Information Processing Systems 28*; Cortes, C., Lawrence, N.D., Lee, D.D., Sugiyama, M., Garnett, R., Eds.; Curran Associates, Inc.: Red Hook, NY, USA, 2015; pp. 91–99.
17. Li, J.; Qu, C.; Shao, J. Ship Detection in SAR Images Based on an Improved Faster R-CNN. In Proceedings of the 2017 SAR in Big Data Era: Models, Methods and Applications (BIGSARDATA), Beijing, China, 13–14 November 2017; pp. 1–6. [CrossRef]
18. Zhang, S.; Wu, R.; Xu, K.; Wang, J.; Sun, W. R-CNN-Based Ship Detection from High Resolution Remote Sensing Imagery. *Remote Sens.* **2019**, *11*, 631. [CrossRef]
19. Zhang, T.; Zhang, X. High-Speed Ship Detection in SAR Images Based on a Grid Convolutional Neural Network. *Remote Sens.* **2019**, *11*, 1206. [CrossRef]
20. Redmon, J.; Divvala, S.; Girshick, R.; Farhadi, A. You Only Look Once: Unified, Real-Time Object Detection. In Proceedings of the 2016 IEEE Conference on Computer Vision and Pattern Recognition (CVPR),: Las Vegas, NV, USA, 27–30 June 2016; pp. 779–788. [CrossRef]
21. Chang, Y.L.; Anagaw, A.; Chang, L.; Wang, Y.; Hsiao, C.Y.; Lee, W.H. Ship Detection Based on YOLOv2 for SAR Imagery. *Remote Sens.* **2019**, *11*, 786. [CrossRef]
22. Rostami, M.; Kolouri, S.; Eaton, E.; Kim, K. Deep Transfer Learning for Few-Shot SAR Image Classification. *Remote Sens.* **2019**, *11*, 1374. [CrossRef]
23. Lanz, P.; Marino, A.; Brinkhoff, T.; Köster, F.; Möller, M. The InflateSAR Campaign: Evaluating SAR Identification Capabilities of Distressed Refugee Boats. *Remote Sens.* **2020**, *12*, 3516. [CrossRef]
24. Lanz, P.; Marino, A.; Brinkhoff, T.; Köster, F.; Möller, M. The InflateSAR Campaign: Testing SAR Vessel Detection Systems for Refugee Rubber Inflatables. *Remote Sens.* **2021**, *13*, 1487. [CrossRef]
25. Marino, A.; Dierking, W.; Wesche, C. A Depolarization Ratio Anomaly Detector to Identify Icebergs in Sea Ice Using Dual-Polarization SAR Images. *IEEE Trans. Geosci. Remote Sens.* **2016**, *54*, 5602–5615. [CrossRef]
26. Holt, B. SAR Imaging of the Ocean Surface. In *Synthetic Aperture Radar Marine User's Manual*; Jackson, C., Apel, J., Eds.; NOAA/NESDIS: Washington, DC, USA, 2004; pp. 25–79.
27. Microwaves101. Miscellaneous Dielectric Constants. Available online: https://www.microwaves101.com/encyclopedias/miscellaneous-dielectric-constants (accessed on 15 January 2023).
28. Agency, E.S. Mediterranean Sea Salinity. Available online: https://www.esa.int/ESA_Multimedia/Images/2017/05/Mediterranean_Sea_salinity (accessed on 17 January 2023).
29. Van Zyl, J.J.; Arii, M.; Kim, Y. Model-Based Decomposition of Polarimetric SAR Covariance Matrices Constrained for Nonnegative Eigenvalues. *IEEE Trans. Geosci. Remote Sens.* **2011**, *49*, 3452–3459. [CrossRef]
30. Klein, L.A.; Swift, C.T. An Improved Model for the Dielectric Constant of Sea Water at Microwave Frequencies. *IEEE Trans. Antennas Propag.* **1977**, *25*, 104–111. [CrossRef]
31. Guo, C.; Ye, H.; Zhou, Y.; Xu, Y.; Wang, L. Scaled Sea Surface Design and RCS Measurement Based on Rough Film Medium. *Sensors* **2022**, *22*, 6290. [CrossRef] [PubMed]

32. Hozhabri, M.; Risman, P.O.; Petrovic, N. Comparison of UWB Radar Backscattering by the Human Torso and a Phantom. In Proceedings of the 2018 IEEE Conference on Antenna Measurements & Applications (CAMA), Västerås, Sweden, 3–6 September 2018; pp. 1–4. [CrossRef]
33. Valenzuela, G.R. Theories for the Interaction of Electromagnetic and Oceanic Waves—A Review. *Bound.-Layer Meteorol.* **1978**, *13*, 61–85. [CrossRef]
34. Robinson, I.S. *Measuring the Oceans from Space: The Principles and Methods of Satellite Oceanography*; Springer Science & Business Media: Berlin/Heidelberg, Germany, 2004.
35. Voronovich, A.G.; Zavorotny, V.U. Theoretical Model for Scattering of Radar Signals in Ku- and C-bands from a Rough Sea Surface with Breaking Waves. *Waves Random Media* **2001**, *11*, 247. [CrossRef]
36. Mouche, A.A.; Chapron, B.; Reul, N. A Simplified Asymptotic Theory for Ocean Surface Electromagnetic Wave Scattering. *Waves Random Complex Media* **2007**, *17*, 321–341. [CrossRef]
37. Kudryavtsev, V.; Hauser, D.; Caudal, G.; Chapron, B. A Semiempirical Model of the Normalized Radar Cross-Section of the Sea Surface 1. Background Model. *J. Geophys. Res. Ocean.* **2003**, *108*, FET 2-1–FET 2-24. [CrossRef]
38. Ren, Y.; Lehner, S.; Brusch, S.; Li, X.; He, M. An Algorithm for the Retrieval of Sea Surface Wind Fields Using X-band TerraSAR-X Data. *Int. J. Remote Sens.* **2012**, *33*, 7310–7336. [CrossRef]
39. Weber, J.E.H. Vertically Varying Eulerian Mean Currents Induced by Internal Coastal Kelvin Waves. *J. Geophys. Res. Ocean.* **2017**, *122*, 1222–1231. [CrossRef]
40. Zhang, L.; Zhang, J.; Zou, B.; Zhang, Y. Comparison of Methods for Target Detection and Applications Using Polarimetric SAR Image. In Proceedings of the Progress in Electromagnetics Research Symposium, Hangzhou, China 24–28 March 2008; Volume 1.
41. Cloude, S.R.; Pottier, E. A Review of Target Decomposition Theorems in Radar Polarimetry. *IEEE Trans. Geosci. Remote Sens.* **1996**, *34*, 498–518. [CrossRef]
42. Cloude, S.; Pottier, E. An Entropy Based Classication Scheme for Land Applications of Polarimetric SAR. *IEEE Trans. Geosci. Remote Sens.* **1997**, *35*, 68–78. [CrossRef]
43. Yamaguchi, Y.; Moriyama, T.; Ishido, M.; Yamada, H. Four-Component Scattering Model for Polarimetric SAR Image Decomposition. *IEEE Trans. Geosci. Remote Sens.* **2005**, *43*, 1699–1706. [CrossRef]
44. Freeman, A.; Durden, S.L. Three-Component Scattering Model to Describe Polarimetric SAR Data. In *Proceedings of the Radar Polarimetry*; International Society for Optics and Photonics: Washington, DC, USA, 1993; Volume 1748, pp. 213–224.
45. Kennaugh, E.M.; Sloan, R.W. *Effects of Type of Polarization On Echo Characteristics*; Technical Report; Ohio State Univ Research Foundation Columbus Antenna Lab: Columbus, OH, USA, 1952.
46. Cameron, W.L.; Leung, L.K. Feature Motivated Polarization Scattering Matrix Decomposition. In Proceedings of the IEEE International Conference on Radar, Arlington, VA, USA, 7–10 May 1990; pp. 549–557.
47. Krogager, E. New Decomposition of the Radar Target Scattering Matrix. *Electron. Lett.* **1990**, *26*, 1525–1527. [CrossRef]
48. Karachristos, K.; Anastassopoulos, V. Land Cover Classification Based on Double Scatterer Model and Neural Networks. *Geomatics* **2022**, *2*, 323–337. [CrossRef]
49. Singh, G.; Mohanty, S.; Yamazaki, Y.; Yamaguchi, Y. Physical Scattering Interpretation of POLSAR Coherency Matrix by Using Compound Scattering Phenomenon. *IEEE Trans. Geosci. Remote Sens.* **2020**, *58*, 2541–2556. [CrossRef]
50. Tello, M.; López-Martínez, C.; Mallorqui, J.J. A Novel Algorithm for Ship Detection in SAR Imagery Based on the Wavelet Transform. *IEEE Geosci. Remote Sens. Lett.* **2005**, *2*, 201–205. [CrossRef]
51. Raney, R.K. Synthetic Aperture Imaging Radar and Moving Targets. *IEEE Trans. Aerosp. Electron. Syst.* **1971**, *AES-7*, 499–505. [CrossRef]

remote sensing

MDPI

Technical Note

A Decade from the Costa Concordia Shipwreck: Lesson Learned on the Contribution of Infrared Thermography during the Maritime Salvage Operations

William Frodella [1,*], Guglielmo Rossi [2], Luca Tanteri [2], Ascanio Rosi [3], Luca Lombardi [1], Francesco Mugnai [4], Riccardo Fanti [1] and Nicola Casagli [1]

1 Department of Earth Sciences, University of Florence, Via G. La Pira 4, 50121 Florence, Italy
2 Civil Protection Centre, University of Florence, Largo Fermi 1, 50142 Florence, Italy
3 Department of Geosciences, University of Padua, Via G. Gradenigo 6, 35131 Padua, Italy
4 Department of Civil and Environmental Engineering, University of Florence, Via di Santa Marta 3, 50139 Florence, Italy
* Correspondence: william.frodella@unifi.it; Tel.: +39-055-2755979

Abstract: On 13 January 2012, the Italian vessel Costa Concordia wrecked on the shore of Giglio Island, about 15 km off the coast of southern Tuscany (Italy), causing the loss of 32 lives. It is considered one of the worst disasters in maritime history. Salvage operations started immediately after the wreck with the coordination of the Italian National Civil Protection Department and the technological support of several Research Centers, which were activated for the management of the consequent emergency phase. A multi-parametric and multiplatform monitoring system was promptly implemented, involving several advanced remote sensing techniques, among which was Infrared Thermography (IRT). In this framework, IRT monitoring was performed during a 35-day period (25 January–1 March 2012), using a terrestrial, hand-held thermal camera. Six different thermal images were acquired daily from the island's coastline in three different positions, both in daylight and night times. The aim was to detect thermal anomalies connected to possible deformations of the vessel and oil spills. Between 3–4 February, IRT successfully revealed on oil spill drifting from the stern of the wreck towards the island harbor. Furthermore, the wreck's thermal dilatation was also analyzed during a 24-day close-range monitoring, providing interesting insights for the interpretation of the deformation monitoring results. This paper presents the outcomes of these innovative and experimental monitoring activities, with the aim of testing the potential of IRT as a versatile and operative tool to be used in maritime and environmental disaster response.

Keywords: shipwreck monitoring; infrared thermography; Costa Concordia; oil spill detection; maritime disaster response

1. Introduction

Shipwrecks can often have major economic and ecological impacts in marine and oceanic waters, as well as in anthropic and natural coastal environments. Oil spills connected to shipwrecks, unfortunately, represent a frequent occurrence and are considered as one of the most vital public issues [1–3]. Regarding oil spills in the sea, remote sensing is playing an increasingly important role as a fundamental tool, since it can provide information on the oil-spill movement through multi-temporal imaging [4,5].

This can then be input to drift-prediction models or used to plan effective countermeasures by the response team [6]. Satellite imagery may provide useful overviews on known and very large spills, using either SAR [7–10] or hyperspectral and multispectral data [11–14]. In this context, due to the coarser spatial resolution, thermal data are less adopted [15–18], while optical sensors offer only marginal capability for oil spill detection [19,20]. In the field of small oil spill detection, mapping and monitoring are common methods often carried out with video or photography from airborne platforms, including Unmanned Aerial Vehicles (UAVs) [5]. In this framework, multispectral [21,22] and thermographic applications [23–25] are present, but still limited, while terrestrial applications, due to logistic and acquisition problems, cannot provide a full coverage of the Areas of Interest (AoIs) [18]. Extensive overviews on the topic of oil spill monitoring are given by [6,9,26–30].

On 13 January 2012, during a cruise in the Mediterranean Sea, the Italian vessel Costa Concordia wrecked offshore of Giglio Island, along the coast of southern Tuscany (Central Italy) (Figure 1a–c). After deviating from the planned route, the ship struck an underwater rock located a few hundred meters east of the island coastline, in front of Giglio Porto locality (Figure 1b). The impact opened a large gash in the ship's portside below the water line, which caused the cutting of the engines' power and the consequent ship drifting towards the island coastline. Luckily, the ship grounded near shore, in a precarious position on a rocky underwater ledge, in front of Gabbianara Point (Figure 1d), where it stood until its removal in July 2014. It is the passenger ship of the largest tonnage to have suffered this consequence; of more than 3229 passengers and 1023 crew, 32 unfortunately died [31].

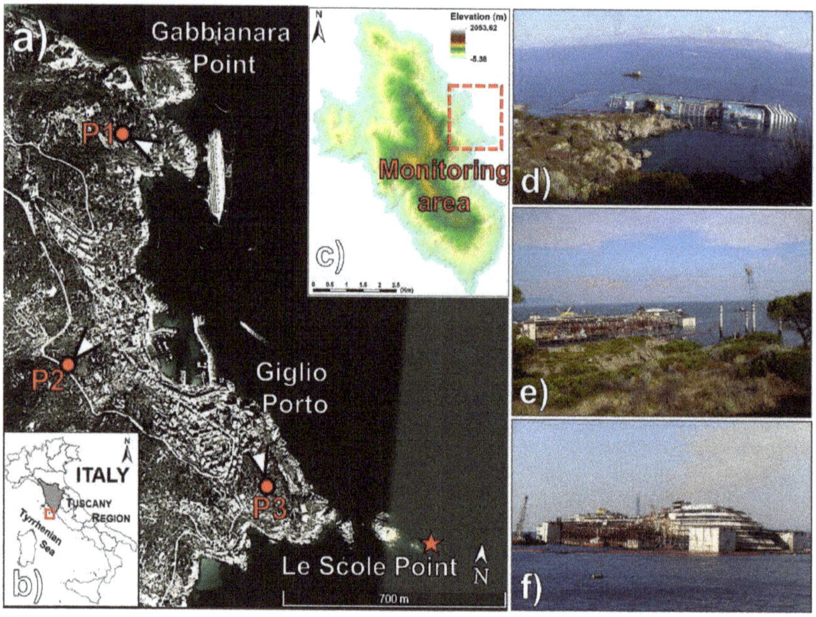

Figure 1. Setting of the study area: (**a**) location of the Costa Concordia after the wreck on Google Earth™ satellite image (March 2012) showing the IRT survey points (P1 = Lazzeretto Point; P2 = Giglio Castello road; P3 = Cannelle); (**b**) geographic setting of the Giglio island (Tuscany/Central Italy); (**c**) elevation map of Giglio Island (red dashed square shows the monitoring area). Pictures of the Costa Concordia shipwreck: (**d**) soon after the wreck in January 2012; (**e**) during the parbuckling operations in September 2013; (**f**) in the hull-removal phase (July 2013).

The Costa Concordia Maritime Disaster Management

Soon after the wreck, salvage operations were set up by the Italian National Civil Protection Department in close coordination with the technical mission structure of the Commissioner delegate, with the technical support of Tuscany Region Central Functional Center, several research centers, and specialized companies. The management of the emergency salvage phases (the largest of this kind to date in maritime disaster history) developed as follows:

- defueling and caretaking (carried out from 18 January 2012 to 15 September 2013): this phase had the primary objective of removing the fuel from the vessel's tanks and engine rooms as quickly as possible, in order to ensure the protection of the marine environment by cleaning of the seabed from the shipwreck debris (Figure 1d);
- parbuckling and post—parbuckling (from 16 September to 31 December 2013): this phase developed with the initial righting of the wreck by rotating the vessel around the longitudinal axis of the ship using cables, giant inflatable buoys, and ballasting with sponsons (Figure 1e);
- hull removal—scrapping (from 31 December 2013 to 14 July 2014): after the hull removal, the Costa Concordia was dragged away from the Giglio Island, to sail for its final cruise to the port of Genoa, where it was finally scrapped [4] (Figure 1f).

During these phases, search and rescue operations were also carried out to extract the victims from the wreck. Especially during the first defueling and caretaking phases, there was a serious concern regarding possible ship displacement and deformations, which in turn could have caused both the sliding of the wreck from the underwater ledge towards the deeper seabed or oil spills. Both hypotheses represented catastrophic scenarios, not only for the rescue teams, but also for the environmental consequences (Giglio Island is in fact part of the protected area of the Tuscan Archipelago National Park). In this framework, in order to ensure the highest safety conditions for the workers and the rescuers operating around and within the wreck, as well as to support the shipwreck removal operations, a multiparametric and multiplatform monitoring system based on-site devices, satellite, and terrestrial remote monitoring techniques was implemented. Amongst traditional seismic [32], topographic [33], and GPS monitoring, several advanced remote sensing techniques were employed, such as space-borne Synthetic Aperture Radar (SAR) satellite [5,34] and ground-based interferometry [35], terrestrial laser scanning, and terrestrial Infrared Thermography (IRT) [36]. Geomechanical and geophysical analysis were also carried out in correspondence with the coastal granite rock mass underlying the wreck in order to assess its general stability [37]. In particular, from 25 January to 1 March 2012, an innovative experimental IRT monitoring campaign was carried out to assess the thermal behavior of the shipwreck and the surrounding sea stretch in order to locate possible thermal anomalies connected to deformations or oil spills. The salvage operations during their entire 30-month duration were a success, because the personnel involved in the rescue could rely on a solid warning in case of possible sliding of the wreck, and few pollutions were produced, even during the defueling. Furthermore, the analysis of the surface temperature variations of the deck contributed to the understanding of the vessel deformation trend. The aim of this paper is to present the lesson learned during the IRT monitoring in this complex emergency management, in order to provide useful outcomes regarding the contribution of thermal sensors analysis in shipwreck monitoring, and in the broader perspective of improving a still not well-developed scientific literature in the field of marine environmental disaster response.

2. Materials and Methods

InfraRed Thermography (IRT) is a versatile and low-cost, non-destructive diagnostic technique, accomplished with calibrated infrared thermal cameras capable of measuring with high spatial and temporal resolution the surface temperature pattern of the monitored scenarios [38].

The product of an IRT survey is a digital image acquired by the camera array detector (called "thermogram" or "thermographic image"), which, following the correction of the sensitive parameters, is converted by the camera's built-in processor into a surface temperature map [39]. The presence of any inhomogeneity within the surveyed material (e.g., an oil coating) highly modifies the material thermal parameters (density, specific heat, and thermal conductivity), reducing the heat transfer and adding its signal to that given by the structure [40]; therefore, the inhomogeneity is displayed in the analyzed surface temperature map as a thermal anomaly (an irregular thermal pattern with respect to the surrounding areas) [41]. Modern high-resolution thermal cameras used both in terrestrial- (hand-held and tripod-mounted) or airborne-based modes, thanks to their portability and easy data collection/processing, represent powerful and cost-effective tools for thermal surveying and monitoring [42]. During the last decades, thanks to the technological development of portable high-resolution and cost-effective devices, IRT has been widely used in the field of civil engineering, cultural heritage, volcanic monitoring, environmental applications, and engineering geology [41–51].

The Costa Concordia Shipwreck IRT Monitoring

The IRT monitoring of the shipwreck was performed in tripod-mounted mode using a FLIR system SC620 thermal camera with a focal plane array (FPA) microbolometer sensor [52] (Figure 2; Table 1). A built-in 3.2 Mpixel digital camera allowed for the comparison between the thermograms and the corresponding optical images, in order to improve the interpretation of the thermal data.

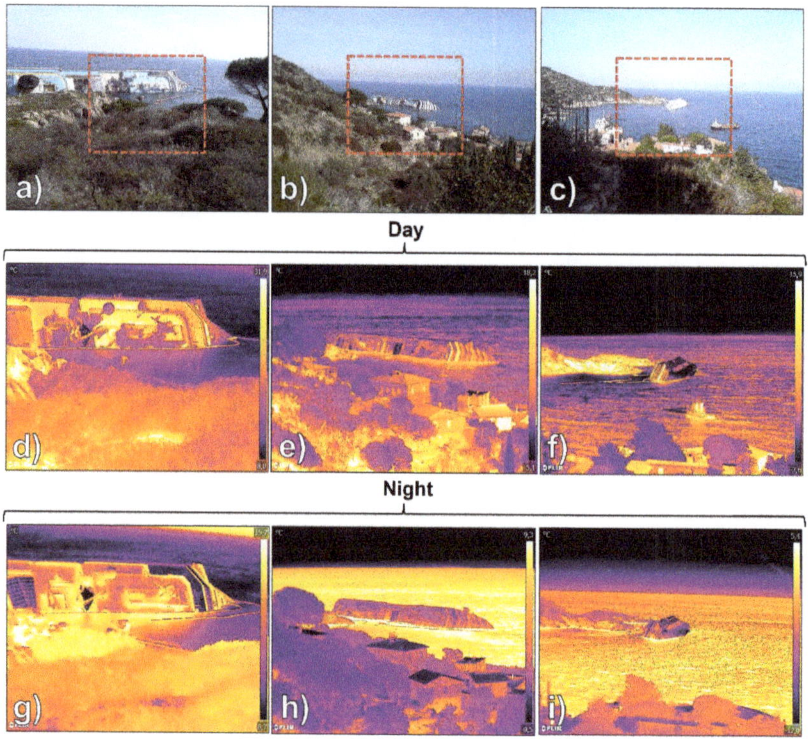

Figure 2. View of the Costa Concordia wreck from the acquiring points, showing the IR field of view (red dashed squares): (**a**) P1; (**b**) P2; (**c**) P3; examples of IRT thermograms during daytime from P1 (**d**), P2 (**e**); P3 (**f**), and during nighttime from P1 (**g**); P2 (**h**); P3 (**i**) (colorbar "iron").

Table 1. Technical features of the adopted thermal camera and IRT survey parameters.

Spectral Range (μm)	Thermal Sensitivity (mK)	Sensor Spatial Resolution (mrad)	IR Resolution (pix)	Field of View (°)	Sensor-Target Distance (m)	Image Resolution (m)
7.5–13	40	0.65	640 × 480	24 × 18	250 (P1); 700 (P2); 800 (P3)	1.6 (P1); 4.55 (P2); 5.20 (P3)

Data correction, thermal focusing, histogram equalization, and mosaicking were performed by means of FLIR Tools + software [53]. Through the thermographic survey, it was possible to measure the thermal radiation emitted by the ship and record the presence of any thermal anomalies in correspondence with the wreck and in the adjacent sea stretch. Initially, two acquisitions were made daily from three different acquisition points (P1–3 in Figure 1a), leading to an image resolution ranging from 1.6 m (P1) to 5.2 m (P3) (Table 1). Depending on their acquiring location, each thermogram was calibrated using the distance values and air temperature-humidity acquired through a pocket thermohygrometer, and a standard emissivity value of 0.95 (due to the complex multi-material scenario, as recommended by FLIR systems). In order to picture the daily temperature variations, two acquisitions were carried out around 12:00 and 00:00. Furthermore, from P1 (acquisition distance of approximately 250 from the shipwreck at a height of 26 m a.s.l.), between 5 to 29 February 2012, a short-term monitoring was also carried out by acquiring a thermogram every two hours approximately during daylight (Figure 2c). The aim was to analyze the surface temperature variations of the Costa Concordia deck, with the aim of assessing the possible effect of thermal dilatation on the vessel deformations (Figure 2d). The average temperatures variations in six distinct areas of the deck were calculated by means of a semi-automated processing procedure in a GIS environment [54]. The thermograms were converted to raster format, co-registered, and analyzed using the ArcGIS "zonal statistics tool". Furthermore, in September 2013 and July 2014, during the parbuckling and hull removal operations (Figure 1e,f), an IR video sequence were also acquired, carried out from a fixed installation in P1 (Figure 2a,b) (Figure 2). By using FLIR ResearchIR software [55], single frames were then extracted and analyzed. The acquisition points P2 and P3 (distance of approximately 700 and 800 m from the shipwreck, and a height of 97 and 40 m a.s.l., respectively) were chosen to monitor the sea stretch surrounding the wreck for the detection of possible oil spills (Figure 2e).

3. Results

The bridge deck temperature measurements acquired in P1 are shown in Figure 3, together with the six control areas on the ship deck (Areas 1–6 in Figure 3a), their average temperature evolution (Figure 3b), and the air temperature recorded by the Giglio Porto air thermometric station (Figure 3c) [56]. The monitoring highlighted daily surface temperature peaks between 2 and 4 p.m. (values ranging approximately from 24.7 to 44.2 °C, in correspondence with Area 2), while the lowest values were recorded in correspondence with Area 4 (from 8.7 to −4.4 °C). Both areas are located on the left deck side; this difference is likely related to the different materials involved. The surface temperature differences plot shows also a higher thermal inertia for Area 2 and 6 (this latter is in the same material with respect to A3, but on the deck's right sector, right above the sea water; Figure 3c). The high temperature range measured in all the control areas (Figure 3a,b) is much higher than the air temperature (Figure 3e); this result is due to the solar radiation of the extensive metal surface of the ship; the high specific heat of aluminum, widely used in the construction deck of the vessel, enhanced the storage of energy radiated by the sun and, thus, reached higher temperatures than air. This may have implied millimetric deformations of the shipwreck due to the linear thermal dilatation of the aluminum (Figure 3d).

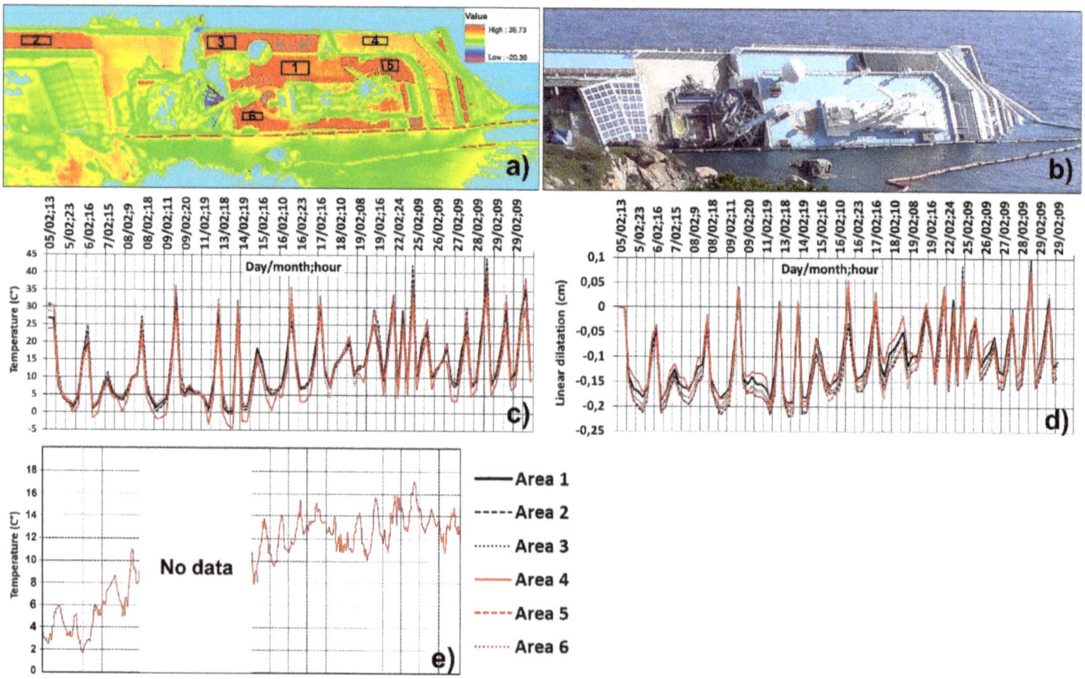

Figure 3. The IRT short-term monitoring campaign from P1: thermographic ΔT Image of the deck sector with the control areas (1–6) created in ArcMap (**a**); corresponding optical image (see the collapsed glass panels of the deck) (**b**); average surface temperature variations in correspondence of the control areas (**c**); linear thermal dilatation in the control sectors (**d**); air temperature data of the marine weather station of Giglio Porto between 5 and 26 February 2012 (**e**). Time scale of panel (**e**) is the same with respect to subfigures (**c**,**d**).

This trend was confirmed by the correlation between the long-term RTS deformation measurements and the air temperature [33,36]. The IRT data collected from P2, and especially from P3, allowed detecting on 3 February 2012 around 00:40 and 12:00, as well as on 4 February 2012 at 00:00, cold thermal anomalies in the stretch of sea between the Costa Concordia shipwreck and Giglio harbor (2–3 °C lower than the surrounding sea water) (Figure 4). Already on 5 February 2012, the spill had ceased. On 1 February 2012, the anti-pollution floating booms surrounding the wreck during the defueling operation had been temporarily cut by the collapse of the deck's glass panels, allowing for oily substances to drift towards the island's harbor, as also confirmed by visual inspections (see Figure 3b). Yet, since there had been no hydrocarbon nor engine oil spills during the defueling operation, considering the extent of the oil slicks (around 30 m in maximum width for around 600 m in maximum length), and since its source point was the vessel's stern, where the kitchens of the restaurant rooms were located, this slick was interpreted as the leak of alimentary oily substances from the vessel's kitchens. Single thermograms, extracted and analyzed from the IR sequence recorded during the parbuckling operations on the night of 3 September 2012, showed that following the very first stages of the wreck righting, thermal anomalies in correspondence with jets of pressurized fluids were coming from the stern-side and the bowside of the chimney (Figure 5). These fluids show temperatures between 22.1 °C (Figure 5a) and 19.9 °C (Figure 5b), suggesting that these could be interpreted as sea water jets developed in the peculiar fluid circulation within the wrecked vessel.

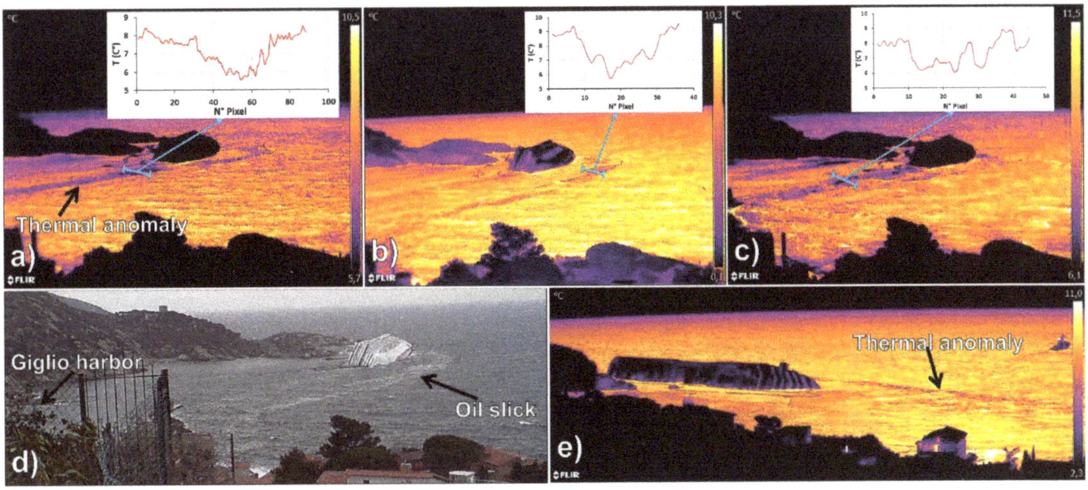

Figure 4. IRT applied for oil slick detection during the defueling operations (colorbar "iron"). Thermograms acquired from P3 on 3 February 2012 at 00:40 (**a**) and 12:30 (**b**); thermogram acquired from P3 on 4 February 2012 at 00:00 (**c**); correspondent optical image of b (**d**); mosaicked thermogram acquired from P2 on 3 February 2012 at 00:40 (**e**).

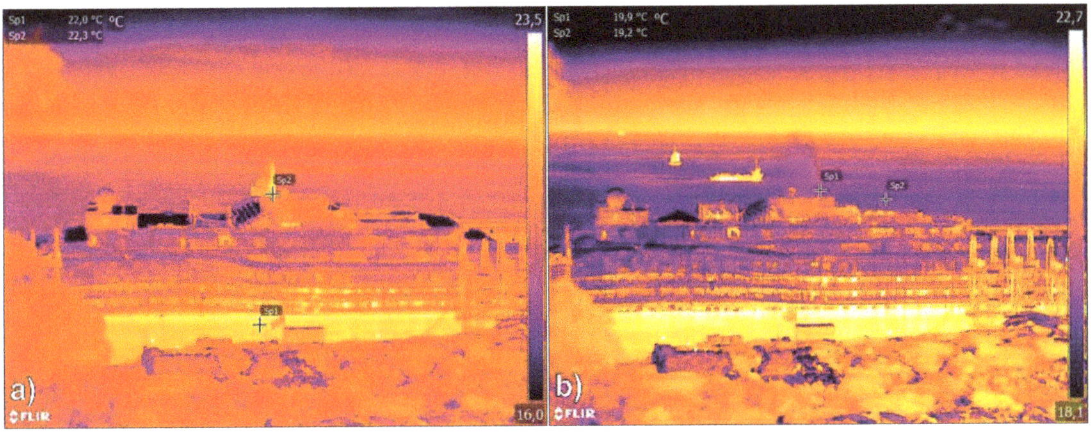

Figure 5. Thermograms selected from the IR video sequence acquired during the parbuckling (colorbar "iron"). Thermogram acquired at 02:42 on 17 September 2013 (**a**); thermogram acquired the same day at 04:12 (**b**) (Sp1–2 represent the spot measurements of single pixels).

In fact, sea water temperature in the warm sea stretch between the wreck and the coastline shows temperatures around 22 °C (the same of the first jet), whereas the slightly colder stretch facing the sea reaches around 19 °C (the same as the second jet) (Figure 5b).

This is probably due to the gradual temperature equilibrium reached between the first and second jet (which occurred with a time interval of around one hour and a half), following the new infilling of sea water in the sunken vessel during the initial phases of the righting. During the hull removal operations in the sea area between the coastline and the ship, no significant thermal anomalies possibly connected to oil spills were detected, as shown by the recorded IR video sequence acquired on 14 July 2014 (see Supplementary materials).

4. Discussion

The Costa Concordia shipwreck salvage operations and its monitoring system represent an extraordinary effort in maritime history, considering the tonnage (114,137 gross registered tonnage) and the size of the ship (290 m in length, 70 in height, and 35 in width height). At the time of the wreckage, 2400 tons of fuel oil were stored inside the tanks, as were several other different pollutants potentially dangerous for the marine environment (e.g., engine oil, alimentary oil, and soap) [5]. For the authorities involved in the emergency management, this was the most alarming threat, together with the possible sliding of the wrecked ship from Punta Gabbianara and subsequent sinking in the deeper seabed, keeping in mind the concern for the security of the operators involved in the rescue and the body recovery. To the complexity of the scenario, it has to be added that several explosive charges were used to penetrate the deformed corridors and rooms of the vessel, which could lead to the potential instability of the wreck. The use of optical satellite remote sensing for oil spills is improving, although its success is generally limited to identifying features at sites where known oil spills have occurred, and it can be profitable used as a base to add other sensors' information (oil shows no spectral characteristics in the visible region) [5]. SAR surely offers the only potential for large area searches and adverse weather remote sensing [57]. Nevertheless, SAR is often costly, its efficiency connected to the revisit time of the satellite, and it is prone to many interferences (false targets can be as high as 95%) [4,5]. In recent years, these limitations can be in part hampered using ESA's Sentinel data, although these are provided only for the European area. Laser fluor sensor is probably the most effective sensor for positively discriminating oil on most backgrounds that include water, soil, ice, and snow; nevertheless, it needs an airborne platform and may be costly [58]. Although there are not still many examples in the literature, IRT from terrestrial and UAV platforms seems to be an ideal tool to detect environmental contamination since, e.g., in oil films on water, it can reduce its surface temperature, therefore representing a thermal anomaly [59]. Furthermore, regarding the monitoring of the hull's thermally induced deformations, the use of wireless sensor networks (WSN) could also have promising results, as shown in several successful applications in the field of disaster risk management [60–62]. In this work, the applicability and potential of IRT as a novel operational tool in a multiparametric and multiplatform shipwreck monitoring system was tested. The results show that although its applications in the field of maritime disaster management are still experimental, IRT can be profitably used in different emergency phases to monitor the vessel thermally induced deformation and small oil spills in the sea stretches surrounding the wreck. In particular, regarding the outcomes of the Costa Concordia monitoring [36], during the first defueling and caretaking phases (up to 1 September 2012), the recorded surface temperature was correlated to the periodic daily thermal dilatations (Figure 3). Temperature did not seem to directly influence the deformations related to accelerations, which in turn were related to tidal, wind, and storm wave effects [36]. IRT showed its potential in detecting between 3 and 4 February 2012 a thermal anomaly of 2 °C with respect to the surrounding sea stretch that was confirmed to be an oil spill coming from the vessel's stern, drifting towards the island harbor (Figure 4). Therefore, a prompt warning was sent to the authorities and personnel involved in the emergency management. During the experimental monitoring, IRT has proven to be a versatile tool, not only for its night vision (e.g., during the parbuckling phase), but also for the capability of the thermal camera to work in a rapid installation-acquisition approach from different points of view and acquiring modes (e.g., from single thermograms to IR video sequences, as in both the parbuckling and hull removal phase) (Figure 5 and Supplementary materials). P1 was the only feasible location for close-range monitoring from a fixed point (the thermal camera was in fact installed together with other instruments of the integrated monitoring system in a cabin-hut with electric power supply) (Figures 1 and 2a,b,e,h). P3 instead granted a favorable commanding view, being on the hilltop just south of Giglio Porto village, overlooking the sea stretch between the wreck and the island harbor (Figures 1 and 2d,g,i).

In this framework the Giglio island's geomorphological setting (relatively steep hilly coastline) granted favorable monitoring point of views with respect to lower flat coastlines. IRT also showed its versatility by detecting after rainfall events cold thermal anomalies connected to seepage along key rock mass discontinuities (granite exfoliation slope-dipping joints) (Figure 6). This represents valuable information, since seepage is an important factor in reducing the wall strength of discontinuities [63], and the most important potential instability phenomena detected in the area by means of rock mass kinematic analysis [51] were plane failures along exfoliation joints, but also because it represents an input for the modelling of the rock mass stability activities [37].

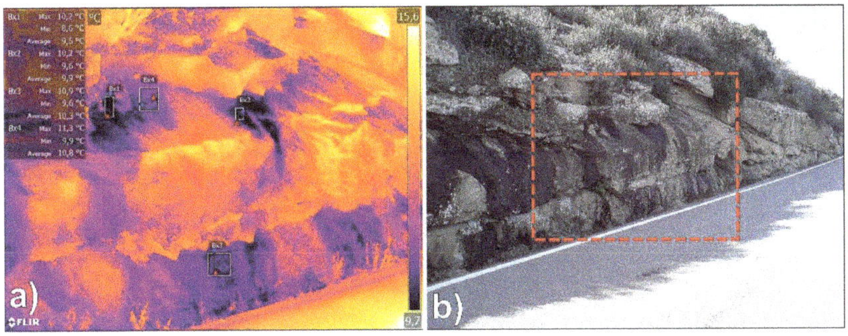

Figure 6. Rock mass IRT survey (colorbar "iron"). Cold thermal anomalies in correspondence of granite exfoliation joints (**a**); correspondent visible image (**b**) (red dashed square highlight the camera field of view).

Furthermore, the possibility to use box plots (Bx1–4 in Figure 6a) for areal surface temperature measurements shows how IRT can not only detect seepage-affected areas (Bx1 and 2), but also potentially discriminate on the discontinuity of the dampness rating and the evidence of water flow, based on the surface temperature (e.g., Bx3–4 in Figure 6a) [50,51]. The timing and the number of daily acquisitions was restricted by the operative and logistic limitations: only a single thermal camera was available, and the dedicated personnel was compelled to assemble daily all the integrated monitoring system data to produce bulletins for the National Civil Protection Department. Furthermore, 24 h IR sequences from fixed installations would have produced a huge amount of data (one minute of IR-video with 60 frames per second is about 1.8 Gigabyte), creating serious data management problems. IRT alone is insufficient for an effective shipwreck monitoring, since it can only detect surface temperature changes; nevertheless, this work showed how this kind of information can be profitably used in integration with other techniques devoted to monitoring the deformations (e.g., RTS and SAR).

5. Conclusions

In conclusion, this work represents an important novel experimental application of IRT in maritime disaster management. Between 3–4 February, IRT successfully detected on oil spill drifting from the stern of the wreck towards the island harbor. Close-range IRT monitoring revealed the wreck thermal dilatation, providing interesting insights for the interpretation of the monitored deformation data. This work shows that IRT can be profitably used in maritime salvage as an ancillary low-cost technique, which can be used operatively in an integrated multiplatform and multi-sensor shipwreck monitoring system with other ground-based remote sensing techniques, such as optical imagery, lidar, and spaceborne and ground-based SAR. Nevertheless, in order to obtain an accurate interpretation of the results, a skilled operator is highly recommended. Future developments in maritime disaster response should include the application of IRT from unmanned UAV platforms and from fixed thermal camera installations for gathering continuous, high-resolution, real-time data.

Supplementary Materials: The following supporting information can be downloaded at: https://www.mdpi.com/article/10.3390/rs15051347/s1, two IR video sequences acquired on 14 July 2014, during the hull removal operations. IR Video S1: 2014-07-14T113854906; IR Video S2: 2014-07-14T150751558.

Author Contributions: W.F. conceived the structure, elaborated the thermal data, wrote the paper, and designed the figures. G.R., L.T. and A.R. conceived the short-term thermal monitoring from P1 position and set the procedure for the data semi-automatic analysis. F.M. and L.L. acquired the IR video sequence during the parbuckling and hull removal operations, respectively. R.F. and N.C. coordinated the research. All authors have read and agreed to the published version of the manuscript.

Funding: This research was carried out in the framework of the agreement with the Department of Earth Sciences of the University of Florence (Italy) "Monitoring of the Costa Concordia (Giglio Island)" on behalf of the Commissioner delegate of the Presidency of the Council of Ministers ex OPCM 3998/2012, funded by the Italian Civil Protection Department.

Data Availability Statement: Not applicable.

Acknowledgments: The authors would like to thank the Italian Civil Protection Department and all the current and former colleagues from the Department of Earth Sciences of the University of Florence (Italy) who were involved in the Costa Concordia shipwreck monitoring (in alphabetical order): Federica Bardi, Alessandro Battistini, Silvia Bianchini, Andrea Ciampalini, Federico Di Traglia, Giulia Dotta, Federica Ferrigno, Francesco Fidolini, Sara Frangioni, Francesca Garfagnoli, Emanuele Intrieri, Daniela Lagomarsino, Alessia Lotti, Stefano Morelli, Massimiliano Nocentini, Teresa Nolesini, Veronica Pazzi, Federico Raspini, Samuele Segoni, Carlo Tacconi Stefanelli and Deodato Tapete.

Conflicts of Interest: The authors declare no conflict of interest.

References

1. Ivanov, A.Y. The oil spill from a shipwreck in Kerch Strait: Radar monitoring and numerical modelling. *Int. J. Remote Sens.* **2010**, *31*, 4853–4868. [CrossRef]
2. Pisano, A.; Bignami, F.; Santoleri, R. Oil spill detection in glint-contaminated near-infrared MODIS imagery. *Remote Sens.* **2015**, *7*, 1112–1134. [CrossRef]
3. Chen, X.; Liu, S.; Liu, R.W.; Wu, H.; Han, B.; Zhao, J. Quantifying Arctic oil spilling event risk by integrating an analytic network process and a fuzzy comprehensive evaluation model. *Ocean Coast. Manag.* **2022**, *228*, 106326. [CrossRef]
4. Lodge, A.E. *The Remote Sensing of Oil Slicks*; John Wiley & Son: Chichester, UK, 1989; p. 147. ISBN -0471920436.
5. Ciampalini, A.; Raspini, F.; Bianchini, S.; Tarchi, D.; Vespe, M.; Moretti, S.; Casagli, N. The Costa Concordia last cruise: The first application of high frequency monitoring based on COSMO-SkyMed constellation for wreck removal. *ISPRS J. Photogramm. Remote Sens.* **2016**, *112*, 37–49. [CrossRef]
6. Fingas, M.; Brown, C. Review of oil spill remote sensing. *Mar. Pollut. Bull.* **2014**, *83*, 9–23. [CrossRef] [PubMed]
7. Martinez, A.; Moreno, V. An oil spill monitoring system based on SAR images. *Spill Sci. Technol. Bull.* **1996**, *3*, 65–71. [CrossRef]
8. Montero, P.; Blanco, J.; Cabanas, J.M.; Maneiro, J.; Pazos, Y.; Moroño, A.; Montero, P.; Balseiro, C.F.; Carracedo, P.; Gomez, B.; et al. Oil spill monitoring and forecasting on the Prestige-Nassau accident. In *Arctic and Marine Oilspill Program Technical Seminar*; Environment Canada: Toronto, ON, Canada, 2003; Volume 2, pp. 1013–1030.
9. Brekke, C.; Solberg, A.H. Oil spill detection by satellite remote sensing. *Remote Sens. Environ.* **2005**, *95*, 1–13. [CrossRef]
10. Zhang, Y.; Lin, H.; Liu, Q.; Hu, J.; Li, X.; Yeung, K. Oil-spill monitoring in the coastal waters of Hong Kong and vicinity. *Mar. Geod.* **2012**, *35*, 93–106. [CrossRef]
11. Rogne, T.; MacDonald, I.; Smith, A.; Kennicutt, M.C.; Giammona, C.S. Multispectral remote sensing and truth data from the Tenyo Maru oil spill. *Photogramm. Eng. Remote Sens.* **1993**, *59*, 3.
12. Grimaldi, C.S.L.; Casciello, D.; Coviello, I.; Lacava, T.; Pergola, N.; Tramutoli, V. An improved RST approach for timely alert and Near Real Time monitoring of oil spill disasters by using AVHRR data. *Nat. Hazards Earth Syst. Sci.* **2011**, *11*, 1281. [CrossRef]
13. Maianti, P.; Rusmini, M.; Tortini, R.; Dalla Via, G.; Frassy, F.; Marchesi, A.; Rota Nodari, F.; Gianinetto, M. Monitoring large oil slick dynamics with moderate resolution multispectral satellite data. *Nat. Hazards* **2014**, *73*, 473–492. [CrossRef]
14. Balogun, A.L.; Yekeen, S.T.; Pradhan, B.; Althuwaynee, O.F. Spatio-Temporal Analysis of Oil Spill Impact and Recovery Pattern of Coastal Vegetation and Wetland Using Multispectral Satellite Landsat 8-OLI Imagery and Machine Learning Models. *Remote Sens.* **2020**, *12*, 1225. [CrossRef]
15. Li, Y.; Liu, Y.; Ma, L.; Li, X. Oil Spill Monitoring Using MODIS Data. In *Second International Conference on Space Information Technology*; International Society for Optics and Photonics: Bellingham, WA, USA, 2007; Volume 6795, p. 67955G.

16. Casciello, D.; Lacava, T.; Pergola, N.; Tramutoli, V. Robust Satellite Techniques for oil spill detection and monitoring using AVHRR thermal infrared bands. *Int. J. Remote Sens.* **2011**, *32*, 4107–4129. [CrossRef]
17. Xing, Q.; Li, L.; Lou, M.; Bing, L.; Zhao, R.; Li, Z. Observation of oil spills through landsat thermal infrared imagery: A case of deepwater horizon. *Aquat. Procedia* **2015**, *3*, 151–156. [CrossRef]
18. Allik, T.H.; Dixon, R.E.; Walters, M. Remote measurement of thick oil spill depth using thermal imagery. In Ocean Sensing and Monitoring X. *Int. Soc. Opt. Photonics* **2018**, *10631*, 106310F.
19. Fingas, M.F.; Brown, C.E. *Oil Spill Remote Sensors: Review, Trends and New Developments*; (No. CONF-970319-); The Environmental Research Institute of Michigan (ERIM): Ann Arbor, MI, USA, 1997.
20. Lehr, W.J.; Simecek-Beatty, D. The relation of Langmuir circulation processes to the standard oil spill spreading, dispersion, and transport algorithms. *Spill Sci. Technol. Bull.* **2000**, *6*, 247–253. [CrossRef]
21. Palmer, D.; Borstad, G.A.; Boxall, S.R.; House, R.; Quay, E. Airborne multispectral remote sensing of the January 1993 Shetlands oil spill. In Proceedings of the Second Thematic Conference on Remote Sensing for Marine and Coastal Environments: Needs, Solutions and Applications, New Orleans, LA, USA, 31 January–2 February 1994; pp. 546–558.
22. Shen, S.S.; Lewis, P.E. Deepwater Horizon oil spill monitoring using airborne multispectral infrared imagery. In Proceedings of the Algorithms and Technologies for Multispectral, Hyperspectral, and Ultraspectral Imagery XVII, Orlando, FL, USA, 25–28 April 2011; Volume 8048, p. 80480H.
23. Grierson, I.T. Use of airborne thermal imagery to detect and monitor inshore oil spill residues during darkness hours. *Environ. Manag.* **1998**, *6*, 905–912. [CrossRef]
24. Jha, M.N.; Levy, J.; Gao, Y. Advances in remote sensing for oil spill disaster management: State-of-the-art sensors technology for oil spill surveillance. *Sensors* **2008**, *8*, 236–255. [CrossRef]
25. De Kerf, T.; Gladines, J.; Sels, S.; Vanlanduit, S. Oil spill detection using machine learning and infrared images. *Remote Sens.* **2020**, *12*, 4090. [CrossRef]
26. Goodman, R. Overview and future trends in oil spill remote sensing. *Spill Sci. Technol. Bull.* **1994**, *1*, 11–21. [CrossRef]
27. Robbe, N.; Hengstermann, T. Remote sensing of marine oil spills from airborne platforms using multi-sensor systems. *Water Pollut. VIII Model. Monit. Manag.* **2006**, *1*, 347–355.
28. Al-Ruzouq, R.; Gibril, M.B.A.; Shanableh, A.; Kais, A.; Hamed, O.; Al-Mansoori, S.; Khalil, M.A. Sensors, features, and machine learning for oil spill detection and monitoring: A review. *Remote Sens.* **2020**, *12*, 3338. [CrossRef]
29. Leifer, I.; Lehr, W.J.; Simecek-Beatty, D.; Bradley, E.; Clark, R.; Dennison, P.; Hu, Y.; Matheson, S.; Jones, C.E.; Holt, B.; et al. State of the art satellite and airborne marine oil spill remote sensing: Application to the BP Deepwater Horizon oil spill. *Remote Sens. Environ.* **2012**, *124*, 185–209. [CrossRef]
30. Zhao, J.; Temimi, M.; Ghedira, H.; Hu, C. Exploring the potential of optical remote sensing for oil spill detection in shallow coastal waters-a case study in the Arabian Gulf. *Opt. Express* **2014**, *22*, 13755–13772. [CrossRef] [PubMed]
31. Schröder-Hinrichs, J.U.; Hollnagel, E.; Baldauf, M. From Titanic to Costa Concordia—A century of lessons not learned. *WMU J. Marit. Aff.* **2012**, *11*, 151–167. [CrossRef]
32. Fiaschi, A.; Matassoni, L.; Lotti, A.; Saccorotti, G. Micro-seismic monitoring after the shipwreck of the Costa Concordia at Giglio Island (Italy). *Acta Geophys.* **2017**, *65*, 1019–1027. [CrossRef]
33. Manconi, A.; Allasia, P.; Giordan, D.; Baldo, M.; Lollino, G. Monitoring the Stability of Infrastructures in An Emergency Scenario: The "Costa Concordia" Vessel Wreck. In *Global View of Engineering Geology and the Environment*; Taylor & Francis Group: London, UK, 2013.
34. Raspini, F.; Moretti, S.; Fumagalli, A.; Rucci, A.; Novali, F.; Ferretti, A.; Claudio Prati, C.; Casagli, N. The COSMO-SkyMed constellation monitors the Costa Concordia wreck. *Remote Sens.* **2014**, *6*, 3988–4002. [CrossRef]
35. Broussolle, J.; Kyovtorov, V.; Basso, M.; Castiglione, G.F.D.S.E.; Morgado, J.F.; Giuliani, R.; Olivero, F.; Sammartino, P.F.; Tarchi, D. MELISSA, a new class of ground based InSAR system. An example of application in support to the Costa Concordia emergency. *ISPRS J. Photogramm. Remote Sens.* **2014**, *91*, 50–58. [CrossRef]
36. Casagli, N.; Moretti, S.; Gabbani, G.; Catani, F.; Fanti, R.; Agostini, A.; Bardi, F.; Battistini, A.; Bianchini, S.; Ciampalini, A.; et al. *Monitoraggio Della Costa Concordia (Isola Del Giglio)—Rapporto Finale. Presidenza Del Consiglio Dei Ministri—Dipartimento Della Protezione Civile*; Internal Report; Earth Science Department, University of Florence: Florence, Italy, 2014. (In Italian)
37. Dotta, G.; Gigli, G.; Ferrigno, F.; Gabbani, G.; Nocentini, M.; Lombardi, L.; Agostini, A.; Nolesini, T.; Casagli, N. Geomechanical characterization and stability analysis of the bedrock underlying the Costa Concordia cruise ship. *Rock Mech. Rock Eng.* **2017**, *50*, 2397–2412. [CrossRef]
38. Maldague, X. *Theory and Practice of Infrared Technology for Non-Destructive Testing*; John-Wiley & Sons: Hoboken, NJ, USA, 2001; 684p.
39. Spampinato, L.; Calvari, S.; Oppenheimer, C.; Boschi, E. Volcano surveillance using infrared cameras. *Earth-Sci. Rev.* **2011**, *106*, 63–91. [CrossRef]
40. Ludwig, N.; Redaelli, V.; Rosina, E.; Augelli, F. Moisture detection in wood and plaster by IR thermography. *Infrared Phys. Technol.* **2004**, *46*, 161–166. [CrossRef]
41. Teza, G.; Marcato, G.; Castelli, E.; Galgaro, A. IRTROCK: A MATLAB toolbox for contactless recognition of surface and shallow weakness of a rock cliff by infrared thermography. *Comput. Geosci.* **2012**, *45*, 109–118. [CrossRef]

42. Frodella, W.; Gigli, G.; Morelli, S.; Lombardi, L.; Casagli, N. Landslide mapping and characterization through infrared thermography (IRT): Suggestions for a methodological approach from some case studies. *Remote Sens.* **2017**, *9*, 1281. [CrossRef]
43. Wu, J.H.; Lin, H.M.; Lee, D.H.; Fang, S.C. Integrity assessment of rock mass behind the shotcreted slope using thermography. *Eng. Geol.* **2005**, *80*, 164–173.
44. Hardgrove, C.; Moersch, J.; Whisner, S. Thermal imaging of sedimentary features on alluvial fans. *Planet. Space Sci.* **2010**, *58*, 482–508. [CrossRef]
45. Frodella, W.; Morelli, S.; Fidolini, F.; Pazzi, V.; Fanti, R. Geomorphology of the Rotolon landslide (Veneto region, Italy). *J. Maps* **2014**, *10*, 394–401. [CrossRef]
46. Pappalardo, G.; Mineo, S.; Zampelli, S.P.; Cubito, A.; Calcaterra, D. InfraRed Thermography proposed for the estimation of the Cooling Rate Index in the remote survey of rock masses. *Int. J. Rock Mech. Min. Sci.* **2016**, *83*, 182–196. [CrossRef]
47. Di Traglia, F.; Nolesini, T.; Solari, L.; Ciampalini, A.; Frodella, W.; Steri, D.; Allotta, B.; Rindi, A.; Marini, L.; Monni, N.; et al. Lava delta deformation as a proxy for submarine slope instability. *Earth Planet. Sci. Lett.* **2018**, *488*, 46–58. [CrossRef]
48. Guerin, A.; Jaboyedoff, M.; Collins, B.D.; Derron, M.H.; Stock, G.M.; Matasci, B.; Boesiger, M.; Lefeuvre, C.; Podladchikov, Y.Y. Detection of rock bridges by infrared thermal imaging and modeling. *Sci. Rep.* **2019**, *9*, 1–19. [CrossRef]
49. Frodella, W.; Elashvili, M.; Spizzichino, D.; Gigli, G.; Adikashvili, L.; Vacheishvili, N.; Kirkitadze, G.; Nadaraia, A.; Margottini, C.; Casagli, N. Combining InfraRed Thermography and UAV Digital Photogrammetry for the Protection and Conservation of Rupestrian Cultural Heritage Sites in Georgia: A Methodological Application. *Remote Sens.* **2020**, *12*, 892. [CrossRef]
50. Frodella, W.; Lazzeri, G.; Moretti, S.; Keizer, J.; Verheijen, F.G. Applying Infrared Thermography to Soil Surface Temperature Monitoring: Case Study of a High-Resolution 48 h Survey in a Vineyard (Anadia, Portugal). *Sensors* **2020**, *20*, 2444. [CrossRef] [PubMed]
51. Frodella, W.; Elashvili, M.; Spizzichino, D.; Gigli, G.; Nadaraia, A.; Kirkitadze, G.; Adikashvili, L.; Margottini, C.; Antidze, N.; Casagli, N. Applying Close Range Non-Destructive Techniques for the Detection of Conservation Problems in Rock-Carved Cultural Heritage Sites. *Remote Sens.* **2021**, *13*, 1040. [CrossRef]
52. FLIR Systems Inc. FLIR ThermaCAM SC620 Technical Specifications. 2009. Available online: www.flir.com/cs/emea/en/view/?id=41965 (accessed on 22 December 2022).
53. FLIR Systems Inc. FLIR Tools + Datasheet. 2015. Available online: https://www.infraredcamerawarehouse.com/content/FLIR%20Datasheets/FLIR%20ToolsPlus%20Datasheet.pdf (accessed on 22 December 2022).
54. ESRI Inc. ArcMap 10.1 Datasheet. 2012. Available online: https://www.esri.com/news/arcnews/spring12articles/introducing-arcgis-101.html (accessed on 1 October 2019).
55. FLIR Systems Inc. ResearchIR 3.4. sp3. 2013. Available online: http://support.flir.com/DsDownload/Assets/T198206_en_40.pdf (accessed on 1 June 2022).
56. Centro Funzionale Regione Toscana (CFR)—Termometria. Available online: https://www.cfr.toscana.it/index.php?IDS=42&IDSS=278 (accessed on 1 December 2022).
57. Dassenakis, M.; Paraskevopoulou, V.; Cartalis, C.; Adaktilou, N.; Katsiabani, K. Remote sensing in coastal water monitoring: Applications in the eastern Mediterranean Sea (IUPAC Technical Report). *Pure Appl. Chem.* **2011**, *84*, 335–375. [CrossRef]
58. Tebeau, P.A.; Etkin, D.S.; French-McCay, D.P. *Cost-Benefit Analysis for Using Laser Fluorosensor for Detecting Heavy Oil*; Coast Guard Research and Development Center: Groton, CT, USA, 2006.
59. Lega, M.; Kosmatka, J.; Ferrara, C.; Russo, F.; Napoli, R.M.A.; Persechino, G. Using advanced aerial platforms and infrared thermography to track environmental contamination. *Environ. Forensics* **2012**, *13*, 332–338. [CrossRef]
60. Aziz, N.A.A.; Aziz, K.A. Managing disaster with wireless sensor networks. In Proceedings of the 13th International Conference on Advanced Communication Technology (ICACT2011), Gangwon-Do, Republic of Korea, 13–16 February 2011; pp. 202–207.
61. Erdelj, M.; Król, M.; Natalizio, E. Wireless sensor networks and multi-UAV systems for natural disaster management. *Comput. Netw.* **2017**, *124*, 72–86. [CrossRef]
62. Intrieri, E.; Gigli, G.; Gracchi, T.; Nocentini, M.; Lombardi, L.; Mugnai, F.; Frodella, W.; Bertolini, G.; Carnevale, E.; Favalli, M.; et al. Application of an ultra-wide band sensor-free wireless network for ground monitoring. *Eng. Geol.* **2018**, *238*, 1–14. [CrossRef]
63. ISRM. Suggested methods for the quantitative description of discontinuities in rock masses. *Int. J. Rock Mech. Min. Sci. Geomech. Abstr.* **1978**, *15*, 319–368.

remote sensing

Technical Note

Evaluation of Assimilation in the MASNUM Wave Model Based on Jason-3 and CFOSAT

Meng Sun [1,2,3], Jianting Du [1,2,3], Yongzeng Yang [1,2,3,*] and Xunqiang Yin [1,2,3]

1 First Institute of Oceanography, Ministry of Natural Resources, Qingdao 266061, China; sunm@fio.org.cn (M.S.); dujt@fio.org.cn (J.D.); yinxq@fio.org.cn (X.Y.)
2 Laboratory for Regional Oceanography and Numerical Modeling, Pilot National Laboratory for Marine Science and Technology, Qingdao 266071, China
3 Key Laboratory of Marine Science and Numerical Modeling (MASNUM), Ministry of Natural Resources, Qingdao 266061, China
* Correspondence: yangyz@fio.org.cn

Abstract: Accurate numerical simulation of ocean waves is one of the most important measures to ensure shipping safety, offshore engineering construction, etc. The use of wave observations from satellite is an efficient way to correct model results. The goal of this paper is to assess the performance of assimilation in the MASNUM wave model for the Indian Ocean. The assimilation technique is based on Ensemble Adjusted Kalman Filter, with a variable ensemble constructed by the dynamic sampling method rather than ensemble members of wave model. Observations of significant wave height from satellites Jason-3 and CFOSAT are regarded as assimilation data and independent validation data, respectively. The results indicate good performance in terms of absolute mean error for significant wave height. Model error decreases by roughly 20–40% in high-sea conditions.

Keywords: wave assimilation; MASNUM; Jason-3; CFOSAT; dynamic sampling

1. Introduction

Ocean shipping and offshore engineering are becoming increasingly frequent. Therefore, refined numerical simulation of ocean waves is one of the most important measures to ensure safety in open oceans. Especially under strong wind conditions, accurate wave prediction of sea states is crucial for guaranteeing safety. The assimilation of satellite wave data plays an important role in improving wave model results.

The increasing number of remote observations has prompted the development of wave data assimilation. Altimeter, synthetic aperture radar (SAR), and surface wave investigation measurements (SWIM), which are loaded on satellites, can provide global surface wave observations. Observations of significant wave height (SWH) from satellites are consecutive, abundant, and reliable, and are extensively used in wave forecasting systems and wave reanalysis [1]. By succeeding Topex/Poseidon, Jason-1, and Jason-2, Jason-3 extends the high-precision ocean altimetry data record, which was launched in January 2016. The China–France Oceanography Satellite (CFOSAT), which was launched in October 2018, carries the SWIM sensor [2]. SWIM provides a significant wave height at nadir-look and directional wave spectra with a wavelength cut-off of 70 m. This work aims to investigate the impact of the assimilation of SWH from Jason-3 in the MASNUM wave model on the Indian Ocean. Validation is performed with independent data from CFOSAT.

The purpose of wave data assimilation is to improve the wave model and to produce better performance. There are two main classes of assimilation methods: one is sequential methods, such as optimal interpolation (OI) [3] and ensemble Kalman filter (EnKF) [4], and the other is variational methods. Sequential methods implement independent corrections at different times, and variational methods aim to find the model solution that minimizes the differences between observations over the whole analysis period [5]. Sequential methods, which are relatively simple and require relatively low computing costs, confront the problem of specifying correlation scales in models and observations. At present, this issue is still one of the focuses in wave data assimilation.

There is no consolidated technique to spread altimeter data on a model grid. (1) Without spreading, only observations on the model grid replace model value [6,7]. (2) Exponential spreading function [8]. (3) Covariance of prediction error/background error/model error, which requires statistical determination and determines the information spread of observations. For the OI method, the prediction error correlation matrix, referred to as P, is assumed to be a function of correlation scale L_{max} [5]. For the EnKF method, background-error covariance, referred as B, is constructed by ensemble members of model states. Apparently, B is flow-dependent, while P is not. It requires tremendous cost to run ensemble members of a wave model.

To reduce the computation cost, Sun et al. [9,10], proposed the dynamic sampling method and static sampling method to construct background-error covariance by the differences between 24-h-interval significant wave heights. Therefore, construction of the model states requires only one model run instead of an ensemble of model runs. In this study, we will assimilate SWH of Jason-3 using a combination of ensemble adjust Kalman filter (EAKF) [11,12] and dynamic sampling method.

The paper is organized as follows. Section 2 briefly describes the wave model MASNUM, assimilation scheme, and satellite observations. Section 3.1 concerns the validation of the results with independent wave observations. The analysis of the assimilation impact in the case of high-sea state is presented in Section 3.2. Discussion and conclusions are given in Sections 4 and 5, respectively.

2. Materials and Methods

2.1. Wave Model and Set-Up

The wave model used in this study is a third-generation global wave model Marine Sciences and Numerical Modeling (MASNUM) [13–15]. The complicated characteristic inlaid method is applied to integrate the wave energy spectrum balance equation. All governing equations have a physical meaning of the wave packet propagation along a ray. Thus, Yuan et al. [14] designed the characteristic inlaid computation scheme to reflect this property. A matching problem is solved between time and space steps using this scheme. In the model, the wave energy spectrum balance equation and its complicated characteristic equations are derived in a wave number space. The breaking dissipation source function adopts a theoretical result based on a statistical study of breaking waves [16].

MASNUM is implemented on a 15 $\times$ 15 km latitude–longitude grid that covers 60.7°E to 99.3°E and 10.4°S to 25.6°N; water depth of the domain is shown in Figure 1. The directional wave spectrum is discretized into 24 directions by a step of 15° and 25 frequencies ranging from 0.042 to 0.413 Hz. The propagation time step is 10 min. Output fields of integrated wave parameters are archived at 1-h intervals. The wave model MASNUM is driven by 1-h analyzed winds with 0.25 $\times$ 0.25° spatial resolution (ERA5) from the European Center for Medium-Range Weather Forecast (ECMWF) atmospheric model. The model topography is interpolated from a gridded bathymetric dataset (GEBCO) [17]. The wave spectrum of the open boundaries for the Indian Ocean domain are interpolated from the hourly output of an outer domain, which covers an area of 20°E to 145.5°E, 70°S to 45.5°N with a spatial resolution of 0.5 Deg.

Two runs of the wave model MASNUM were performed for the period from 1 July 2020 to 31 January 2021. The first run used the assimilation of Jason-3 altimeter wave heights (referred to as Run A). The second run is a baseline run of the wave model MASNUM without assimilation (referred to as Run B). For independent validation, we use observations of significant wave height from CFOSAT which are excluded from assimilation.

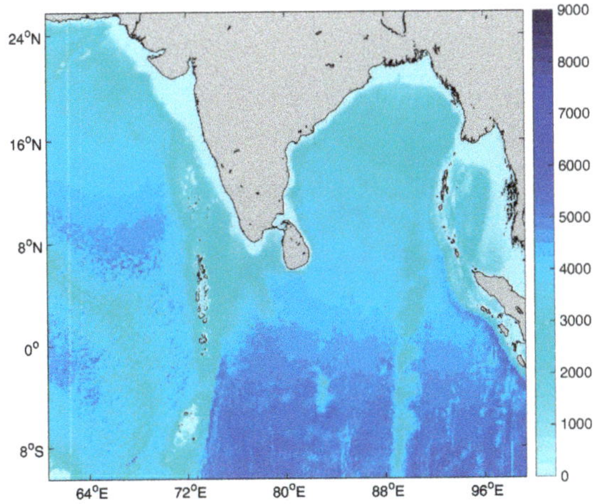

Figure 1. Study domain with water depth. Unit is meters.

2.2. Assimilation Scheme

The assimilation procedure is split into two parts: first, the significant wave height of altimeter from the Jason-3 satellite is used to construct an analyzed field of SWH by assimilation technique using a combination of EAKF and dynamic sampling [10]. The second part consists of the reconstruction of the spectrum from the analyzed SWH. The information contained in the first-guess spectrum $E^b(K,\theta)$ is used to produce an analyzed spectrum $E^a(K,\theta)$ as $E^a(K,\theta) = \left(\frac{SWH^a}{SWH^b}\right)^2 E^b(K,\theta)$ [7]. Here, we use a limitation for the factor $\left(\frac{SWH^a}{SWH^b}\right)^2$ to avoid introducing shocks in the model, i.e., if $\frac{SWH^a}{SWH^b} < 0.5$, we set the factor to the square of 0.5; if $\frac{SWH^a}{SWH^b} > 1.5$, we set the factor to the square of 1.5.

Flowchart of assimilation procedure is shown in Figure 2. In the process of assimilating, only one wave model member is implemented. The assimilation is performed sequentially every 6 h. SWH observations from Jason-3 are assimilated into the MASNUM wave model with the following scheme: (1) Dynamic ensemble, which is an approximation to background error, is constructed using the dynamic sampling method, i.e., collecting wave model results without assimilation on the interval (t − 7d, t + 7d) and then calculating the difference between 24 h-interval SWH fields. Here t represents assimilation time. Thus, dynamic ensemble changes with time. Specifically, dynamic ensemble is flow-dependent. There are 24 samples for each day. (2) The dynamic ensemble is superposed to the SWH field at time t to obtain ensemble states with 360 members. (3) The two-part filter method [11,12] is applied to update ensemble states using observations at time $\tau \in (t - 6\text{h}, t + 6\text{h})$. (4) The analyzed SWH is the mean of the updated ensemble states. (5) An analyzed wave spectrum is reconstructed by multiplying the first-guess spectrum $E^b(K,\theta)$ by the factor $\left(\frac{SWH^a}{SWH^b}\right)^2$.

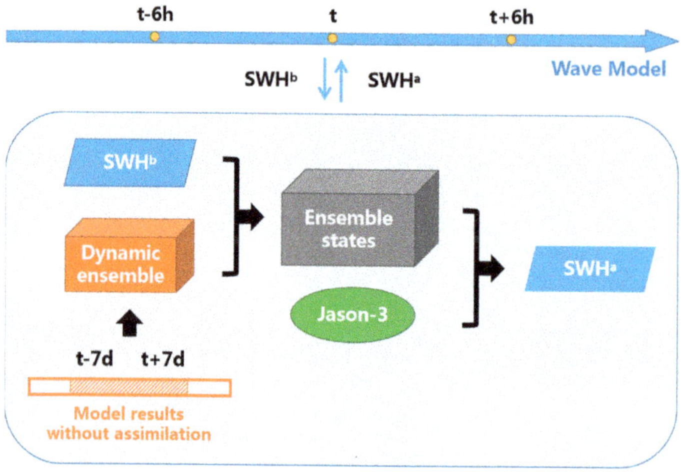

Figure 2. Flowchart of wave assimilation.

2.3. Satellite Data

Jason-3 launched in January 2016 as the result of a joint effort by CNES, NASA, EUMETSAT, and NOAA. The satellite carries a typical suite of altimetry mission instruments that will acquire highly accurate measurements of ocean surface height to extend the data record compiled by Topex/Poseidon, Jason-1, and Jason-2 [18]. Jason-3 has an orbital repetition cycle of 10 days.

CFOSAT launched in October 2018 as a joint mission from the Chinese and French Space Agencies, devoted to the observation ocean surface wind and waves [19,20]. The mission contains two main instruments: a wave scatterometer called SWIM (surface wave investigation and monitoring), which provides significant wave height at nadir-look direction and directional wave spectra with a wavelength cut-off 70 m, and the wind scatterometer (SCAT). CFOSAT has an orbital repetition cycle of 13 days.

The data from Jason-3 and CFOSAT have been downloaded from the CMEMS web portal (Copernicus Marine Environment Monitoring Service, https://marine.copernicus.eu/ accessed on 17 July 2021). The domain is well covered by Jason-3 and CFOSAT. SWH from the satellites have been generated on a grid size that is $0.2 \times 0.2^{\circ}$. There are 300,000 points collected from Jason-3, and 260,000 points collected from CFOSAT from 1 July 2020 to 31 January 2021. The spatial distribution of SWH based on the observations of Jason-3 and CFOSAT is shown in Figure 3. The distance between adjacent orbit tracks of CFOSAT is closer compared to that of Jason-3. There are two high-value regions of significant wave height in the domain, which are in the northwest and the south Indian Ocean.

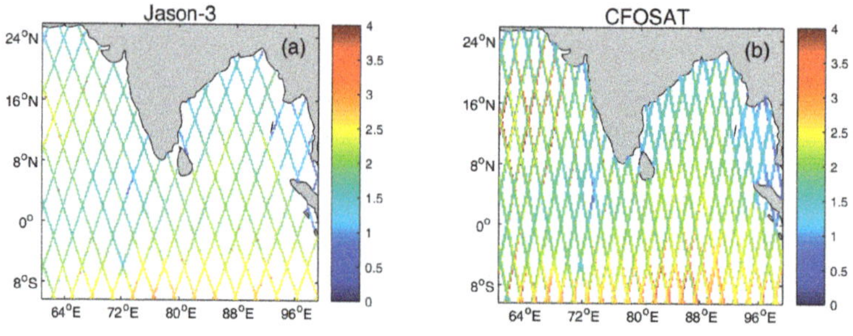

Figure 3. Significant wave height of satellite Jason-3 and CFOSAT from 1 July 2020 to 31 January 2021. Unit is meters. (**a**) SWH from satellite Jason-3; (**b**) SWH from satellite CFOSAT.

3. Results

Two runs of the wave model MASNUM were performed for the period from 1 July 2020 to 31 January 2021. The first run used the assimilation of Jason-3 altimeter wave heights (referred to as Run A). The second run is a baseline run of the wave model MASNUM without assimilation (referred to as Run B).

First, we analyzed the impact, which consists of comparing model outputs from Runs A and B with independent wave observations of SWH from CFOSAT. This impact evaluates the efficiency of the assimilation system. The performance of these two runs is discussed by computing the statistical parameters for significant wave heights. Moreover, the impact was analyzed based on a comparison between model outputs with and without assimilation. This indicates how large the difference of wave parameters induced by the assimilation is. Another impact is considered and this consists of comparing model outputs with observations along orbit tracks from CFOSAT, especially in high-sea conditions.

3.1. Validation

In this section, we use the significant wave heights from CFOSAT as independent wave observations. SWH from the MASNUM wave model was first collocated with CFOSAT satellite data, which is along the orbit track using temporal–spatial interpolation. Statistical parameters on the difference between the observed and model wave variables were computed with the following definitions:

$$\text{Mean error (ME)} = \overline{\mathrm{M_i} - \mathrm{O_i}} \tag{1}$$

$$\text{Root Mean Square Error (RMSE)} = \sqrt{\overline{(\mathrm{M_i} - \mathrm{O_i})^2}} \tag{2}$$

$$\text{Mean Absolute Error (MAE)} = \overline{|\mathrm{M_i} - \mathrm{O_i}|} \tag{3}$$

$$\text{Relative Error (RE)} = \overline{|\mathrm{M_i} - \mathrm{O_i}|/\mathrm{O_i}} \tag{4}$$

$$\text{Improvement Rate (IR)} = \frac{\mathrm{Error_{noasm}} - \mathrm{Eerror_{asm}}}{\mathrm{Error_{noasm}}}\% \tag{5}$$

$$\text{Correlation Coefficient(Corr)} = \frac{\overline{\mathrm{M_i}\,\mathrm{O_i}} - \overline{\mathrm{M}}\,\overline{\mathrm{O}}}{\sqrt{\overline{\mathrm{M_i}^2} - \overline{\mathrm{M}}^2}\sqrt{\overline{\mathrm{O_i}^2} - \overline{\mathrm{O}}^2}} \tag{6}$$

$$\text{Scatter index (SI)} = \frac{\sqrt{\overline{\left[(\mathrm{M_i} - \overline{\mathrm{M}}) - (\mathrm{O_i} - \overline{\mathrm{O}})\right]^2}}}{\overline{\mathrm{O}}} \tag{7}$$

Table 1 shows the statistical parameters for Runs A and B. First, this indicates that the wave model MASNUM generally underestimates the significant wave height. After the assimilation of Jason-3 SWH data, the root mean square errors are reduced from 0.36 m to 0.32 m. Improvement rates of simulation error are more than 10%. This indicates that the MASNUM wave model reproduces relatively well the observed wave regimes, which are recorded by CFOSAT. The impact of assimilation is statistically weak given the applied high number of observations (number is roughly 260,000). Therefore, further detailed analysis is necessary.

Statistical parameters for each month from 1 July 2020 to 31 January 2021 are shown in Figure 4. There is a similar tendency of RMSE and MAE, with maximum value in July and minimum value in January. However, there is little difference of relative error between each month. This is correlated with seasonal variation of surface waves in the Indian Ocean. As can be seen, SWH from CFOSAT in Figure 5a,d, monthly mean wave height in winter is less than 3 m, while in summer it is in the 2–5 m range. The comparison with CFOSAT indicates a correction of assimilation in July is more noticeable, i.e., the impact of assimilation is more remarkable in high-sea conditions. In Figure 4c, the correlation coefficient increased after data assimilation. Figure 4f indicates that scatter index decreased after data assimilation.

Table 1. Statistical analysis of significant wave heights from runs with and without assimilation at orbit tracks of CFOSAT. The period of analysis is from 1 July 2020 to 31 January 2021.

Error	Run B: Noasm	Run A: Asm	Improvement Rate
Mean error	−0.15	−0.13	13%
RMSE	0.36	0.32	11%
Mean absolute error	0.26	0.23	12%
Relative error	0.13	0.11	15%
Correlation coefficient	0.91	0.92	-
Scatter index	0.16	0.15	-

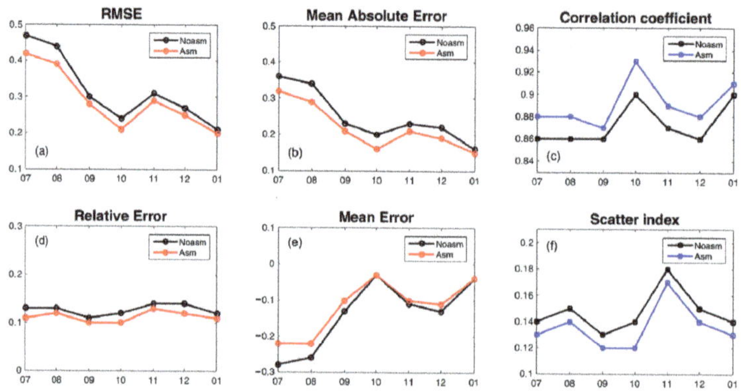

Figure 4. Statistics of model results against observations from CFOSAT satellite computed every month during the period from 1 July 2020 to 31 January 2021. (**a**) Root mean square error; (**b**) Mean absolute error; (**c**) Correlation coefficient; (**d**) Relative error; (**e**) Mean error; (**f**) Scatter index.

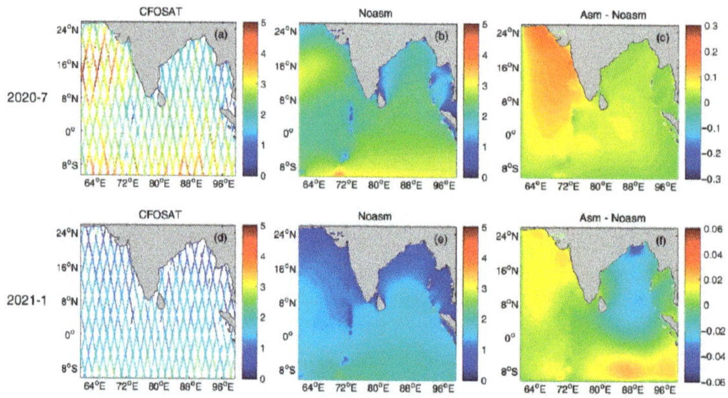

Figure 5. Spatial distribution of SWH (in meters) in July 2020 and January 2021. (**a**) SWH of CFOSAT in July 2020; (**b**) SWH of MASNUM wave model without assimilation in July 2020; (**c**) Difference of SWH with and without assimilation of Jason-3 wave data in July 2020; (**d**) SWH of CFOSAT in January 2021; (**e**) SWH of MASNUM wave model without assimilation in January 2021; (**f**) Difference of SWH with and without assimilation of Jason-3 wave data in January 2021.

Comparisons between model outputs with and without assimilation indicate the difference of wave parameters induced by the assimilation. Figure 5c,f show the difference of SWH between Run A (with assimilation) and Run B (without assimilation) in July 2020 and January 2021, respectively. This clearly indicates that the monthly mean impact of the assimilation on significant wave height can reach roughly 0.2 m in July and ±0.03 m in January. SWH of CFOSAT and of wave model without assimilation in July are shown in Figure 5a,b. It is easy to see the strong impact of assimilation in the northwest Indian Ocean where the model is underestimating the significant wave height.

3.2. Analysis in High-Sea Conditions

The assimilation of Jason-3 altimeter wave data was analyzed for high-sea conditions in this section. Several cases are selected to evaluate the impact of assimilation. Generally, due to the propagation effect of ocean surface waves, wave data from Jason-3 are assimilated in one place. The impact of assimilation could be observed by CFOSAT at another place later. We attempt to detect this relationship using wave parameters and investigate the impact of assimilation.

3.2.1. Case A

Group velocity Cg, which describes the energy propagation of ocean surface waves, is given as follows.

$$\omega = \frac{2\pi}{T},\ \omega^2 = gk,\ \lambda = \frac{2\pi}{k},\ C = \frac{\lambda}{T},\ C_g = \frac{1}{2}C \tag{8}$$

where ω and T represent frequency and period respectively, k and λ indicate wave number and wavelength separately, g represents gravitational acceleration. C is wave phase velocity.

SWH time series at orbit tracks on 19–20 July 2020 are shown in Figure 6, referred to as Case A. In Figure 6c, black dot lines indicate observations of Jason-3, and blue lines and green lines represent SWH of the model with and without assimilation of Jason-3. Locations of orbit tracks J1 and J2 are shown in Figure 6a,d. Mean absolute error of SWH of model outputs decreased 29% and 31% at orbit tracks J1 and J2 compared to measurements of Jason-3.

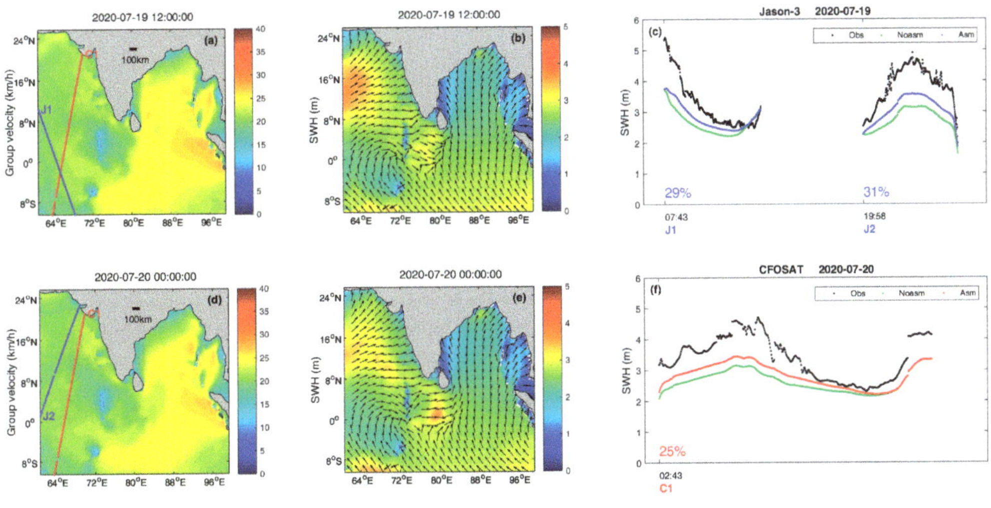

Figure 6. Case A: SWH time series at orbit tracks on 19–20 July 2020. (**a**,**d**): Color represents group velocity in km/h. Blue and red dotted lines indicate orbit tracks of Jason-3 and CFOSAT, respectively. Location of character J1/J2/C1 indicates the start of the orbit track. (**b**,**e**): Color and arrows represent SWH and mean wave direction of model outputs without assimilation. (**c**) Black dotted lines indicate Jason-3 measurements; blue and green lines indicate SWH of model outputs with and without assimilation of Jason-3; percentage represents improvement rate of mean absolute error of SWH with assimilation compared to without assimilation. (**f**) Black dotted lines indicate CFOSAT measurements; red lines and green lines indicate SWH of model outputs with and without assimilation of Jason-3; percentage represents improvement rate of mean absolute error of SWH with assimilation compared to without assimilation.

We found that there is one orbit track of CFOSAT near orbit track J1 and J2, which is referred to as C1. The location of orbit track C1 is shown in Figure 6a,d. In Figure 6f, black dot lines represent the measurements of CFOSAT, and red lines and green lines indicate SWH of the wave model with and without assimilation of Jason-3. Improvement rate of mean absolute error of SWH is 25% at orbit track C1 compared with independent observations of CFOSAT.

Based on a mean wave period of the model without assimilation, group velocity Cg on 20 July 2020 at 0:00 UTC are shown in Figure 6d. SWH and mean wave direction are given in Figure 6e. Orbit track J2 and C1 differ by 6–7 h. At orbit track J2, group velocity is roughly 20 km/h and ocean surface waves propagated northeast. The first half of orbit track C1, where correction of assimilation is pronounced, is approximately 120–140 km from orbit track J2. This clearly indicates the efficiency of assimilation.

Group velocity Cg on 19 July 2020 at 12:00 UTC is shown in Figure 6a. SWH and mean wave direction are given in Figure 6b. Orbit track J1 and C1 differ by 19 h. Group velocity is roughly 20 km/h at orbit track J1. Ocean surface waves propagated northeast to the north of the crossing of J1 and C1, and propagated northwest at the south of the crossing simultaneously. Propagation distance is roughly 380 km, which covers orbit track C1. There is little influence of assimilation on the latter half of orbit track C1. One reason for that is the large time difference between the two orbits. The other reason is that the correction of assimilation in the latter half of orbit track J1 is relatively small.

3.2.2. Case B

SWH time series at orbit tracks on 25–26 July 2020 are shown in Figure 7, referred to as Case B. In Figure 7c,f, black dotted lines indicate observations of Jason-3, and blue lines and green lines represent SWH of model with and without assimilation of Jason-3. Locations of orbit tracks J1/J2/J3 are shown in Figure 7a,d,g. Mean absolute error of SWH of the model outputs decreased 54%, 56%, and 6% at orbit tracks J1/J2/J3 compared to measurements of Jason-3.

We found that there are two orbit tracks of CFOSAT near orbit tracks J1/J2/J3, which are referred to as C1/C2. Locations of orbit tracks C1/C2 are shown in Figure 7a,d,g. In Figure 7i, black dotted lines represent measurements of CFOSAT, and red lines and green lines indicate SWH of the wave model with and without assimilation of Jason-3. Improvement rate of mean absolute error of SWH is 42% and 24% at orbit tracks C1/C2 compared with independent observations of CFOSAT.

Based on mean wave period of model without assimilation, group velocity Cg on 26 July 2020 at 12:00 UTC is shown in Figure 7g. SWH and mean wave direction are given in Figure 7h. Orbit track J3 and C2 differ by 9 h. In the latter half of orbit track J3, group velocity is roughly 23 km/h and ocean surface wave propagated towards northwest. Propagation distance is roughly 210 km, which covers orbit track C2, and the performance of assimilation is obvious at the first half of orbit track C2. Figure 7b,e,h show that significant wave height near 8°S increased gradually from 25 July 12:00 to 26 July 12:00. This indicates that ocean waves are potentially growing in the latter half of orbit track J3. Despite a small correction at orbit track J3, the impact of assimilation gradually amplifies with the propagation of surface waves.

Group velocity Cg on 26 July 2020 at 0:00 UTC is shown in Figure 7d. SWH and mean wave direction are given in Figure 7e. Orbit track J2 and C1 differ by 8 h. Surface waves in the latter half of orbit track J2 propagated northeast and away from the orbit track C1, therefore there is no influence of the former on the latter. At the first half of orbit track J2, group velocity is roughly 23 km/h and ocean surface wave propagated northwest. Propagation distance is roughly 190 km, which covers orbit track C1, and the performance of assimilation is remarkable at the latter half of orbit track C1.

Group velocity Cg on 25 July 2020 at 12:00 UTC is shown in Figure 7a. SWH and mean wave direction are given in Figure 7b. Orbit track J1 and C1 differ by 21 h. Surface wave at orbit track J1 propagated towards northeast mostly, thus there is little impact of assimilation on orbit track C1 at the south of the crossing of J1 and C1. At the north of the crossing, group velocity at the first half of orbit track J1 is roughly 20 km/h. Propagation distance of influence of assimilation is roughly 420 km, which covers orbit track C1. The impact of assimilation gradually weakens with long-range propagation of surface waves, therefore correction in the first half of the orbit track C1 is not pronounced.

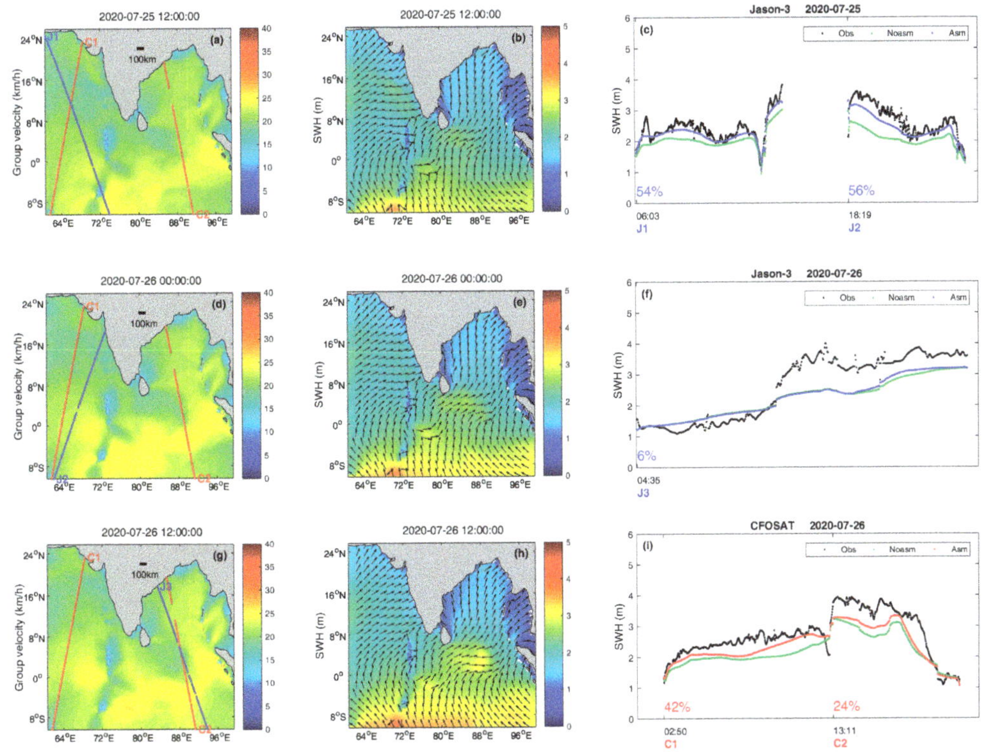

Figure 7. Case B: SWH time series at orbit tracks on 25–26 July 2020. (**a**,**d**,**g**): Color represents group velocity in km/h. Blue and red dotted lines indicate orbit tracks of Jason-3 and CFOSAT, respectively. Location of character J1/J2/J3/C1/C2 indicates the start of the orbit track. (**b**,**e**,**h**): Color and arrows represent SWH and mean wave direction of model outputs without assimilation. (**c**,**f**): Black dotted lines indicate Jason-3 measurements; blue and green lines indicate SWH of model outputs with and without assimilation of Jason-3; percentage represents improvement rate of mean absolute error of SWH with assimilation compared to without assimilation. (**i**) Black dotted lines indicate CFOSAT measurements; red lines and green lines indicate SWH of model outputs with and without assimilation of Jason-3; percentage represents improvement rate of mean absolute error of SWH with assimilation compared to without assimilation.

3.2.3. Case C

SWH time series at orbit tracks on 28 July 2020 are shown in Figure 8, referred to as Case C. In Figure 8b, black dotted lines indicate observations of Jason-3, and blue lines and green lines represent SWH of the model with and without assimilation of Jason-3. Location of orbit tracks J1 is shown in Figure 8a. Mean absolute error of SWH of the model outputs decreased 41% at orbit tracks J1 compared to measurements of Jason-3.

We found that there is one orbit track of CFOSAT near orbit track J1, which is referred to as C1. The location of orbit track C1 is shown in Figure 8a. In Figure 8d, black dotted lines represent measurements of CFOSAT, and red lines and green lines indicate SWH of the wave model with and without assimilation of Jason-3. Improvement rate of mean absolute error of SWH is 39% at orbit track C1 compared with independent observations of CFOSAT.

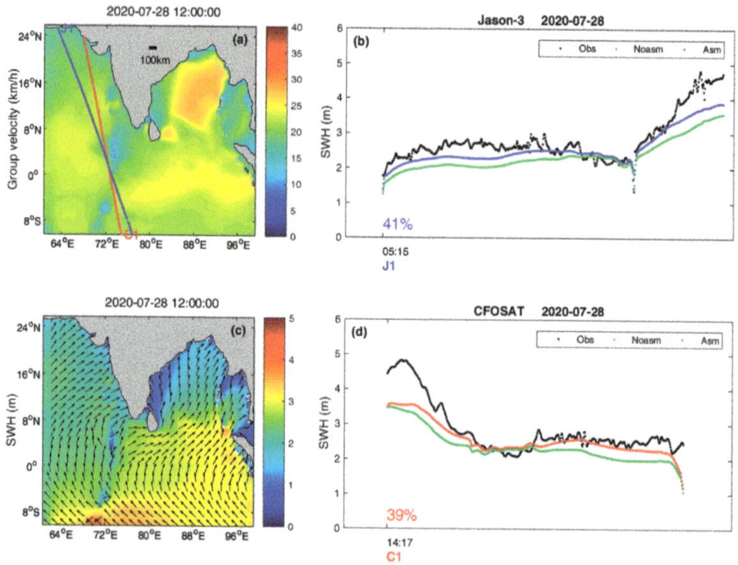

Figure 8. Case C: SWH time series at orbit tracks on 28 July 2020. (**a**) Color represents group velocity in km/h. Blue and red dotted lines indicate orbit tracks of Jason-3 and CFOSAT, respectively. Location of character J1/C1 indicates the start of the orbit track. (**b**) Black dotted lines indicate Jason-3 measurements; blue and green lines indicate SWH of model outputs with and without assimilation of Jason-3; percentage represents improvement rate of mean absolute error of SWH with assimilation compared to without assimilation. (**c**) Color and arrows represent SWH and mean wave direction of model outputs without assimilation. (**d**) Black dotted lines indicate CFOSAT measurements; red lines and green lines indicate SWH of model outputs with and without assimilation of Jason-3; percentage represents improvement rate of mean absolute error of SWH with assimilation compared to without assimilation.

Based on mean wave period of model without assimilation, group velocity Cg on 28 July 2020 at 12:00 UTC is shown in Figure 8a. SWH and mean wave direction are given in Figure 8c. Orbit track J1 and C1 differ by 9 h. At the north of the crossing of J1 and C1, ocean surface waves propagated northeast at group velocity of roughly 23 km/h. At the south of the crossing, they propagated northwest at group velocity of roughly 25 km/h. SWH at the whole orbit track C1 are corrected by assimilation. This illustrates that mean wave direction is one of the indispensable elements in analysis of propagation of assimilation.

3.2.4. Case D

SWH time series at orbit tracks on 6 October 2020 are shown in Figure 9, referred to as Case D. In Figure 9b, black dotted lines indicate observations of Jason-3, blue lines and green lines represent SWH of model with and without assimilation of Jason-3. The location of orbit track J1 is shown in Figure 9a. Mean absolute error of SWH of model outputs decreased 36% at orbit track J1 compared to measurements of Jason-3.

We found that there is one orbit track of CFOSAT near orbit track J1, which is referred to as C1. The location of orbit track C1 is shown in Figure 9a. In Figure 9d, black dotted lines represent measurements of CFOSAT, and red lines and green lines indicate SWH of the wave model with and without assimilation of Jason-3. The improvement rate of mean absolute error of SWH is 32% at orbit track C1 compared with independent observations of CFOSAT.

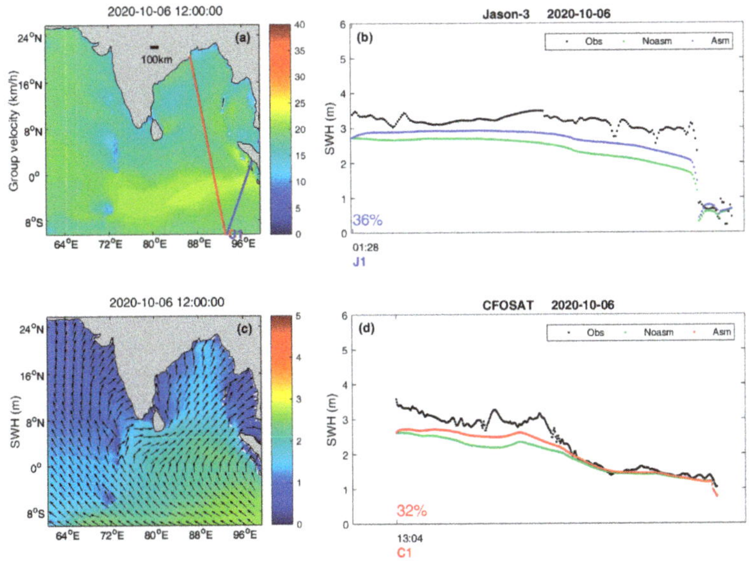

Figure 9. Case D: SWH time series at orbit tracks on 6 October 2020. (**a**) Color represents group velocity in km/h. Blue and red dotted lines indicate orbit tracks of Jason-3 and CFOSAT, respectively. Location of character J1/C1 indicates the start of the orbit track. (**b**) Black dotted lines indicate Jason-3 measurements; blue and green lines indicate SWH of model outputs with and without assimilation of Jason-3; percentage represents improvement rate of mean absolute error of SWH with assimilation compared to without assimilation. (**c**) Color and arrows represent SWH and mean wave direction of model outputs without assimilation. (**d**) Black dotted lines indicate CFOSAT measurements; red lines and green lines indicate SWH of model outputs with and without assimilation of Jason-3; percentage represents improvement rate of mean absolute error of SWH with assimilation compared to without assimilation.

Based on mean wave period of the model without assimilation, group velocity Cg on 6 October 2020 at 12:00 UTC is shown in Figure 9a. SWH and mean wave direction is given in Figure 9c. Orbit track J1 and C1 differ by 12 h. At orbit track J1, group velocity is roughly 20 km/h and ocean surface waves propagated northwest. The first half of orbit track C1, where correction of assimilation is considerable, is approximately 240 km from orbit track J1. This clearly indicates the efficiency of assimilation. This also illustrates that length of orbit track is one of decisive factors of the influence range of assimilation.

4. Discussion

The assimilation of observations from satellites is an efficient way to correct numerical model errors. In this study, measurements of SWH from Jason-3 are assimilated to improve MASNUM wave model outputs in the Indian Ocean. Validation is constructed by comparing significant wave height from model runs with and without assimilation against observations from CFOSAT. The period of assimilation experiments is from 1 July 2020 to 31 January 2021. Background errors are constructed by dynamic sampling instead of ensemble members of model runs. Variable state ensemble is updated using the EAKF method. During the assimilating procedure, only one member of a wave model is needed, which reduces the computation resource requirements.

Validation of model results indicates that mean absolute error of significant wave height reduces roughly 10% after assimilation. The monthly statistic of model error indicates that the wave model underestimates wave height in high-sea conditions. Significant wave height of model results with assimilation in July 2020 increased by roughly 0.1–0.3 m in the northeast of the Indian Ocean compared with that of the baseline run. In this study, several cases in high-sea conditions were analyzed. The improvement rate of mean absolute error reached roughly 20–40% at a single orbit track. Analysis confirms that observations along orbit tracks from Jason-3 are assimilated in one place, and the impact of assimilation propagates with ocean waves and then could be observed by CFOSAT in another place.

The wave scatterometer SWIM carried by CFOSAT provides a combined directional wave spectra from several incidence angles with an azimuthal cut-off of 70 m for wavelength. Based on the dynamic sampling method, preliminary experiments with assimilation of wave spectrum from CFOSAT will be developed in further study.

5. Conclusions

Based on observations of significant wave height from the satellites CFOSAT and Jason-3, the effect of wave data assimilation in the MASNUM wave model for the Indian Ocean was analyzed and evaluated. Ensemble states of SWH were constructed by dynamic sampling method rather than by an ensemble of wave model members. This approach reduces the computational cost dramatically. Comparison between model outputs and measurements along orbit tracks indicated that assimilation improves model results remarkably in high-sea conditions. This study also analyzed several factors that affected independent validation of assimilation.

Author Contributions: Conceptualization, M.S. and Y.Y.; methodology, M.S., X.Y.; software, M.S., J.D.; validation, M.S., J.D.; formal analysis, M.S., Y.Y.; investigation, M.S.; resources, Y.Y.; data curation, M.S., J.D.; writing—original draft preparation, M.S.; writing—review and editing, Y.Y., J.D.; visualization, M.S.; supervision, Y.Y., X.Y.; project administration, Y.Y.; funding acquisition, M.S., Y.Y., and J.D. All authors have read and agreed to the published version of the manuscript.

Funding: This research was jointly funded by the National Key Research Program of China, grant number 2017YFC1404201, Basic Scientific Fund for National Public Research Institutes of China, grant number 2019Q07 and 2019Q06, and National Natural Science Foundation of China, grant number 41906186 and 41906028, Qingdao Postdoctoral Science Foundation QDBSH2019005.

Acknowledgments: We acknowledge the data support from the European Center for Medium-Range Weather Forecast (ECMWF, http://www.ecmwf.int/, accessed on 17 July 2021), the Copernicus Marine Environment Monitoring Service (CMEMS, https://marine.copernicus.eu/, accessed on 17 July 2021), AVISO (https://www.aviso.altimetry.fr/en/home.html, accessed on 17 July 2021), GEBCO (https://www.gebco.net/, accessed on 17 July 2021). We thank the reviewers for their careful reading and valuable comments, which helped us to improve the manuscript.

Conflicts of Interest: The authors declare no conflict of interest.

References

1. Law-Chune, S.; Aouf, L.; Dalphinet, A.; Levier, B.; Drillet, Y.; Drevillon, M. WAVERYS: A CMEMS global wave reanalysis during the altimetry period. *Ocean Dyn.* **2021**, *71*, 357–378. [CrossRef]
2. Hasselmann, K.; Chapron, B.; Aouf, L.; Ardhuin, F.; Collard, F.; Engen, G.; Schulz-Stellenfleth, J. The ERS SAR wave mode: A breakthrough in global ocean wave observations. *Eur. Space Agency ESASP* **2012**, *1326*, 167–197.
3. Hollingsworth, A. Objective analysis for numerical weather prediction. *J.M.S.J.* **1986**, *64*, 11–59. [CrossRef]
4. Evensen, G. Sequential data assimilation with a nonlinear quasi-geostrophic model using Monte Carlo methods to forecast error statistics. *J. Geophys. Res. Space Phys.* **1994**, *99*, 10143–10162. [CrossRef]
5. Lionello, P.; Günther, H.; Janssen, P.A.E.M. Assimilation of altimeter data in a global third-generation wave model. *J. Geophys. Res. Space Phys.* **1992**, *97*, 14453–14474. [CrossRef]
6. Janssen, P.A.E.M.; Lionello, P.; Reistad, M.; Hollingsworth, A. Hindcasts and data assimilation studies with the WAM model during the Seasat period. *J. Geophys. Res. Space Phys.* **1989**, *94*, 973–993. [CrossRef]
7. Esteva, D.C. Evaluation of preliminary experiments assimilating Seasat significant wave heights into a spectral wave model. *J. Geophys. Res. Space Phys.* **1988**, *93*, 14099. [CrossRef]

8. Francis, P.E.; Stratton, R.A. Some experiments to investigate the assimilation of SEASAT altimeter wave height data into a global wave model. *Q. J. R. Meteorol. Soc.* **1990**, *116*, 1225–1251. [CrossRef]
9. Sun, M.; Yin, X.; Yang, Y. Construction and application in global wave data assimilation of static sample set. *Oceanologia Lim-nologia Sinica* **2014**, *45*, 918–927. [CrossRef]
10. Sun, M.; Yin, X.; Yang, Y.; Wu, K. An effective method based on dynamic sampling for data assimilation in a global wave model. *Ocean Dyn.* **2017**, *67*, 433–449. [CrossRef]
11. Anderson, J.L. An ensemble adjustment Kalman filter for data assimilation. *Mon. Weather Rev.* **2001**, *129*, 2884–2903. [CrossRef]
12. Anderson, J.L. A local least squares framework for ensemble filtering. *Mon. Weather Rev.* **2003**, *131*, 634–642. [CrossRef]
13. Yuan, Y.; Pan, Z.; Hua, F.; Sun, L. LAGFD-WAM wave numerical model (I), the basic physical model. *Acta Oceanol. Sin.* **1992**, *14*, 1–7.
14. Yuan, Y.; Hua, F.; Pan, Z.; Sun, L. LAGFD-WAM numerical wave model-II. Characteristics inlaid scheme and its application. *Acta Oceanol. Sin.* **1992**, *14*, 12–24.
15. Yang, Y.; Qiao, F.; Zhao, W.; Teng, Y.; Yuan, Y. MASNUM ocean wave numerical model in spherical coordinates and its application. *Acta Oceanol. Sin.* **2005**, *27*, 1–7. [CrossRef]
16. Yuan, Y.; Tung, C.C.; Huang, N.E. Statistical Characteristics of Breaking Waves. In *Wave Dynamics and Radio Probing of the Ocean Surface*; Phillips, O.M., Hasselmann, K., Eds.; Springer: Boston, MA, USA, 1986; pp. 265–272. [CrossRef]
17. GEBCO. Available online: https://www.gebco.net/ (accessed on 17 July 2021).
18. AVISO. Available online: https://www.aviso.altimetry.fr/en/home.html (accessed on 17 July 2021).
19. Hauser, D.; Tison, C.; Amiot, T.; Delaye, L.; Mouche, A.; Guitton, G.; Aouf, L.; Castillan, P. CFOSAT: A new Chinese-French satellite for joint observations of ocean wind vector and directional spectra of ocean waves. *Remote Sens. Oceans Inland Waters Tech. Appl. Chall.* **2016**, *9878*, 98780. [CrossRef]
20. Hauser, D.; Tison, C.; Amiot, T.; Delaye, L.; Corcoral, N.; Castillan, P. SWIM: The First Spaceborne Wave Scatterometer. *IEEE Trans. Geosci. Remote Sens.* **2017**, *55*, 3000–3014. [CrossRef]

remote sensing

MDPI

Letter

Satellite Altimetry and Tide Gauge Observed Teleconnections between Long-Term Sea Level Variability in the U.S. East Coast and the North Atlantic Ocean

Qing Xu [1,2], Kai Tu [1], Yongcun Cheng [3,*], Weiping Wang [4], Yongjun Jia [5] and Xiaomin Ye [5]

1. College of Oceanography, Hohai University, Nanjing 210098, China; maggiexu@hhu.edu.cn (Q.X.); 161331010008@hhu.edu.cn (K.T.)
2. Key Laboratory of Coastal Disaster and Defence (Hohai University), Ministry of Education, Nanjing 210098, China
3. Shenzhen AeroImgInfo Technology Co. Ltd., Shenzhen 518000, China
4. South China Sea Marine Survey and Technology Center, Ministry of Natural Resources of China, Guangzhou 510310, China; wangweiping@smst.gz.cn
5. National Satellite Ocean Application Service, Beijing 100081, China; jiayongjun@mail.nsoas.org.cn (Y.J.); yxm@mail.nsoas.org.cn (X.Y.)

* Correspondence: chengyongcun@piesat.cn

Received: 30 October 2019; Accepted: 26 November 2019; Published: 28 November 2019

Abstract: Rising sea levels amplify the threat and magnitude of storm surges in coastal areas. The U.S. east coast region north of Cape Hatteras has shown a significant sea level rise acceleration and is believed to be a "hot-spot" for accelerating tidal flooding. To better understand the forcing mechanism of long-term regional sea level change, in order to more efficiently implement local sea level rise adaptation and mitigation measures, this work investigated the teleconnections between low-frequency sea level variability in the coastal region north of Cape Hatteras and the subpolar/tropical North Atlantic Ocean by using tide gauge measurements, satellite altimetry data and a sea level reconstruction dataset. The correlation analysis demonstrates that the tide-gauge measured sea level variability in the area north of Cape Hatteras is highly and positively correlated with that observed by satellite altimetry in the subpolar and tropical North Atlantic between 1993 and 2002. Over the following decade (2003–2012), the phase of the teleconnection in the subpolar region was reversed and the spatio-temporal correlation in the tropical North Atlantic was enhanced. Furthermore, the positive correlation in the region north of Cape Hatteras's near shore area is strengthened, while the negative correlation in the Gulf Stream front region is weakened. The North Atlantic Oscillation and Atlantic Multidecadal Oscillation, which affect variations of the Atlantic Meridional Overturning Circulation and Gulf Stream, were shown to have significant impacts on the decadal changes of the teleconnections. Coherent with satellite altimetry data, the reconstructed sea level dataset in the 20th century exhibits similar spatial correlation patterns with the Atlantic Meridional Overturning Circulation, North Atlantic Oscillation and Atlantic Multidecadal Oscillation indices.

Keywords: sea level variability; satellite altimetry; teleconnection; Atlantic Meridional Overturning Circulation; North Atlantic Oscillation; Atlantic Multidecadal Oscillation

1. Introduction

Recently, a significant sea level rise acceleration has been detected along the U.S. east coast north of Cape Hatteras (NCH) using tide gauge data [1–3]. The region is also believed to be a "hot-spot" for accelerating tidal flooding [4]. Some studies have revealed that the climate-related changes in the Atlantic Meridional Overturning Circulation (AMOC) and Gulf Stream, as well as the subsistence of the coastal zones are potential drivers behind this [3,5].

Rising sea levels amplify the threat and magnitude of the storm surge-induced urban flooding in coastal areas. Therefore, it is important to combine tide gauge measurements, satellite altimetry and sea level data on longer time series (e.g., the sea level reconstruction) together to better understand the forcing mechanism of the regional sea level variability and help to implement the sea level rise adaptation and mitigation strategy. Yearly averaged tide gauge data have been found to be highly and positively correlated with satellite altimetry data in NCH near shore regions [6]. But are there teleconnections between low-frequency sea level variability in the U.S. east coast and deep-water regions of the North Atlantic (e.g., [7])? If so, how do they link to each other? Do the correlations exist on long time scales? To answer these questions, we investigated the relationships of tide-gauge measured long-term sea level variability to satellite altimetry observations and sea level reconstruction data (1900–2012, [8]) in the North Atlantic.

The paper is organized as follows. The data and methods used in this work are described in Section 2. Section 3 presents the variations of correlations between long-term sea level variability along the U.S. east coast observed by tide gauges and that in the North Atlantic observed by satellite altimeters. Analysis of the factors contributing to the variations are given in Section 4. The results from the sea level reconstruction are shown in Section 5. The summary is presented in Section 6.

2. Data and Methodology

2.1. Satellite Altimetry Data

In this study, we used daily absolute dynamic topography (ADT) data from the SSALTO/Duacs gridded delayed-time products, which were merged from all available satellite altimetry data with spatial resolution of 0.25° × 0.25°. The ADT, known as the sea surface height with respect to the geoid, is a sum of sea level anomaly (SLA) and mean dynamic topography. The daily ADT maps are monthly averaged for further analysis.

2.2. Tide Gauge Data

The monthly sea level data recorded by 32 tide gauge stations (Figure 1) along the U.S. east coast were downloaded from the Permanent Service for Mean Sea Level. The data completeness over the time span of 1993–2012 exceeds 90% at all the tide gauges. The classical inverted barometer correction was applied at each tide gauge station using NCEP (National Centers for Environmental Prediction)/NCAR (National Center for Atmospheric Research) reanalyzed (2.5° × 2.5° grids) monthly sea level pressure data. Considering the high coherent regional sea level variability along the U.S. east coast and to fill the data gaps in the tide gauge measurements, the tide gauges are divided into two groups, i.e., 26 sites in NCH (blue dots in Figure 1) and 6 (black dots in Figure 1) in the region south of Cape Hatteras (SCH) [9]. We averaged the detrended tide gauge data in the two defined regions and the mean value in each composite sea level time series were removed.

To compare with altimetry data and the sea level reconstruction dataset, four representative regions in the North Atlantic Ocean were selected and shown in Figure 1. Significant positive and negative correlations between tide gauge and altimetry data have been found in the NCH near shore area (Region 1) and Gulf Stream front region (Region 2) [6], respectively. Regions 3 and 4 denote the subpolar (e.g., the Labrador Sea) and subtropical regions, respectively.

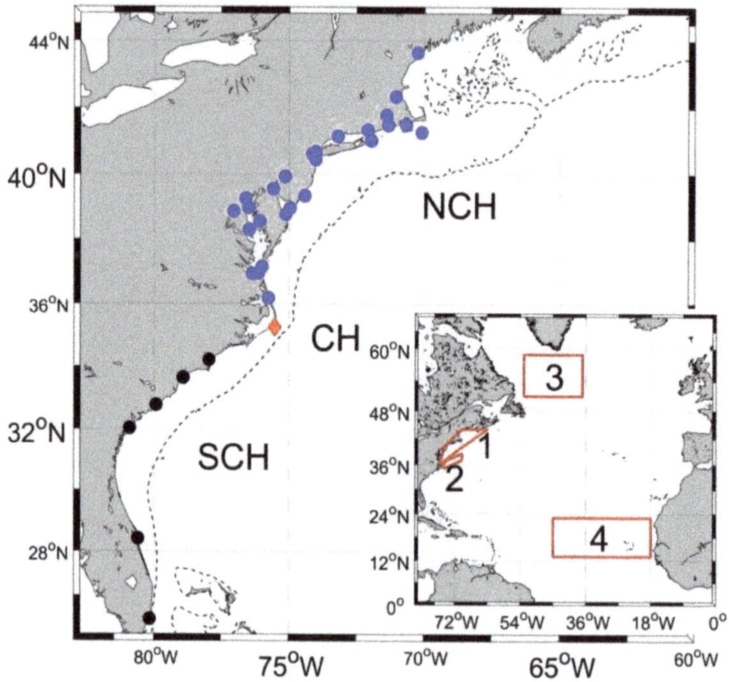

Figure 1. Locations of the tide gauge stations in each group: North (blue dots) and south of Cape Hatteras (black dots). The location of Cape Hatteras (CH) is marked as the red diamond. A water depth of 200 m isobath is overlaid (dashed curve). The red boxes 1–4 denote the four representative regions: the NCH near shore area (1), Gulf Stream front region (2), subpolar (3) and subtropical (4) North Atlantic Ocean.

2.3. Sea Level Reconstruction Data

The reconstructed global sea level by combining sea surface temperature (SST) with tide gauge and altimetry data could represent the contribution of climate variability to sea level change. The data are available over the time period from 1900 to 2012 (see Reference [10] for more details about the dataset). Particularly, the reconstructed sea level variability is highly correlated with the Atlantic Multidecadal Oscillation (AMO) in the North Atlantic Ocean.

3. Spatio-Temporal Correlations between Sea Level Variability in the U.S. East Coast and the North Atlantic Ocean

In order to show the teleconnections of long-term sea level variability in the U.S. east coast with that in the North Atlantic Ocean, the spatio-temporal correlation coefficients between tide gauge records in SCH (Figure 2a–c)/NCH (Figure 2d–f) and altimeter data at each grid are shown in Figure 2. All the time series were low-pass filtered with a filter half amplitude period of 1 year. To exhibit the changes of the teleconnection patterns on a decadal time scale, the calculations were made over 1993–2012 (Figure 2a,d), 1993–2002 (Figure 2b,e) and 2003–2012 (Figure 2c,f), respectively. Only the correlations significant at 95% confidence level are shown in Figure 2.

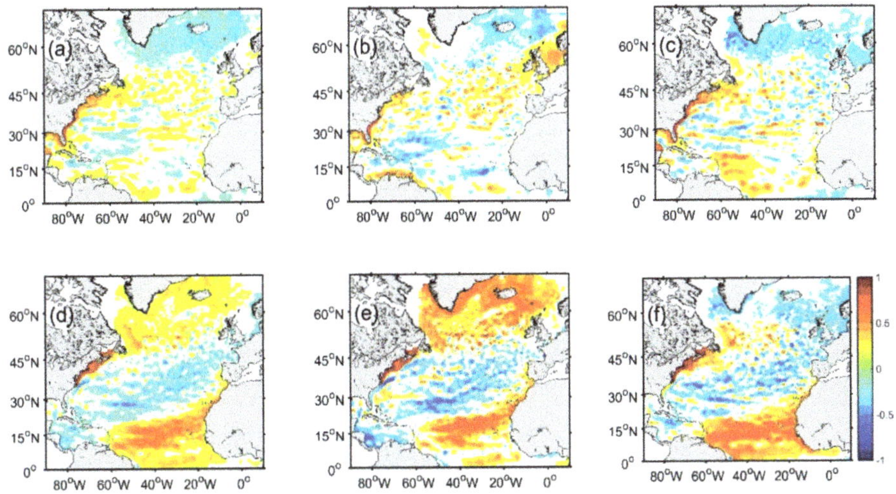

Figure 2. Geographical distribution of the correlation coefficient between tide gauge measurements south of Cape Hatteras (**a**–**c**), north of Cape Hatteras (**d**–**f**) and satellite altimeter data over 1993–2012 (left panels), 1993–2002 (middle panels) and 2003–2012 (right panels). The white regions denote correlations that are not significant at the 95% confidence level.

In Figure 2a, the SCH averaged sea level variability observed by tide gauges are highly correlated with the altimeter observations in the U.S. near shore areas. Higher correlations are found over these areas in 2003–2012 (Figure 2c) than a decade ago (Figure 2b, 1993–2002). In addition, there is an obvious negative correlation in the sub-polar gyre (e.g., the Labrador Sea and Irminger Sea) in 2003–2012 (Figure 2c), which is not observed in the first decade. This might be due to the possible change in the forcing factors of the regional sea level variability.

Compared with the region SCH, the patterns of the spatio-temporal correlations between altimetry data and tide gauge measurements in NCH (Figure 2d–f) are distinct, indicating the forcing mechanisms are possibly different in these two regions. High correlations are shown from the coasts to the outer shelf and sub-tropical regions (10° N–20° N) in Figure 2d. This has not been investigated in previous works [6]. Compared with the first decade of 1993–2002 (Figure 2e), there is a correlation reversal in the northern North Atlantic and coastal regions of Greenland in 2003–2012 (Figure 2f) and an enhanced correlation in NCH near shore and sub-tropical regions. The change of the high correlations implies a shift from subpolar-related to tropical-related regional sea level variability along the NCH coast in the past two decades. In the following sections, considering that the change of correlation in NCH is more significant than that in SCH and it is a "hot-spot" for accelerating flooding, we focus on the sea level variability in this area.

As shown in Figure 2e, over 1993–2002, the tide gauge data in NCH are negatively correlated with altimeter data near 36° N and 73° W, the southeast of the Gulf Stream's mean position. This was also pointed out in Reference [6]. However, the correlations disappear in the following decade and a wider spatial coverage of higher correlations exhibits on the NCH shelves, which may be related to the eastward shift of the Gulf Stream path [5].

4. Potential Drivers of the Teleconnection

4.1. Atlantic Meridional Overturning Circulation (AMOC)

In this work, we use a monthly AMOC proxy [11] to investigate the relationship of AMOC to satellite altimetry observations. Figure 3a–c show the spatial pattern of correlations between the AMOC proxy and altimeter data over the time span of 1993–2012, 1993–2002 and 2003–2012, respectively. The white regions denote the correlations that are not significant at the 95% confidence level. In Figure 3a, significant positive correlations appear in the mid-Atlantic Ocean and along the path of the Gulf Stream, while negative correlations are observed in the tropical Atlantic and NCH near shore regions. The comparison of Figure 3c with Figure 3b demonstrates a strengthening of the negative correlations in the U.S. northeast coast, Labrador Sea and tropical North Atlantic in 2003–2012, while the Gulf Stream front (Region 2 in Figure 1) exhibits a weakening of the positive correlation. In the northeastern Atlantic, the correlations reversed in the last two decades.

Compared with Figure 2, one can see the changes of the spatio-temporal correlation patterns agree fairly well with that between the composite tide gauge data in NCH and altimetry data. This indicates the variation in AMOC in recent years is more closely related to sea level variation in the Gulf Stream. Consistent with the other studies (e.g., [12]), the recent weakening of AMOC will increase coastal sea levels but decrease offshore sea levels and lead to higher correlations between tide gauge observed sea level variability in the U.S east coast and altimetry data in the North Atlantic.

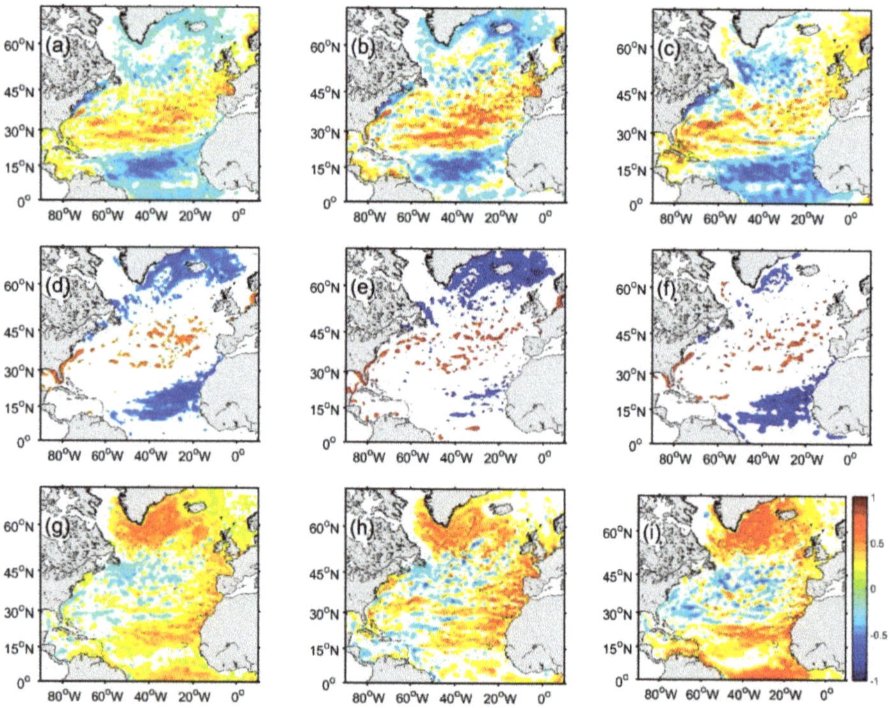

Figure 3. Geographical distribution of the correlation coefficient between the reconstructed monthly AMOC proxy and altimeter observed sea level anomaly (SLA) over the time span of 1993–2012 (**a**), 1993–2002 (**b**) and 2003–2012 (**c**). (**d–f**) are similar to (**a–c**) but for correlations between the North Atlantic Oscillation (NAO) winter index and annually averaged altimeter data. (**g–i**) are similar to (**d–f**) but for correlations between monthly Atlantic Multidecadal Oscillation (AMO) index and altimeter data. The white regions denote correlations that are not significant at the 95% confidence level.

4.2. North Atlantic Oscillation (NAO)

The sea level variability in the North Atlantic is highly correlated with NAO [7]. Long term variations of AMOC and Gulf Stream are linked to the NAO index through Ekman transport [13]. Furthermore, the decline of the AMOC proxy in 2009–2010 is linked to the drop in the NAO winter index [14]. Hereby we tested the variations of spatio-temporal correlations between altimeter data and the NAO winter index in the past two decades.

The correlation coefficients between annually averaged altimetry SLA time series and the NAO winter index is shown in Figure 3d–f. The patterns of the correlations in the northern Atlantic Ocean between 1993 and 2002 (Figure 3e) are consistent with that found by [7]. Moreover, negative correlations are found in the NCH near shore region and tropical North Atlantic. However, a significant decrease and an increase in negative correlations are shown in the northern and tropical North Atlantic, respectively, in the following decade (Figure 3f). This is highly consistent with the variations of the correlations between tide gauge and altimeter data over the years (Figure 2e–f).

The similarity between the general patterns of the SLA–AMOC correlation (Figure 3a) and the SLA–NAO correlation (Figure 3d) confirms the close relation between AMOC and NAO as indicated in numerous studies mentioned before. That is also in line with the findings that the sea level variations in the North Atlantic closely represent the AMOC variability driven by the NAO forcing at interannual to decadal time scales [15]. However, some differences in the above correlation patterns in the subpolar region, especially in recent years near the Greenland Sea (Figure 3c), may reveal the influence of some other factors, such as the increase in Greenland ice sheet melt (see Section 2).

4.3. Atlantic Multidecadal Oscillation (AMO)

Figure 3g–i show the variations of spatio-temporal correlations between the SLA and AMO index in the past two decades. An enhancement of the positive correlation is found in the tropical North Atlantic and Labrador Sea in the last decade.

The AMO and NAO indices are not independent of each other. Figure 4a,b show the correlations between a high NAO/AMO index (i.e., outside ± σ of the index, where σ denotes the standard deviation of the index) and altimeter data over 1993–2012. One can see the variability of high station-based NAO is negatively correlated with long-term sea level variability in the northeast Atlantic (Figure 4a). Significant positive correlations between the high AMO index and altimeter data are observed in the Subpolar Gyre and tropical Atlantic Ocean (Figure 4b). Moreover, the correlation coefficients exceed 0.5 in NCH coastal regions. That implies the AMO has an important impact on the variation of long-term sea level variability in the NCH near shore area, whereas a combination of low NAO and high AMO can contribute to higher coastal water levels in this region in the last decade.

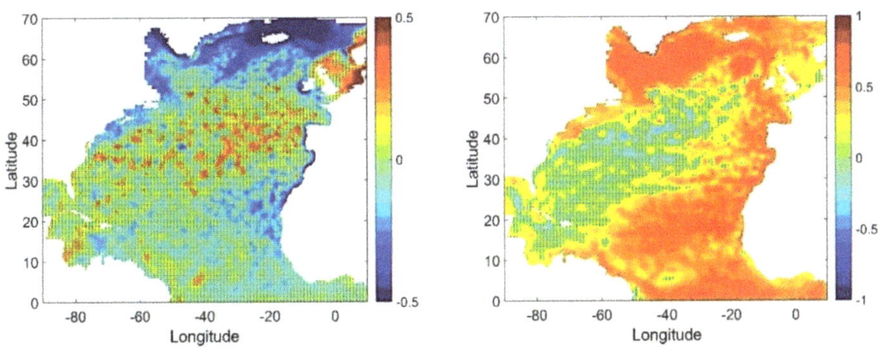

Figure 4. Geographical distribution of the correlation coefficient between high NAO (**left**)/AMO (**right**) index (outside ± σ of the NAO/AMO index) and altimetry sea level time series during 1993 to 2012.

To further explain the mechanism of the teleconnection between sea level variability and the AMO/NAO/AMOC, the empirical orthogonal function (EOF) analysis is performed to altimetry data with the trend and seasonal signal removed. The leading EOF mode (EOF1) accounts for 5.4% of the total sea level variance (Figure 5a). The subpolar regions exhibit strong sea level variability accompanied with opposite sea level variability in the Gulf Stream and western subtropical Atlantic. Coherent variability along the NCH and subpolar/subtropical Atlantic regions confirms the teleconnection between sea levels in coastal areas and remote regions in Figure 2d–f.

Figure 5b–d compare the principle component time series (PCTS1) of EOF1 with the negative station-based NAO index, AMO index and negative AMOC proxy, respectively. In Figure 5b, the low-passed NAO (13-month running mean) leads PCTS1 by 6 months. In Figure 5c,d, 13-month and 25-month running means were applied to PCTS1 and AMO/AMOC, respectively, to exhibit the low-frequency sea level variability. Coherent variations are shown between the PCTS1 and the indices with correlations of 0.49, 0.49 and 0.55, respectively. Particularly, the extremely low NAO, high AMO and low AMOC are linked to the strong sea level variation in 2010–2011. Combining Figure 5a with Figure 5b–d explains the decadal variation of the correlations presented in Figure 3, which is associated with the variations of the climate indices.

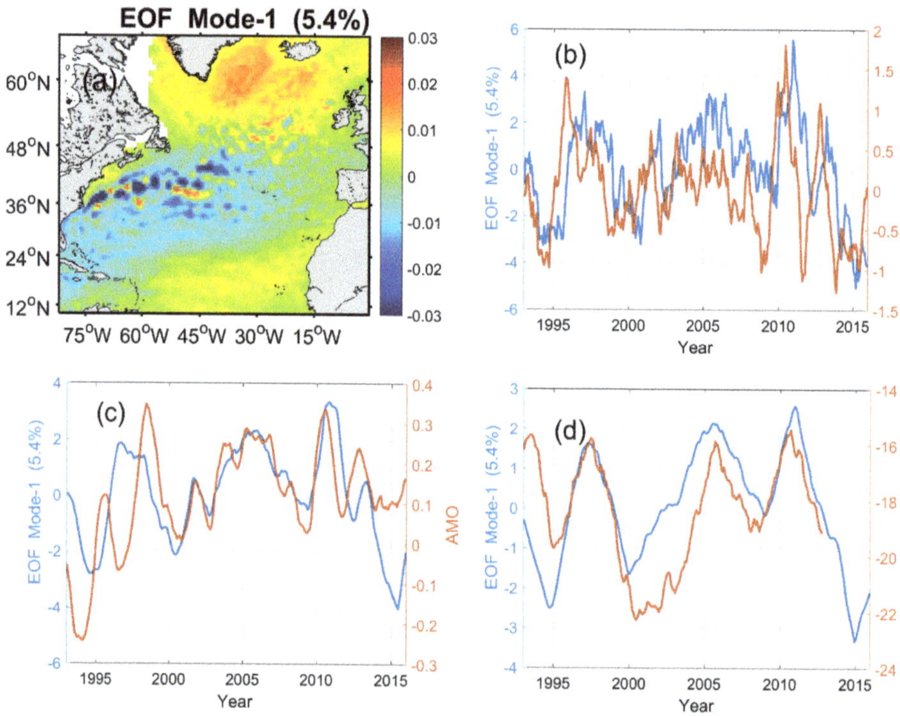

Figure 5. The first empirical orthogonal function (EOF) mode of altimeter SLA in the North Atlantic Ocean (**a**) and comparison of PCTS1 with NAO (**b**), AMO (**c**) and AMOC (**d**) indices.

5. Long-Term Sea Level Variability from Sea Level Reconstruction

In the above sections, we investigated the variations of correlations between tide gauge measurements, altimetry data and climate indices over the period 1993–2012. Does the correlation and teleconnection exist on a longer time scale? Here we try to answer the question based on the reconstructed sea level data over the 20th century (1901–2012).

Before further analysis of the low-frequency sea level variability in the North Atlantic Ocean, we first evaluated the reconstructed sea level data against satellite altimetry data in the four representative regions in Figure 1. The time series of sea level anomaly from the reconstruction dataset and altimeters are shown in Figure 6a–d with trends and seasonal signals removed. All regions exhibit high correlations larger than 0.7. In Figure 6a, the sea level peak in 1998 in Region 1 is captured by the sea level reconstruction dataset. Strong sea level variability in the Gulf Stream is well reproduced (Figure 6b). In the subpolar (Figure 6c) and subtropical (Figure 6d) regions, the correlations exceed 0.8.

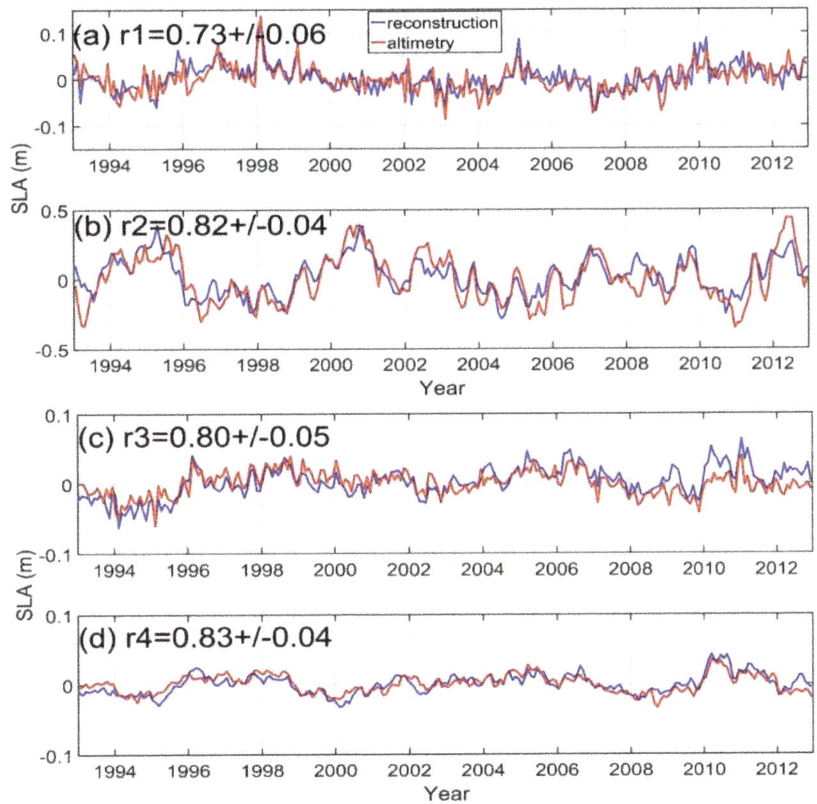

Figure 6. Time series of regional mean sea level anomaly from the reconstruction and satellite altimetry in the four representative regions (in Figure 1): (**a**) the NCH near shore area, (**b**) Gulf Stream front region, (**c**) subpolar and (**d**) subtropical North Atlantic Ocean. The trends and seasonal signals in the time series were removed.

Similar to Figure 3, Figure 7a–c show the correlation coefficient between the filtered sea level reconstruction data and AMOC/NAO winter/AMO indices over 1901–2012 (1935–2012 for AMOC). Compared with the satellite altimetry era (Figure 3a), coherent spatial patterns are observed in the 20th century (Figure 7a) but with a weaker correlation. Comparing Figure 7b with Figure 3d, the reconstructed sea level shows stronger correlation in NCH near shore and subpolar regions. As for the AMO, very similar spatial patterns are presented in altimetry data and the sea level reconstruction dataset (Figures 3g and 7c).

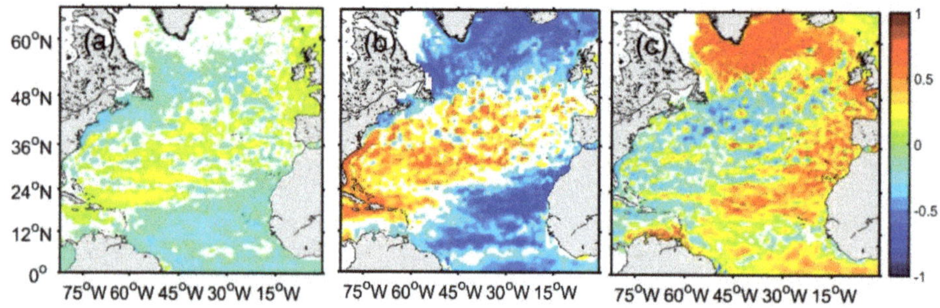

Figure 7. Geographical distribution of the correlation coefficient between the annual averaged reconstructed SLA and monthly AMOC proxy over 1935–2012 (**a**), NAO winter (**b**) and AMO indices (**c**) over the time span of 1901–2012. The white zones denote correlations that are not significant at the 95% confidence level.

Figure 8 shows the time evolution of the reconstructed sea level anomaly in the past 112 years and AMOC/NAO/AMO indices in the four selected regions. In Figure 8a, a higher correlation could be obtained if the calculation started from 1960. The correlations between the sea level anomalies in the four regions are listed in Table 1. Negative correlations are found between sea level in the Gulf Stream front and the other three regions, with a correlation coefficient (CC) of −0.45, −0.53 and −0.73, respectively. Sea level variability in the coastal regions are tele-connected with that in the subpolar and subtropical regions with a CC of 0.38 and 0.51. Similar to the altimetry era, the sea level in the subpolar and subtropical regions are highly correlated with each other with an absolute value of CC over 0.7. Combining Figure 8a with Figure 5, it is obvious that the remarkable high sea level variations in NCH coastal regions in 1998–1999 and 2010–2011 are related to AMOC variations and low NAO/high AMO indices.

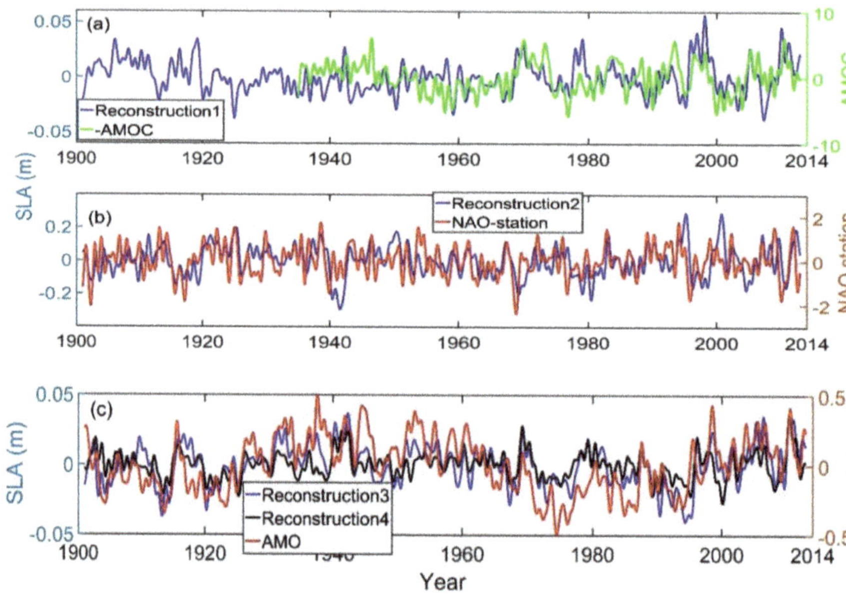

Figure 8. Regional mean reconstructed SLA in (**a**) Region 1: the NCH near shore area; (**b**) Region 2: Gulf Stream front region; and (**c**) Regions 3 and 4: subpolar and subtropical North Atlantic Ocean.

Table 1. The correlation coefficient between the sea level reconstructions in the four regions in Figure 1 over 1901–2012. Regions 1–4 denote the NCH near shore area, Gulf Stream front region, subpolar and subtropical North Atlantic, respectively.

	Region 1	Region 2	Region 3
Region 2	−0.45		
Region 3	0.38	−0.53	
Region 4	0.51	−0.73	0.71

Table 2 summarizes the correlations between regional sea level reconstructions and the three climate indices. NAO is correlated with the reconstructed sea level in the four regions, but AMO is only correlated with that in the subpolar and subtropical regions. Compared with the other regions, a higher correlation between AMOC and sea level is shown in the NCH near shore area.

Table 2. The correlation coefficient between the regional sea level reconstruction and NAO/AMO/AMOC indices over 1901–2012 (1935–2012 for AMOC). Regions 1–4 denote the NCH near shore area, Gulf Stream front region, subpolar and subtropical North Atlantic, respectively.

	NAO	AMO	AMOC
Region 1	−0.30	0.01	−0.37
Region 2	0.38	−0.07	0.21
Region 3	−0.37	0.59	−0.15
Region 4	−0.51	0.38	−0.22

6. Summary

In this study, we found a high teleconnection between sea level variability in NCH observed by tide gauges and that captured by satellite altimetry in the subpolar and tropical North Atlantic. The correlation varies on a decadal time scale. In the Labrador Sea, the positive correlations during 1993 to 2002 reverses to negative in the following decade (2003–2012). Compared with the subpolar North Atlantic, satellite data in the tropical region exhibit a stronger connection with tide gauge data in NCH in 2003–2012, which implies a shift of local sea level variability from subpolar-related to tropical-related in the past two decades. Moreover, the negative correlations between tide gauge measurements and altimeter data near 36° N and 73° W disappear in 2003–2012, which could be related to the shift in the Gulf Stream path and has not been reported in previous studies.

At basin scales, surface heat flux changes contribute a lot to the low-frequency sea level variations [16]. Hence, we further investigated the variability of ocean heat content (OHC) to explain the remote connections between tide gauge and satellite altimeter data in the last 20 years. Figure 9a,b show the distribution of the standard deviation of the detrended OHC over 1993–2002 and 2003–2012, respectively. The AMOC transports the heat from the South Atlantic and tropical North Atlantic to high latitudes. In Figure 9, note the decrease in OHC variability along the Gulf Stream main path, which is possibly related to the slowdown of the AMOC accompanied with less transportation of warm water to the subpolar Atlantic Ocean.

The climatic warming and the melting of the Greenland ice sheet have a great impact on the deep (dense) water formation in the northern North Atlantic and Labrador Sea. The reversed correlations in the Labrador Sea in the last two decades (Figure 2e,f) are consistent with the cooling regions caused by the slowdown of AMOC due to the freshwater anomaly added to the northern North Atlantic [17]. Therefore, the decrease in deep-water formation in the Labrador Sea and the associated reduction of AMOC may partly lead to the correlation variations. In the tropical regions, the areas with positive correlations agree well with the warm westward limb of the AMOC at 10°–20° N. The weakening of AMOC decreases the westward transportation of the warm water. That explains the variability of the sea level here, particularly in 1998, 2005 and 2010 (see Figure 5d).

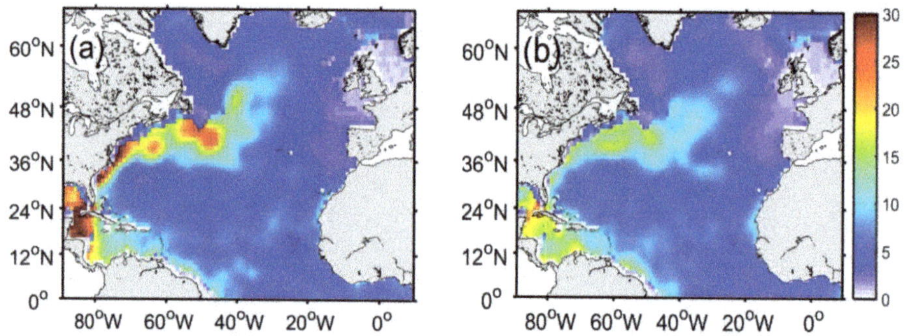

Figure 9. Geographical distribution of the standard deviation of the detrended seasonal ocean heat content (10×10^{18} J, 0–700 m) over the time span of 1993–2002 (**a**) and 2003–2012 (**b**).

The variations of the correlations between tide gauge observations in NCH and altimeter data are accompanied with lower or higher negative correlations between NAO winter and annual sea level data in the subpolar and tropical Atlantic Ocean (Figure 3d–f), suggesting that the NAO may be an important factor remotely influencing the local sea level variability.

Another factor linked to the teleconnection of NHC–North Atlantic sea level variability is AMO (Figure 3g–i). A positive AMO phase can increase sea level by thermal expansion, and the associated enhanced AMOC tends to weaken the sea level rise rate in the northeast coastal areas of the U.S. and Subpolar Gyre regions (e.g., [18,19]). The two effects combine to yield the spatial patterns of sea level change associated with the AMO (Figures 4b and 5c). The sea level variability in the NCH near shore area may be linked with that in the subtropical North Atlantic through the observed warming of the tropical Atlantic Ocean (Figures 2h and 4b), and remotely connected with that in the northern North Atlantic through NAO (Figure 4a) and its relationship with AMOC variability.

The climate indices AMOC, AMO and NAO are related to each other. Multidecadal variation of the NAO can induce multidecadal variation of the AMOC and poleward ocean heat transport in the Atlantic (e.g., AMO). The NAO variation could change the AMOC. The sea level reconstruction (1900–2012) demonstrates remarkable decadal sea level variability in the North Atlantic Ocean. Coherent with altimetry data, the reconstructed dataset in the 20th century exhibits similar spatial correlation patterns with the AMOC, NAO and AMO indices (Figure 7 and Table 2). It confirms the teleconnection between sea level variability in NCH and the subpolar/subtropical North Atlantic, which may be attributed to variations in AMOC and NAO/AMO.

Author Contributions: Conceptualization and methodology, Q.X. and Y.C.; data processing and teleconnection analysis, Q.X., K.T. and X.Y.; mechanism analysis and discussion, Q.X., Y.C., W.W. and Y.J.; writing–original draft preparation, Q.X. and K.T.; writing–review and editing, Y.C., W.W., Y.J. and X.Y.

Funding: This research was funded by the National Key Project of Research and Development Plan of China (Grant No. 2016YFC1401905), the Guangdong Special Fund Program for Marine Economy Development (Grant No. GDME-2018B001), National Natural Science Foundation of China (Grant No. 41976163), Dayu Scholar Program of Hohai University (for Q. X.) and the Priority Academic Program Development of Jiangsu Higher Education Institutions.

Acknowledgments: We thank the constructive suggestions from Benjamin D. Hamlington, Tal Ezer and Hans-Peter Plag at Old Dominion University. The altimeter products were produced by Ssalto/Duacs and distributed by Aviso with support from CNES and available at http://www.aviso.altimetry.fr. The monthly tide gauge data were downloaded from Permanent Service for Mean Sea Level (PSMSL, http://www.psmsl.org/data/obtaining/). The AMO proxy data were obtained from: http://www.esrl.noaa.gov/psd/data/timeseries/AMO/.

Conflicts of Interest: The authors declare no conflict of interest.

References

1. Boon, J.D. Evidence of sea level acceleration at us and Canadian tide stations, Atlantic coast, north America. *J. Coast. Res.* **2012**, *28*, 1437–1445. [CrossRef]
2. Boon, J.D.; Mitchell, M. Nonlinear change in sea level observed at north American tide stations. *J. Coast. Res.* **2015**, *31*, 1295–1305. [CrossRef]
3. Sallenger, A.H.; Doran, K.S.; Howd, P.A. Hotspot of accelerated sea-level rise on the atlantic coast of north america. *Nat. Clim. Chang.* **2012**, *2*, 884–888. [CrossRef]
4. Ezer, T.; Atkinson, L.P. Accelerated flooding along the us east coast: On the impact of sea-level rise, tides, storms, the gulf stream, and the north Atlantic oscillations. *Earth Future* **2014**, *2*, 362–382. [CrossRef]
5. Ezer, T.; Atkinson, L.P.; Corlett, W.B.; Blanco, J.L. Gulf stream's induced sea level rise and variability along the us mid-Atlantic coast. *J. Geophys. Res. Ocean.* **2013**, *118*, 685–697. [CrossRef]
6. Andres, M.; Gawarkiewicz, G.G.; Toole, J.M. Interannual sea level variability in the western north Atlantic: Regional forcing and remote response. *Geophys. Res. Lett.* **2013**, *40*, 5915–5919. [CrossRef]
7. Volkov, D.L.; van Aken, H.M. Annual and interannual variability of sea level in the northern north Atlantic ocean. *J. Geophys. Res. Ocean.* **2003**, *108*. [CrossRef]
8. Hamlington, B.; Leben, R.; Strassburg, M.; Kim, K.Y. Cyclostationary empirical orthogonal function sea-level reconstruction. *Geosci. Data J.* **2014**, *1*, 13–19. [CrossRef]
9. Thompson, P.; Mitchum, G. Coherent sea level variability on the north Atlantic western boundary. *J. Geophys. Res. Ocean.* **2014**, *119*, 5676–5689. [CrossRef]
10. Hamlington, B.; Leben, R.; Kim, K.Y. Improving sea level reconstructions using non-sea level measurements. *J. Geophys. Res. Ocean.* **2012**, *117*. [CrossRef]
11. Stepanov, V.N.; Iovino, D.; Masina, S.; Storto, A.; Cipollone, A. Observed and simulated variability of the Atlantic meridional overturning circulation at 41° N. *J. Mar. Syst.* **2016**, *164*, 42–52. [CrossRef]
12. Ezer, T. Detecting changes in the transport of the gulf stream and the Atlantic overturning circulation from coastal sea level data: The extreme decline in 2009–2010 and estimated variations for 1935–2012. *Glob. Planet. Chang.* **2015**, *129*, 23–36. [CrossRef]
13. Zhao, J.; Johns, W. Wind-forced interannual variability of the Atlantic meridional overturning circulation at 26.5° N. *J. Geophys. Res. Ocean.* **2014**, *119*, 2403–2419. [CrossRef]
14. Srokosz, M.A.; Bryden, H. Observing the Atlantic meridional overturning circulation yields a decade of inevitable surprises. *Science* **2015**, *348*, 1255575. [CrossRef] [PubMed]
15. Li, F.; Jo, Y.-H.; Yan, X.-H.; Liu, W.T. Climate signals in the mid to high latitude north Atlantic from altimeter observations. *J. Clim.* **2015**, *29*, 4905–4925. [CrossRef]
16. Wang, Z.; Lu, Y.; Dupont, F.; Loder, J.W.; Hannah, C.; Wright, D.G. Variability of sea surface height and circulation in the north Atlantic: Forcing mechanisms and linkages. *Prog. Oceanogr.* **2015**, *132*, 273–286. [CrossRef]
17. Yin, J.; Goddard, P.B. Oceanic control of sea level rise patterns along the east coast of the United States. *Geophys. Res. Lett.* **2013**, *40*, 5514–5520. [CrossRef]
18. Hu, A.; Meehl, G.A.; Han, W.; Yin, J. Transient response of the MOC and climate to potential melting of the Greenland ice sheet in the 21st century. *Geophys. Res. Lett.* **2009**, *36*. [CrossRef]
19. Liu, Z. Dynamics of interdecadal climate variability: A historical perspective. *J. Clim.* **2012**, *25*, 1963–1995. [CrossRef]

remote sensing

MDPI

Correction

Correction: Lanz et al. The InflateSAR Campaign: Developing Refugee Vessel Detection Capabilities with Polarimetric SAR. *Remote Sens.* 2023, *15*, 2008

Peter Lanz [1,2,*], Armando Marino [3], Morgan David Simpson [3], Thomas Brinkhoff [2], Frank Köster [1,4] and Matthias Möller [5]

1 Department of Computing Science, Carl von Ossietzky University of Oldenburg, Ammerländer Heerstraße 114-118, 26129 Oldenburg, Germany; frank.koester@dlr.de
2 Institute for Applied Photogrammetry and Geoinformatics, Jade University Oldenburg, Ofener Str. 16/19, 26121 Oldenburg, Germany; thomas.brinkhoff@jade-hs.de
3 Department of Biological and Environmental Sciences, University of Stirling, Stirling FK9 4LA, UK; armando.marino@stir.ac.uk (A.M.); m.d.simpson@stir.ac.uk (M.D.S.)
4 Institute for AI Safety and Security, German Aerospace Center (DLR), Rathausallee 12, 53757 Sankt Augustin, Germany
5 Faculty for Humanities and Cultural Sciences, Otto-Friedrich-University of Bamberg, Am Kranen, 96045 Bamberg, Germany; matthias.moeller@uni-bamberg.de
* Correspondence: peter.lanz@uni-oldenburg.de

Error in Figure

In the original publication [1], there was a mistake in the legend for Figures 11a–h, 12a–d and 13a–d. All 16 Figures have the headline "Incidence Angle" which is wrong and was removed in the corrected figures below. The correct Figures 11–13 appears below.

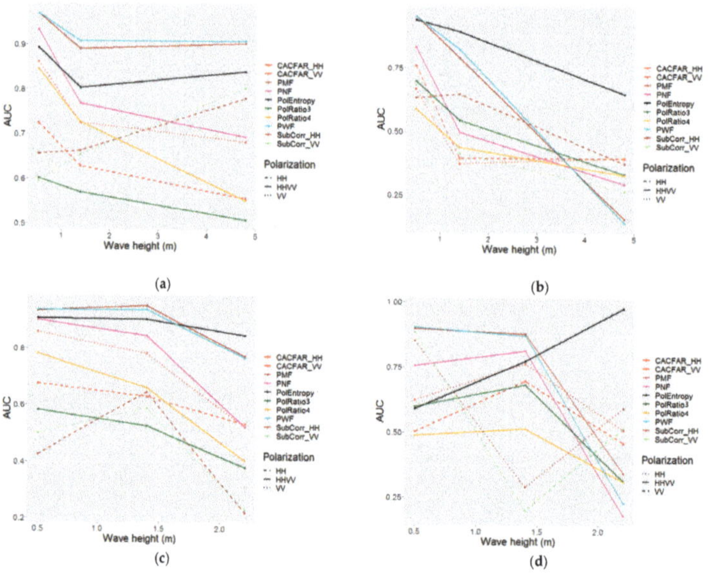

Figure 11. *Cont.*

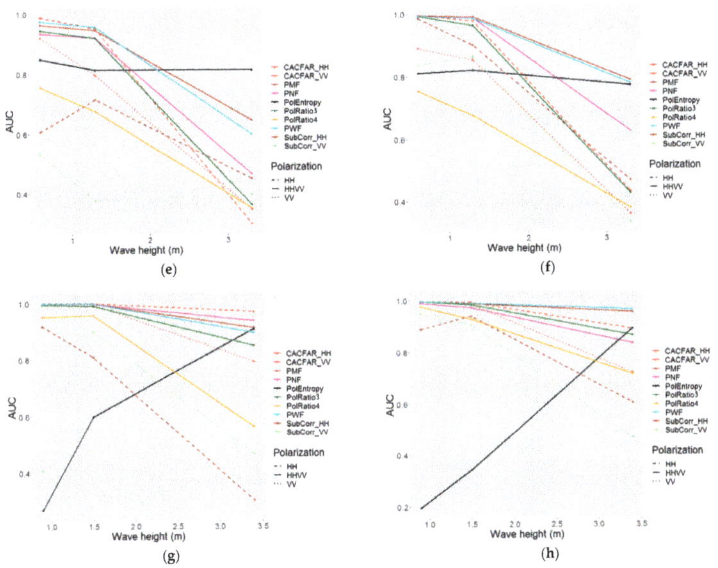

Figure 11. Comparison of the detector AUCs for different wave heights for HH VV. (**a**) Wind: cross, incidence angle: low and boat orientation: 45°. (**b**) Wind: cross, incidence angle: low and boat orientation: 90°. (**c**) Wind: up/down, incidence angle: low and boat orientation: 45°. (**d**) Wind: up/down, incidence angle: low and boat orientation: 90°. (**e**) Wind: up/down, incidence angle: medium and boat orientation: 45°. (**f**) Wind: up/down, incidence angle: medium and boat orientation: 90°. (**g**) Wind: up/down, incidence angle: high and boat orientation: 45°. (**h**) Wind: up/down, incidence angle: high and boat orientation: 90°.

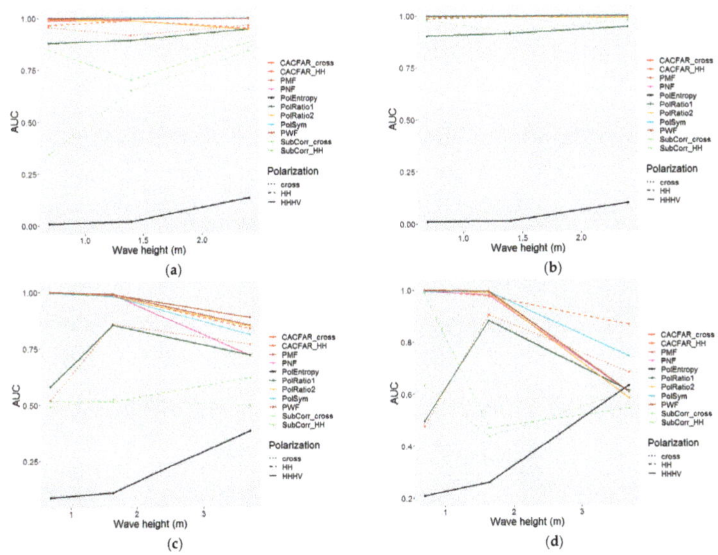

Figure 12. Comparison of the detector AUCs for different wave height for HV HH. (**a**) Wind: up/down, incidence angle: medium and boat orientation: 45°. (**b**) Wind: up/down, incidence angle: medium and boat orientation: 90°. (**c**) Wind: up/down, incidence angle: high and boat orientation: 45°. (**d**) Wind: up/down, incidence angle: high and boat orientation: 90°.

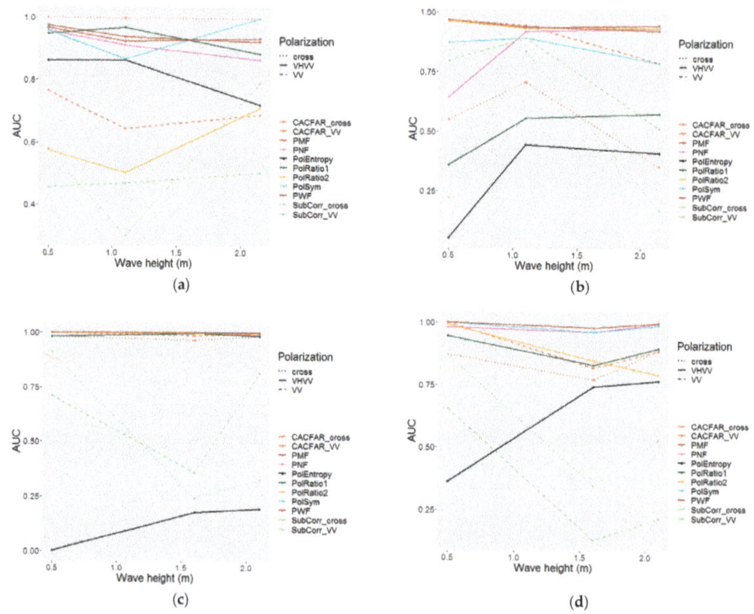

Figure 13. Comparison of the detector AUCs for different wave heights for VH VV. (**a**) Wind: up/down, incidence angle: medium and boat orientation: 45°. (**b**) Wind: up/down, incidence angle: medium and boat orientation: 90°. (**c**) Wind: up/down, incidence angle: high and boat orientation: 45°. (**d**) Wind: up/down, incidence angle: high and boat orientation: 90°.

Error in Table

In the original publication, there was a mistake in Tables 11–13 as published. The names of the detectors in row 5 ("PolRatio1") and in row 6 ("PolRatio2") are incomplete. The corrected Tables 11–13 appears below.

Table 11. The AUCs of the vessel-detection algorithms with different sensor parameters. The background colours indicate different AUCs from white (low AUCs) to dark green (very high AUCs).

Polarization	**HH VV**			**HV HH**		**VH VV**		
Incidence Angle	**Low**	**Medium**	**High**	**Medium**	**High**	**Medium**	**High**	**Avg**
PMF	0.787	0.888	0.976	0.996	0.906	0.945	0.995	**0.928**
PWF	0.779	0.882	0.975	0.997	0.912	0.941	0.995	**0.926**
PNF	0.67	0.822	0.956	0.994	0.881	0.871	0.986	**0.883**
PolEntropy	0.834	0.813	0.534	0.046	0.279	0.559	0.373	**0.491**
PolRatio1/3	0.529	0.768	0.947	0.913	0.689	0.714	0.939	**0.786**
PolRatio2/4	0.554	0.599	0.849	0.983	0.9	0.768	0.936	**0.799**
SubCorr_HH	0.565	0.688	0.744	0.9	0.608			**0.701**
SubCorr_VV	0.528	0.557	0.684			0.604	0.481	**0.571**
SubCorr_cross				0.785	0.531	0.483	0.537	**0.584**
CACFAR_HH	0.6	0.775	0.975	0.98	0.943			**0.854**
CACFAR_VV	0.628	0.695	0.915			0.798	0.946	**0.797**
CACFAR_cross				0.972	0.701	0.766	0.909	**0.837**
PolSym				0.999	0.92	0.895	0.991	**0.951**
avg	**0.647**	**0.749**	**0.856**	**0.869**	**0.752**	**0.759**	**0.826**	

Table 12. The AUCs of the vessel-detection algorithms with high sea states (SPAN > −17 dB). The background colours indicate different AUCs from white (low AUCs) to dark green (very high AUCs).

Polarization		**HH VV**		**HV HH**		**VH VV**		
Incidence Angle	**Low**	**Medium**	**High**	**Medium**	**High**	**Medium**	**High**	**Avg**
PMF	0.787	0.843	0.935		0.728	0.945		**0.848**
PWF	0.779	0.83	0.932		0.745	0.941		**0.846**
PNF	0.67	0.75	0.887		0.662	0.871		**0.768**
PolEntropy	0.834	0.804	0.9		0.506	0.559		**0.721**
PolRatio1/3	0.529	0.669	0.858		0.664	0.714		**0.687**
PolRatio2/4	0.554	0.521	0.64		0.714	0.768		**0.639**
SubCorr_HH	0.565	0.634	0.455		0.58			**0.558**
SubCorr_VV	0.528	0.494	0.468			0.604		**0.523**
SubCorr_cross					0.551	0.483		**0.517**
CACFAR_HH	0.6	0.666	0.931		0.85			**0.762**
CACFAR_VV	0.628	0.59	0.758			0.798		**0.694**
CACFAR_cross					0.723	0.766		**0.745**
PolSym					0.776	0.895		**0.835**
avg	**0.647**	**0.68**	**0.776**		**0.682**	**0.759**		

Table 13. The AUCs of the vessel-detection algorithms with different orientations of the rubber vessel. The blue background colours indicate better AUCs for the 45° inclined vessel, the orange colours show higher AUCs for the vessel oriented at 90°.

Polarization		**HH VV**		**HV HH**		**VH VV**		
Incidence Angle	**Low**	**Medium**	**High**	**Medium**	**High**	**Medium**	**High**	**Avg**
PMF	0.23	−0.07	−0.01	0	0.08	0	0.01	**0.03**
PWF	0.25	−0.08	−0.02	0	0.09	0	0.01	**0.04**
PNF	0.22	−0.10	0.05	−0.01	0.04	0.08	0.02	**0.04**
PolEntropy	0.06	0.02	0.12	0.01	−0.17	0.51	−0.50	**0.01**
PolRatio1/3	0	−0.05	0	−0.02	0.05	0.44	0.1	**0.07**
PolRatio2/4	0.22	−0.01	−0.05	−0.02	0.09	−0.35	0.12	**0**
SubCorr_HH	0	−0.19	−0.13	−0.18	−0.12			**−0.13**
SubCorr_VV	0.07	−0.24	−0.18			−0.26	0.3	**−0.06**
SubCorr_cross				−0.35	−0.06	0.15	−0.09	**−0.09**
CACFAR_HH	0.05	−0.06	0.03	−0.03	−0.01			**0**
CACFAR_VV	0.22	−0.01	0.03			−0.20	0.09	**0.03**
CACFAR_cross				−0.06	0.03	0.46	0.14	**0.14**
PolSym				0	0.02	0.09	0.02	**0.03**
avg	**0.13**	**−0.08**	**−0.02**	**−0.06**	**0**	**0.08**	**0.02**	**0.01**

The authors state that the scientific conclusions are unaffected. This correction was approved by the Academic Editor. The original publication has also been updated.

Reference

1. Lanz, P.; Marino, A.; Simpson, M.D.; Brinkhoff, T.; Köster, F.; Möller, M. The InflateSAR Campaign: Developing Refugee Vessel Detection Capabilities with Polarimetric SAR. *Remote Sens.* **2023**, *15*, 2008. [CrossRef]

MDPI AG
Grosspeteranlage 5
4052 Basel
Switzerland
Tel.: +41 61 683 77 34

Remote Sensing Editorial Office
E-mail: remotesensing@mdpi.com
www.mdpi.com/journal/remotesensing

www.ingramcontent.com/pod-product-compliance
Lightning Source LLC
Chambersburg PA
CBHW061822210926
51850CB00006B/783